Practical Graph Analytics with Apache Giraph

Claudio Martella

Roman Shaposhnik

Dionysios Logothetis

Apress®

Practical Graph Analytics with Apache Giraph

ISBN-13 (pbk): 978-1-4842-1252-3

ISBN-13 (electronic): 978-1-4842-1251-6

Managing Director: Welmoed Spahr
Lead Editor: Jonathan Gennick
Development Editor: Douglas Pundick
Technical Reviewer: Steven Harenberg
Editorial Board: Steve Anglin, Mark Beckner, Gary Cornell, Louise Corrigan, Jim DeWolf,
    Jonathan Gennick, Robert Hutchinson, Michelle Lowman, James Markham, Susan McDermott,
    Matthew Moodie, Jeffrey Pepper, Douglas Pundick, Ben Renow-Clarke, Gwenan Spearing,
    Matt Wade, Steve Weiss
Coordinating Editor: Jill Balzano
Copy Editor: Tiffany Taylor
Compositor: SPi Global
Indexer: SPi Global
Artist: SPi Global
Cover Designer: Anna Ishchenko

Distributed to the book trade worldwide by Springer Science+Business Media New York, 233 Spring Street, 6th Floor, New York, NY 10013. Phone 1-800-SPRINGER, fax (201) 348-4505, e-mail orders-ny@springer-sbm.com, or visit www.springeronline.com. Apress Media, LLC is a California LLC and the sole member (owner) is Springer Science + Business Media Finance Inc (SSBM Finance Inc). SSBM Finance Inc is a Delaware corporation.

For information on translations, please e-mail rights@apress.com, or visit www.apress.com.

Apress and friends of ED books may be purchased in bulk for academic, corporate, or promotional use. eBook versions and licenses are also available for most titles. For more information, reference our Special Bulk Sales–eBook Licensing web page at www.apress.com/bulk-sales.

Any source code or other supplementary material referenced by the author in this text is available to readers at www.apress.com. For detailed information about how to locate your book's source code, go to www.apress.com/source-code/.

Contents at a Glance

Contents

About the Authors

Claudio Martella is passionate about graphs. He is a member of the Large-Scale Distributed Systems group at the VU University Amsterdam. His topics of interest are large-scale distributed systems, graph processing, and complex networks. He has been a contributor to Apache Giraph since its incubation; he is a committer and a member of the project's Podling Project Management Committee (PPMC).

Dionysios Logothetis is a software engineer at Facebook. He is interested in building systems and tools for large-scale data management with a focus on graph mining. He has experience developing analysis systems built around Giraph and Hadoop. Dionysios holds a PhD in Computer Science from the University of California, San Diego, and also a degree in electrical and computer engineering from the National Technical University of Athens.

Roman Shaposhnik is a vice president and one of the lead developers of Apache Bigtop, a 100% open source and community-driven Big Data management distribution built on top of Apache Hadoop. He has been working on making Hadoop ecosystem components more accessible and easier to use, and he has contributed to a wide array of Apache projects, from Avro to ZooKeeper. In addition to his day job building Data Fabric APIs at Pivotal Inc., Roman currently serves as a vice president of Apache Incubator, helping exciting and new open source projects join the Apache family.

About the Technical Reviewer

Steve Harenberg has a BSc in Mathematics from the University of North Carolina at Chapel Hill. He is currently a PhD student in the Department of Computer Science at North Carolina State University, where his research focuses on the design and analysis of graph mining algorithms and their application to real-world problems. In his spare time, he enjoys cross-skating, juggling, and spending time with his dog, Aipa, and cat, Lily.

Introduction

We live in an age of so-called Big Data. We hear terms like *data scientist,* and there is much talk about analytics and the mining of large amounts of corporate data for tidbits of business value. There are even apocryphal stories involving diapers and beer selling together in the same store aisle. The common theme is the problem of having large amounts of data and somehow converting that data into actionable information.

Enter graph theory. It's a branch of mathematics concerned with pairwise relationships between objects. Graph theory can be taught abstractly, and probably often is. It's very practical though. Imagine mapping all the link relationships in a web site. One page might turn out to be in more relationships than all the others, and perhaps that page is an important one. Likewise, one can examine relationships between people in a group, and perhaps the person having the largest number of connections could also be seen as having the widest influence. Certainly, you'd want that well-connected person if your goal were to spread a piece of news or gossip quickly.

The book you're holding is about Apache Giraph and its use in creating graph structures used in analyzing large data sets. Apache Giraph is a graph processing engine designed to be scalable and to quickly answer business and scientific questions based upon connections between people and objects. It is used at Facebook, for example, as the basis for mining and selling information derived from the many random posts that you make throughout the day.

If you have large amounts of data to be analyzed, and especially if there is information to be derived from the relationships between data points, then this book *Practical Graph Analytics with Apache Giraph* can unleash tremendous value. Buy the book today. Read it. And reap the business and scientific value that's hidden away in your data.

Annotation Conventions

Many of the code examples in this book are annotated using markers in the form #1, #2, #3 and so forth. Whenever you see such a marker in a code listing, know that it is not part of the code per se. It is instead a reference to an annotation resolved at the bottom of the listing.

For example, following are some lines from Listing 5-4 in Chapter 5:

```
public class GiraphHelloWorld extends                              #2
    BasicComputation<IntWritable, IntWritable,
                     NullWritable, NullWritable> {
  @Override
  public void compute(Vertex<IntWritable,                          #3
                      IntWritable, NullWritable> vertex,
                 Iterable<NullWritable> messages) {
```

Look at that listing, and you'll see the references #2 and #3 resolved following the listing in body text. Here's how that looks:

```
...
    System.exit(ToolRunner.run(new GiraphRunner(), args));
  }
}
```

#1 Required important statements. We will omit those in all future listings.

#2 Extending the most basic abstract superclass for defining our graph computation.

#3 Defining compute method that would be called for every vertex.

You'll see the foregoing annotation convention used throughout the book. Annotations in code in the form #1, #2, #3 correspond to comments given immediately following the code listing.

PART I

Giraph Building Blocks

PART II

Clojure Scripting Basics

CHAPTER 1

■ ■ ■

Introducing Giraph

This chapter covers

   Using large amounts of graph data to obtain deeper insights

   Understanding the specific objectives and design goals of Giraph

   Positioning Giraph in the Hadoop ecosystem

   Identifying the differences between Giraph and the other graph-processing tools

   This chapter discusses the importance of data-driven decision making and how to execute it with Big Data. You see examples of how using large amounts of data can bring added value. You also learn how Giraph fits among the myriad of tools you can use to analyze large datasets. The chapter positions Giraph in the Hadoop ecosystem and introduces how it plays with its teammates. In addition, the chapter positions Giraph in contrast with other graph-based technologies, such as graph databases, and explains when and why you should use each technology. This is an introductory chapter, and as such it introduces concepts and notions that are explored more deeply in the following chapters.

Data, Data, Data

A couple of years ago, the *New York Times* published a fascinating article about data analytics.[1] According to the story, a father near Minneapolis complained to a Target employee about his teenage daughter receiving coupons for pregnancy-related items. "My daughter got this in the mail!" he said, pointing out the printed coupon. "She's still in high school, and you're sending her coupons for baby clothes and cribs? Are you trying to encourage her to get pregnant?" The Target representative apologized, and then called the father a few days later to apologize further—but the conversation took an unexpected turn. "I had a talk with my daughter," the father said. "It turns out there have been some activities in my house I haven't been completely aware of. She's due in August. I owe you an apology."

   Target, like other retailer shops, assigns each customer a unique ID (such as their credit card number) and uses these IDs to record the customer purchases in stores and online. By analyzing this historical data, companies are able to profile customers and shopping behavior. One of the things they can recognize is pregnancy. Not only that, but they can estimate the due date within a small time window. Apparently lotions are good indicators; many people buy lotion, but pregnant women tend to buy large quantities of unscented lotions around their second trimester. Moreover, in their first trimester, they buy supplements like calcium, magnesium, and zinc. These products are not specific to pregnancy, but a sudden spike in consumption,

---

[1]Charles Duhigg, "How Companies Learn Your Secrets," *New York Times*, Feb. 16, 2012, www.nytimes.com/2012/02/19/magazine/shopping-habits.html.

along with the purchase of other products such as hand sanitizers and washcloths, acts as a signal for pregnancy and closeness to the delivery date. Target's statisticians analyzed customer purchases to figure out these patterns.

It is likely that in the near future, when you walk through shops and malls, your mobile phone will notify you of offers that fit your current needs and tastes, like those shoes you always wanted or the spaghetti you are running out of. But purchase data is only one example of data that can be used to learn more about customer behavior. We are surrounded by new devices and sensors that empower us to measure and record the world around us with increasing precision. As Mark Weiser put it, "The most profound technologies are those that disappear. They weave themselves into the fabric of everyday life until they are indistinguishable from it."[2] It is probably time for our ubiquitous computing devices to disappear into the fabric of everyday life, with their Internet connections and their set of full-fledged sensors. When this happens, we will be provided with a volume of data that we have not had to manage before. To allow these technologies to effectively integrate and disappear, the infrastructure that manages the data they produce must evolve with them.

The abilities provided by these technologies affect us deeply and broadly, including the way we communicate and socialize. The more we can connect and share, the more we do. As illustrated in Figure 1-1, the total amount of data created and shared—from documents to pictures to tweets—in 2011 was of 2 zettabytes,[3] around nine times the amount in 2005; and the predicted amount for 2015 is 8 zettabytes. This data includes information about social behavior, finances, weather, biology, and so on. It is not only valuable for the sciences but can also be used effectively by enterprises to understand their customers and provide better services.

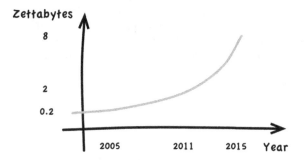

Figure 1-1. *Past and predicted amount of data created and shared*

Now consider this. We act differently depending on our mood. When we feel good and happy, we tend to open up to the world, try new things, and take risks. On the other hand, a negative mood can make us pessimistic and less prone to new adventures. Recently, scientists have taken this idea to an extreme and have tried to see if there is a connection between people's moods and their behavior on the stock market. Do they invest differently, buying or selling certain stocks, depending on how they feel? Data exists about stock-market trends, but it is much more difficult to collect data about the mood of millions of people every second. So the scientists looked at Twitter. People post on Twitter about all sorts of things: what they are doing, a movie they liked, something that made them happy or angry, jokes and quotes, and so on. Tweets are simple and contain more emotional content than blog posts or articles. As such, the text in tweets can be quite representative of the mood of the people who wrote them—and they write thousands every second.

---

[2]Mark Weiser, "The Computer for the Twenty-First Century," *Scientific American*, 1991.
[3]A zettabyte is 1 trillion gigabytes. Source: John Gantz and David Reinsel, IDC, "Extracting Value from Chaos," June 2011, `www.emc.com/collateral/analyst-reports/idc-extracting-value-from-chaos-ar.pdf`.

Sentiment analysis is an application of *natural language processing (NLP)* techniques to determine the attitude of a writer toward a particular topic. Simply put, by looking at the words and characters used, such as adjectives, verbs, emoticons, and punctuation, it is possible to assign that text a positive or negative classification. Sometimes it is possible to assign an evaluation of the affective state of the writer, such as "happy," "sad," or "angry." These techniques are applied to social media to measure customer attitudes toward certain brands, or voter attitudes regarding candidates during political campaigns.

Returning to the connection between moods and the stock market, scientists were able to show that after a number of days of overall calm, as measured on Twitter, the Dow Jones index rose, whereas the reverse happened after an anxious period. Although the researchers claimed accuracy of around 87%, it is still under discussion whether the Twitter data alone can really produce such precise predictions; but it is generally accepted that it can act reliably as an indicator. Whether for predicting the behavior of financial markets or trends in social media, this is a great example of how small pieces of information, such as 140-character tweets, can be very valuable when looked at together.

Practitioners of business intelligence are looking at data analytics to discover patterns in their data and support their decision-making. Looking at how customers use products allows designers and engineers to understand how to improve those products. Data about user behavior provides insights into how to adapt products to users with new or better features. The study of product usage has been traditionally performed via surveys and usability studies. Users are asked to describe how they perform certain actions and tasks— How do you search for a lost e-mail in your client software, choose a hiking backpack on an e-commerce site, organize your appointments in your calendar?—and to list features that are missing. They may also be asked to use a product in a controlled environment, such as a lab, giving the product designers the opportunity to measure the effectiveness of their design decisions. But these approaches have one big drawback: they cannot be performed on a large scale. How many users can you test, and how realistic are these evaluations? Or, how much data can be collected, and how reliable is it?

Analyzing data is not only useful for analysts and scientists. Machine-learning algorithms can discover patterns and classify and categorize data as part of an application. For example, consider users and books. If you find out that Mark and Sharon have similar tastes in books, you can advise Mark to read the books that Sharon has read but Mark has not (yet), and vice versa. Or, to turn it around, if you know that two books are similar, you can advise the readers of the first to read the second, and so on. What these algorithms learn automatically from the data can be integrated directed into an application, making it more personalized and proactive. These concepts are at the core of *recommender systems*, such as those used by Amazon and Facebook to recommend books and friends. Or think of a search engine like Google. Google Search receives queries from users asking for specific content. It also collects which entries the users click on the results page, not to mention clicks on content provided by other Google products such as Gmail, Google Maps, and so on. Putting this information together, Google Search can provide different search results for the same query, depending on the user who performs it. If I search for a music shop, Google will probably give more relevance to shops that sell jazz music in Amsterdam. This is called *personalized search.*

One interesting fact about data is that the more you have, the better you can understand and model the phenomena you are looking at. In fact, studies show something even more striking. Recently, scientists have tested "naive" machine-learning algorithms against more sophisticated ones. They have discovered that often, the naive algorithms outperform the more sophisticated ones if they are fed large enough volumes of data. Consider the Netflix challenge. Netflix provided a large dataset containing the ratings of nearly half a million users for 18,000 movies. The challenge was to predict user ratings for movies they have not seen yet. These studies show that a simpler algorithm that takes into account additional data about the movies, such as information from the Internet Movie Database (IMDb), can outperform a more sophisticated algorithm that uses only the Netflix data. This example is tricky, because IMDb is a different (independent) dataset; but in general, machine-learning algorithms benefit from more data because it reduces noise and helps build better models.

Another counterintuitive point is that, given a large enough dataset, a sophisticated algorithm may perform more poorly than a naive one. Basically, more sophisticated heuristics do not help when algorithms are given a bigger picture; rather, they may achieve worse results. It is as if the additional complexity of the newer algorithms compensates for absence of evidence, and hence they work better with small datasets, but the complexity biases the results that can be achieved if more evidence is available. One way to put it is that the sophisticated algorithms can be more *opinionated*, instead of basing their conclusions on facts. This is good news. It means you can mine data with a fairly simple and well-understood toolset (the basic algorithms that have been studied extensively over the last 30 years), provided you can execute them at scale.

■ **Important** The ability to create models based on data is the basis of creating better products and services. The volume of data being produced is already huge and is constantly growing. Unfortunately, there is no shortcut that will let you avoid facing these large volumes of data. You have to design algorithms and build platforms that can manage data at this scale. Giraph is one of these platforms.

From Big Data to Big Graphs

The previous section discussed the importance of looking at large volumes of data to extract useful intelligence. The examples presented use many small pieces of data that are aggregated to extract trends and whatnot. What is the average age of people buying a product? In what month of the year does that product sell best? Which other product has a similar trend? The more data you have, the more accurate the answers you can give to questions like these. However, for certain kinds of data, you can also take advantage of the *connections* between the entities in your data. In those cases, more important than the pieces of information (and their aggregations) is the way those pieces are connected—more precisely, the structure of the network created by these connections. Unfortunately, the power that comes from analyzing connected data has a cost, because analyzing graphs is usually computationally expensive and introduces new challenges and specific tools. Giraph is a framework you can use to run graph analytics on very large graphs across many commodity machines. It was specifically designed to solve problems that take the connectedness of data into account. In other words, it was designed to process graphs.

A *graph* is a structure that represents abstract entities (or *vertices*) and their relationships (or *edges*). Graphs are used in computer science in a variety of domains to represent data and express problems. Many common data structures like trees and linked lists are graphs: their nodes are connected with links according to the criteria that define each specific data structure. Graphs are also used in other disciplines; for example, in the social sciences, a graph is usually called a *network*. Networks are used to represent relationships between individuals, such as in a social network, as illustrated in Figure 1-2. As in the case of graphs in computer science, social networks were used in the social sciences long before social networking sites like Facebook became popular. And in biology, graphs are used to map interactions between proteins. Having such a connected structural model lets researchers look at data in a more cohesive way than as just a collection of single items. Chapter 2 looks at graphs in more depth and explains how to model data and problems through them.

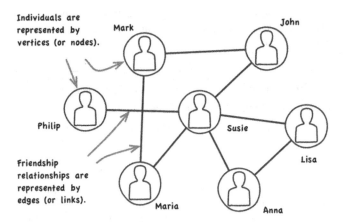

Figure 1-2. *A social network of individuals represented with a graph*

To give you a better idea of how looking at data connections as a graph can give you added value, and the types of problems for which Giraph may be helpful, let's consider some examples. Recent research has shown that by looking at the social network of 1.3 million individuals on Facebook, it was possible to predict which people were partners and whether they were going to break up in the next two months (of course, without looking at their relationship status). The researchers used only the network structure—the friendship relationships between individuals. The concept behind this research is *social dispersion*. Basically, just looking at the number of friends two individuals have in common—called *embeddedness*—does not provide enough information and is a low predictor metric. Dispersion, on the other hand, measures how much these common friends are *not* connected. In other words, high dispersion between two people means they have friends in common but only a few of those friends are friends with each other. According to the data, couples with long-lasting relationships tend to present high dispersion. Intuitively, the results suggest that strong romantic relationships are those in which people participate in different social groups, which they share with their partners but which remain separate. Looking at one individual and selecting from her social network individuals with whom she has high dispersion generates a list of possible partners for that individual; about 60% of the time, the person at the top of this list is indeed the correct partner. Moreover, couples without this particular social structure are more likely to split in the near future.

Another example of how looking at data as a graph can bring added value is the Web. As you see in the following chapters, one of the reasons for Google's initial success was that it looked at the Web as a graph. Web pages point to other pages through hyperlinks. Early on, Google's competitors were crawling the Web by jumping from page to page and following the links on each page, but the graph structure was not used to provide search results. Results were provided solely based on the presence of query terms on web pages, which led to the spamming of popular keywords on unrelated pages. Google, on the other hand, decided to look at the structure of the graph—which pages linked to which other pages—to rank pages based on a notion of popularity. According to Google's PageRank algorithm, the popularity of a page depends on the number of links pointing to it and how popular those pages are. This means the entire structure of the graph is taken into account to define the ranking of vertices. This approach was more sophisticated, but the results were impressive, and you know how the story continued.

A final example relates to the temporal aspect of networks. A couple of years ago, researchers used a network representing sex buyers and escorts to study the spread of diseases. This dataset was a graph in which connections between individuals represented sexual intercourse; it also contained a temporal aspect, because connections had a timestamp. By studying the structure of the graph and how it developed over time, researchers were able to simulate epidemics and study how they developed. If they simulated some of the individuals being affected by a sexually transmitted infection and looked at the connections, how many people might be eventually infected, and how did the structure of the graph affect this outcome? Interestingly, similar models have been applied to online social networks to study how information is spread via networks like Twitter and Facebook—a process that by no coincidence is often called *viral*.

■ **Important**   The Web, online social networks, the relationships among users who rate movies in a movie database, and biological networks are only a few examples of graphs. Whether in the context of business or science, viewing data as connected adds value to it. Analyzing these connections can help you extract intelligence in ways that are not possible otherwise.

Unfortunately, looking at data as graphs often requires computationally expensive algorithms as the larger, interleaved nature of data is taken into account. Moreover, graphs are becoming larger and larger. For example, the Facebook social graph contains more than 1 billion users, each with around 200 friendship relationships. The Web, as indexed by Google, contains more than 50 trillion pages. These graphs are huge. If you want to extract information out of the interleaved connections between entities in your data with a simple but powerful programming paradigm, read on: Giraph is probably the right tool for your use case.

Why Giraph?

Some of the things mentioned in the previous sections are not exactly new. Big Data is a movement that has grown during the last decade from a problem faced by a few (Google, Yahoo!, Facebook, and so on) to a complete paradigm shift. Initially, Big Data was characterized by the design of new solutions that overcame the limitations of traditional ones when managing large volumes of data. The following three technical challenges come with managing data at a large scale:

- Data is dynamic, and solutions must support high rates of updates and retrievals. Think of a web application with millions of users clicking links and requesting pages every second.

- Data is large, and its management must be distributed across a number of machines. Think of petabytes of data that are too large to be stored and processed by a single machine.

- The computing environment can be dynamic as well, and solutions have to manage data reliably across a number of machines that can and will fail at any moment. Think of a cluster of thousands of commodity machines, possibly distributed across different data centers.

Big Data solutions must face these challenges to support applications at large scale.

Managing Big Graphs requires solving these challenges and more. Graphs add more problem-specific challenges that Giraph was designed to handle:

- Graph algorithms are often expressed as *iterative* computations, and a graph-processing system that supports such algorithms needs to perform iterations efficiently.

- Graph algorithms tend to be *explorative*, which introduces random access to your data that is difficult to predict.

- Usually, a small computation is executed multiple times on each graph entity, and such operations should be simple to express and *parallelize*.

This book looks at how Giraph attacks these specific challenges technically. For now, consider that Giraph was designed with graphs and graph algorithms in mind, and as such it can execute graph algorithms up to *100* times faster than other Big Data frameworks like MapReduce.

■ **Definition**   Giraph is a framework designed to execute *iterative* graph algorithms that can be easily *parallelized* across hundreds of commodity machines. Giraph is fault-tolerant and easy to program, and it can process graphs of *massive* scale.

Although it is important to execute graph algorithms efficiently and reliably, it is just as important that the programming paradigm be simple and not require time-consuming and error-prone operations. Giraph offers what is called a *vertex-centric* programming model. This model requires you to put yourself in the shoes of a vertex that can exchange messages with the other vertices in a sequence of iterations. None of the programming complexity of a large-scale parallel and distributed system is exposed to the user. The programmer concentrates on the problem-specific aspects of the algorithm, and Giraph executes the algorithm in parallel across the available machines transparently. In practice, this means extending a class and implementing a single function. Often, Giraph algorithms are fewer than 50 lines and do not require *any* concurrent primitives such as locks, semaphores, or atomic operations.

■ **Important**   Giraph provides a vertex-centric programming model that requires the developer to think like a vertex that can exchange messages with other vertices. The programming model hides the complexity of programming a parallel and distributed system. Giraph executes your code transparently in parallel across the available machines.

Giraph was designed to compute graph analytics and social-network analysis. As you see in the next chapter, these are often executed as *offline computations*. This means Giraph applications usually sit in the back office, are run periodically over large datasets, and take minutes or hours. Giraph is designed for heavy-lifting tasks, not for quick, interactive queries.

Let's take a quick look at an application architecture (in the broadest sense) and where Giraph would fit. Chapter 2 looks at this in more depth. Figure 1-3 shows such an architecture. As you can see, the architecture is divided into the three typical macro components: the front end, the back end, and the back office. The *front end* is where the application's client-related components are running and includes the mobile application, the code that runs in the desktop browser, and so on. The *back end* is where the application servers, the databases, and all the (distributed) infrastructure run, in cooperation with the front end, to give the user a unified experience when they interact with the application. In the case of an application like Facebook, these two components are responsible for providing the data and the logic when you click the buttons of social-networking features like surfing your newsfeed, looking at your friends' activity and so on. In the back end are your pictures, your comments and likes, and the social graph. In the back office resides the logic that is executed periodically to compute and materialize offline the content that is used online by the application. Giraph lives in the back office and is executed to compute application logic like friend recommendations, ranking of activity items, and so forth. It is also used on demand by the data-science team to run analytics on user-activity data collected by the back end. For example, Giraph is used to execute analyses like that mentioned earlier, seeing whether partnership relationships can be inferred from the social graph through the social-dispersion metric.

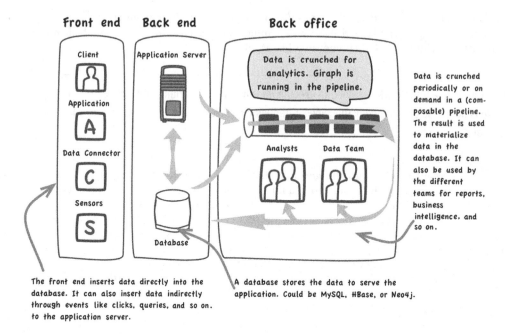

Figure 1-3. *An application's architecture and where Giraph fits in*

To summarize, Giraph is used to run expensive computations that are executed asynchronously with respect to the interaction between the user and the application, and for this reason it is said to live in the back office. The following sections explore how Giraph integrates with technology that resides in this component of the application architecture and how it differs from other tools that are positioned both in in the back office and in the back end.

■ **Important**   Giraph resides in the back office. It is used to compute *offline* graph analytics that are run periodically on data collected from interactions between the front end and the back end, along with all the additional data that may be available in the back office.

GOOGLE PREGEL AND APACHE GIRAPH

Many Apache projects under the Hadoop umbrella are heavily inspired by Google technologies. Hadoop originally consisted of the Hadoop Distributed File System and the MapReduce framework. These two systems were both inspired by two Google technologies: the Google File System (GFS) and Google MapReduce, described in two articles published in 2003–2004.[4] In 2010, Google published an article about a large-scale graph-processing system called Pregel.[5] Apache Giraph is heavily inspired by Pregel.

---

[4]Sanjay Ghemawat, Howard Gobioff, and Shun-Tak Leung, "The Google File System," ACM SIGOPS Operating Systems Review. Vol. 37. No. 5. ACM, 2003.

Jeffrey Dean and Sanjay Ghemawat, "MapReduce: Simplified Data Processing on Large Clusters." Communications of the ACM 51.1 (2008): 107–113.

[5]Grzegorz Malewicz, Matthew H. Austern, et. al., "Pregel: A System for Large-Scale Graph Processing." Proceedings of the 2010 ACM SIGMOD International Conference on Management of Data. ACM, 2010.

Giraph was initially developed at Yahoo! and was incubated at the Apache Foundation during the summer of 2011. In 2012, Giraph was promoted to an Apache Top-Level Project. Giraph is an (open source) loose implementation of Google Pregel, and it is released under the Apache License. In 2013, version 1.0 was released as proof of its stability and the number of features added since its initial release.

Giraph enlists contributors from Facebook, Twitter, and LinkedIn, and it is currently used in production at some of these companies and others. Giraph shares with Pregel its computational model and its programming paradigm, but it extends the Pregel API and architecture by introducing a master *compute* function, out-of-core capabilities, and so on, and removing the single point of failure (SPoF) represented by the master. You see all these functionalities in detail throughout the book.

Giraph and the Hadoop Ecosystem

Hadoop is a popular platform for the management of Big Data. It has an active community and a high adoption rate with yearly growth of around 60%,[6] and a number of companies make supporting these enterprises their mission. The global Hadoop market is predicted to grow to billions of dollars in the next five years, and skill with Hadoop is considered one of the hottest competitive factors in the ICT job market.[7]

Hadoop started as an implementation of the Google distributed filesystem and the MapReduce framework in the Apache Nutch project. Over the years, it has turned into an independent Apache project. Nowadays, it is much more. It has evolved into a full-blown ecosystem. Under its umbrella are projects beyond the Hadoop Distributed File System and the MapReduce framework. All these systems were developed to tackle different challenges related to managing large amounts of data, from storage to processing and more. Giraph is a relatively newcomer to the Hadoop ecosystem.

▪ **Important**   Although Giraph is part of the Hadoop ecosystem and runs on Hadoop, it does not require you to be an expert in Hadoop. You just need to be able to start a Hadoop machine (or cluster).

Here is a selection of projects related to Giraph. Do not be worried if you do not know (or do not use) some of these tools. The aim of this list is to show you that Giraph can cooperate with many popular projects of the Hadoop ecosystem:

- *MapReduce*: A programming model and a system for the processing of large data sets. It is based on scanning files in parallel across multiple machines, modelling data as keys and values. *Giraph can run on Hadoop as a MapReduce job.*

- *Hadoop Distributed File System (HDFS)*: A distributed filesystem to store data across multiple machines. It provides high throughput and fault tolerance. *Giraph can read data from and write data to HDFS.*

- *ZooKeeper*: A centralized service for maintaining configuration information, naming, providing distributed synchronization, and providing group services. *Giraph uses ZooKeeper to coordinate its computations and to provide reliability and fault tolerance.*

---

[6]*Wall Street Journal*, CIO Report, "Hadoop There It Is: Big Data Tech Gaining Traction." April 12, 2014. For WSJ subscribers, http://blogs.wsj.com/cio/2014/04/12/hadoop-there-it-is-big-data-tech-gaining-traction/.
[7]"Job Trends," Indeed, www.indeed.com/jobtrends.

- *HBase*: The Hadoop database. It is a scalable and reliable data store, supporting low-latency random reads and high write throughput. It has a flexible column-based data model. *Giraph can read data from HBase.*

- *Cassandra*: A distributed database that focuses on reliability and scalability. Cassandra is fully decentralized and has no SPoF. *Giraph can read data from Cassandra.*

- *Hive*: A data warehouse for Hadoop. It lets you express ad hoc queries and analytics for large data sets with a high-level programming language similar to SQL. *Giraph can read data from a Hive table.*

- *HCatalog*: A table- and storage-management service for data created using Apache Hadoop. It operates with Hive, Pig, and MapReduce. *Giraph can read data stored through HCatalog.*

- *Gora*: Middleware that provides an in-memory data model and persistence for Big Data. Gora supports persisting to column stores, key-value stores, document stores, and RDBMSs. Through Gora, Giraph can read and write graph data from and to any data store supported by Gora.

- *Hama*: A pure bulk synchronous parallel (BSP) computing framework on top of HDFS for massive scientific computations such as matrix, graph, and network algorithms. *Giraph was designed to solve iterative algorithms like Hama. Unlike Hama, Giraph does not require additional software to be installed on a Hadoop cluster.*

- *Mahout*: A library of scalable machine-learning algorithms on top of Hadoop. It contains algorithms for clustering, classification, recommendations, and mining of frequent items. *Giraph can run machine-learning algorithms designed specifically for graphs.*

- *Nutch*: Software for web search applications. It consists of a web crawler and facilities for ranking and indexing web pages. *Nutch can use Giraph to compute the rankings of web pages.*

As you can see, these projects are very different, from databases to processing tools. But they share a common characteristic: they are designed for a large scale. Giraph fits in the list by providing a programming model and a system for processing large graphs. You can think of Giraph as a MapReduce that is specific to graph algorithms. Giraph uses existing Hadoop installations by running as a MapReduce job or as an Apache Hadoop YARN (Yet Another Resource Negotiator) application, where a YARN-based Hadoop installation is available. Giraph can read data from HDFS, HBase, HCatalog, and Hive, and it uses ZooKeeper to coordinate computation. Collaboration also happens also in the reverse direction: as mentioned, Nutch can use Giraph to compute the rankings of web pages. You see more about how to integrate Giraph with the rest of the ecosystem throughout this book. Figure 1-4 shows the high-level architecture of Giraph and how it uses different Hadoop projects to work.

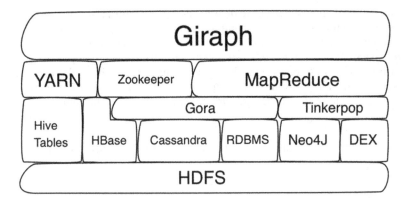

Figure 1-4. *Architecture of Giraph and how it integrates with other Hadoop projects*

■ **Important**   If you already have a Hadoop infrastructure in place or plan to deploy more in the future, Giraph will play well with most of it.

Giraph and Other Graph-Processing Tools

It is important to understand what distinguishes Giraph from other tools on the graph-processing scene in order to make the right decision about when to use it. Here's a simple heuristic. If your computation requires touching a small portion of the graph (for example, a few hundred vertices) to return results within milliseconds, maybe because you need to support an interactive application, then you should look at a graph database like Neo4j, OrientDB, or the TinkerPop tools. On the other hand, if you expect your computations to run for minutes or hours because they require exploring the whole graph, Giraph is the right tool for the job. If your computations are aggregating values extracted from scanning text files or computing SQL-like joins between large tables, then you should look at tools like MapReduce, Pig, and Hive.

In particular, we want to stress the difference between a tool like Giraph and a graph database. You can use a proportion to understand the relationship between Giraph and a graph database like Neo4j: Giraph is to graph databases as MapReduce is to a NoSQL database like HBase or MongoDB (or a relational database like MySQL). In the same way MapReduce is usually used to run expensive analytics on data represented through tuples, Giraph can run graph-mining algorithms on data represented through a graph. In both cases, MapReduce and Giraph run analytics on the data stored in the databases. In both cases, the databases are used to serve low-latency queries: for example, to support an interactive application. What differs is the data model used to represent data. This analogy works well for presentation purposes, but don't let it mislead you. Giraph can process also data stored in a NoSQL database, and MapReduce can do the same with graph databases.

■ **Tip**   You can think of the relationship between Giraph and a graph database as the following proportion: Giraph : graph database = MapReduce : NoSQL database.

This is a very simple yet effective heuristic. But let's try to make things even clearer. The emergence of Big Data has created a set of new tools. Hadoop is an ecosystem that hosts some of them, but more are available. The landscape of these tools is usually divided into two groups: *databases* for storing data and serving queries with low latency, and tools for long-lasting computations such as *analytics*. These two groups of tools have different use cases and requirements, which lead to different design decisions. Databases focus on serving thousands of queries per second that require milliseconds (or a few seconds) to return results. Think of a database backing the content of a web page after a click. Tools for analytics focus on serving a few computation requests per minute, each requiring minutes or hours to return results. Think of computing communities in a large social network. Giraph positions itself in this second group, because it was designed to run long-lasting computations that require analyzing the entire graph with multiple passes. If you want to learn more about databases for graphs, look at systems like Neo4j, OrientDB, and the TinkerPop community.

■ **Tip**   If you are interested in getting to know more about graph databases like Neo4J and OrientDB, you can check out *Practical Neo4j*, by Greg Jordan (Apress 2014).

All these tools specialize in various data models: key-value pairs, columns, documents, or graphs. Figure 1-5 provides a visual representation of some of them. Of these, graphs are probably the most peculiar. Compared to simple key-value pairs and columns, graphs have a more unstructured and interleaved nature. Moreover, graph algorithms tend to be explorative, where portions of the graph are visited to answer a specific query (compared to a bunch of specific key-value pairs, columns, or documents being retrieved). For these reasons, graph-processing systems require design decisions that take into account the specific characteristics of graphs and their algorithms. Running graph algorithms with general-purpose systems is possible—for example, on systems based on tuples, such as relational databases or MapReduce—but these solutions tend to perform worse.

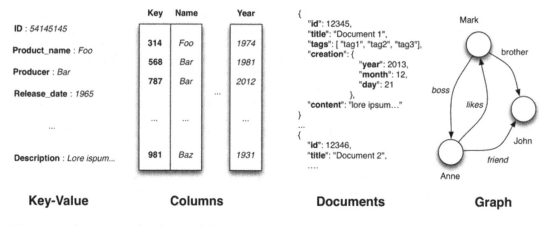

Figure 1-5. *Comparing other data models to a graph*

Giraph was designed for graphs since the beginning. And it is not the only such tool. Other systems similar to Giraph are GraphLab and GraphX. Of these, Giraph is the only one that runs transparently on a Hadoop cluster and has a large open source community. GraphX is perhaps most similar to Giraph; although it offers a simplified API for graph processing, it requires implementing your algorithm in Scala with a different API than that provided by Giraph (which is arguably less simple), it requires Spark, and it is generally slower than Giraph. GraphLab (now known as Dato Core or GraphLab Create) used to be an open source solution, but it is now proprietary. Figure 1-6 gives you an idea of how Giraph compares to other data-management systems.

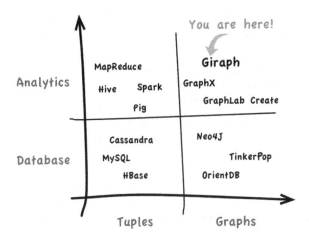

Figure 1-6. *Positioning Giraph with respect to other data-management tools*

To summarize, Giraph is a good tool for these tasks:

- Analyzing a large set of connected data across a cluster of commodity machines (running Hadoop)

- Running an algorithm that is compute-intensive and that processes your graph iteratively, perhaps in multiple passes

- Periodically running an algorithm offline in the back end

And these are the tasks Giraph is *not* good for:

- Running aggregations on a set of unconnected or unrelated pieces of data

- Working with a small graph that can easily fit on a single machine

- Running a computation that is expected to touch a small portion of the graph—for example, to support an interactive workload like a web application

Summary

Big Data and data analytics are opening new ways of collecting traces and understanding user behavior. Mastering techniques to analyze this data allows for better products and services. Looking at aggregations of many small chunks of data provides insights into the data, but examining the way these chunks are *connected* enables you to take a further step into better understanding your data.

In this chapter, you learned the following:

- A graph is a neat and flexible representation of entities and their relationships. Graphs can be used to represent social networks, the Internet, the Web, and the relationships between items such as products and customers.

- Processing large graphs introduces specific challenges that are not tackled successfully by traditional tools for data analytics. Graph analytics requires tools specifically designed for the task.

- Giraph is a framework for analyzing large data sets represented through graphs. It has been designed to run graph computations reliably across a number of commodity machines.

- Unlike graph databases, Giraph is a batch-processing system. It was designed to run computationally expensive computations that analyze the entire graph for analytics, not to run queries expected to be computed within milliseconds.

- Giraph is part of the Hadoop ecosystem. Giraph jobs can run on existing Hadoop clusters without installing additional software. Giraph plays well with its teammates, and it can read and write data from a number of data stores that are part of the Hadoop ecosystem.

Now that you have been introduced the general problem of processing large volumes of connected data and the ecosystems of Big Data and graph processing, you are ready to look at how to model data as graphs, how graphs can fit in real-world use cases, and how Giraph can help you solve the related graph-analytical problems.

CHAPTER 2

■ ■ ■

Modeling Graph Processing Use Cases

This chapter covers

- Modeling your data as a graph
- Differences between offline and online computations
- Fitting Giraph in an application architecture
- Use case for Giraph at a web-search company
- Use case for Giraph at an e-commerce site
- Use case for Giraph at an online social networking site

Giraph goes beyond a basic library to facilitate graph processing. It provides a programming model and a framework to solve graph problems through offline computations at large scale. To use Giraph, you need to understand how to model data with graphs and how to fit Giraph in a system architecture to process graph data as part of an application. For this reason, this chapter is dedicated to three topics. First it introduces graphs and shows how you can use graphs to model data in a variety of domains. Then you learn the difference between offline and online computations to help you identify the types of processing for which Giraph is a good solution. The remainder of the chapter focuses on fitting Giraph into a system architecture. To achieve this goal, the chapter presents three uses cases based on real-world scenarios.

Graphs Are Everywhere

A graph is a neat, flexible structure to represent entities and their relationships. You can represent different things through a graph: a computer network, a social network, the Internet, interactions between proteins, a transportation network—in general terms, data. What do these examples have in common? They are all composed of entities *connected* by some kind of relationship. A computer network is composed of connected devices, a social network is a network of people connected by social relationships (friends, family members, co-workers, and so on), cities are connected by roads, and neurons are connected by synapses. The World Wide Web is probably one of the clearest examples: pages, images, and other resources are connected through the href attribute of links.

■ **Definition** A *graph* is a structure to represent entities that are *connected* through relationships.

A graph can be used to represent these connected structures. It consists of *vertices* (or nodes) and the *edges* (or arcs, or links) that connect them. The two vertices connected by an edge are usually called the edge *endpoints*. The vertices connected to a specific vertex, through the respective edges, are called the vertex *neighbors*. An edge that connects a vertex to itself is called a *loop*. In diagrams, vertices are usually represented by circles and edges are represented by lines. Figure 2-1 shows these fundamental concepts; in its most basic form, there is nothing more to a graph.

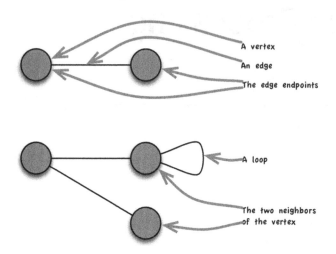

Figure 2-1. *Basic concepts of a graph*

This section guides you through several examples of how to model different data using a graph. Each example introduces a number of graph concepts. Let's start with a network of computers.

Modeling a Computer Network with a Simple, Undirected Graph

A computer network comprises a collection of computing devices such as desktops, servers, workstations, routers, mobile phones, and so on, connected by means of a communication medium. Figure 2-2 shows an example of a computer network. Ignore the technology or protocol that enables the devices to communicate (such as Ethernet, WiFi, or a cellular network), and focus on the fact that they are connected and can exchange data. This is the fundamental aspect that makes a computer network a network and a graph.

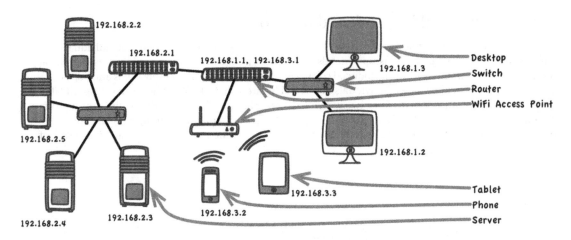

Figure 2-2. *A computer network*

It is natural to model a computer network with a graph by representing computers as vertices and by connecting two vertices through an edge if the two computers they stand for are connected. Clearly, there is not one single correct way to model a computer network through a graph; it is up to the modeler to decide, for example, whether hubs and switches should be represented in the graph, or which layer of the TCP/IP stack should be considered. To identify each device in a graph, you assign each vertex a *unique identifier* (ID); in the case of a computer network, you can use the device hostname or an IP address. Each data domain has a natural data element that can be used as an ID for vertices, such as a passport number for a social network, a URL for a web page, and so on.

Figure 2-3 shows a graph representing the computer network from Figure 2-2. In this example, the IP address of each computer is used as the ID of the vertex, and two computers are connected by an edge if they can communicate directly at the network layer (hence ignoring switches, hubs, access points, and so on).

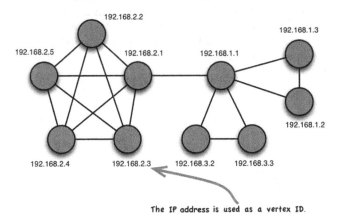

Figure 2-3. *A graph representing a computer network*

Modeling a Social Network and Relationships

Let's move to another example: a social network. A *social network* consists of a number of individuals connected by some kind of relationship—friends, co-workers, and so on. You can easily represent a social network using the graph concepts introduced so far, with vertices representing individuals and edges connecting two vertices if the two individuals they stand for are socially tied. But how can you specify the particular type of relationship between two individuals? You can use a *label*. A label identifies the type of relationship, and it can be attached to an edge. Figure 2-4 shows a graph representing a social network of five individuals. The individual's first name is the ID, and a label is attached to each edge to qualify the type of relationship.

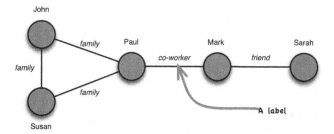

Figure 2-4. *A graph representing a social network*

Until now, you have only considered symmetric relationships, such as friendship. However, edges can be extended with *direction*. The direction defines the *source* vertex and the *destination* vertex of an edge. An edge with direction is called a *directed* edge. All edges that have a specific vertex as a source are called its *outgoing* edges, and those that have that vertex as a destination are called its *incoming* edges. One of the differences between the Twitter and Facebook social graphs is the result ot this aspect. For Facebook, a friendship relationship is symmetric (it requires confirmation by both parties), and it is better represented through an undirected edge. On the other hand, Twitter lets you specify a follower relationship that does not have to be reciprocated; hence a directed edge is more appropriate. Again, whether a relationship should be modeled through direction depends on the relationship, and the decision is in the hands of the modeler. In plots, directed edges are usually represented by an arrow. A graph with directed edges is called a *directed graph*, and a graph with undirected edges is called an *undirected graph*. Figure 2-5 shows the same network of five individuals, this time modeled using a directed graph.

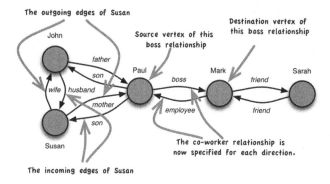

Figure 2-5. *A directed graph representing a social network*

Modeling Semantic Graphs with Multigraphs

All the graphs so far allow only one edge between a specific pair of vertices in the undirected case (only one friend edge between Mark and Sarah), and two in the directed case (one edge for John father Paul and one edge for Paul son John). *Multigraphs*, on the other hand, allow multiple edges between the same pair of vertices, in both the directed and undirected cases, and can also support labels attached to each edge.

A good example of a labeled directed multigraph is a resource description framework (RDF) graph. According to the RDF data model, information can be described through a series of triples, each composed of a subject, a predicate, and an object. Ignoring the specifics such as syntax notations, serialization, and so on, RDF can be used to represent knowledge about different domains. Imagine that you want to represent the fact that Rome is the capital of Italy. ("Rome", "is capital of", "Italy") is a valid RDF triple to represent such information, where "Rome" is the subject, "is capital of" is the predicate, and "Italy" is the object of the triple. ("Resource Description Framework", "has abbreviation", "RDF") is another example of a triple. If you think about it, each of these triples is nothing more than a labeled directed edge, where the subject is the source vertex, the predicate is the label, and the object is the destination vertex. Having such a general and flexible way of describing concepts unleashes your ability to represent pretty much anything that can be expressed through a triple. RDF graphs are often referred to as *semantic networks or graphs* (also making RDF one of the core components of the Semantic Web), because they are frequently used to describe the semantics of things through their relationships.

DBpedia is an example of such a semantic graph. DBpedia is an effort to represent the structured information in Wikipedia—for example, in the info boxes—in the form of a graph. Figure 2-6 lists a number of (simplified) triples from DBpedia. Constructing a graph from this table is straightforward: each triple can be represented by an outgoing edge, leaving from the vertex representing the subject and ending in the vertex representing the object. Subject and object labels can be used for vertices IDs, and predicate labels can be represented by the edge labels. Figure 2-7 shows such a graph.

Subject	Predicate	Object
United States	areaTotal	9826675.0
United States	anthem	The Star Spangled Banner
United States	leaderName	Barack Obama
United States	leaderName	Joe Biden
United States	leaderName	John Boehner
Barack Obama	residence	White House
Barack Obama	tenantOf	White House
Barack Obama	birthPlace	Hawaii
Barack Obama	orderInOffice	President of the United States
Hawaii	areaTotal	28311.0
Hawaii	country	United States

Figure 2-6. *A set of triples from DBpedia*

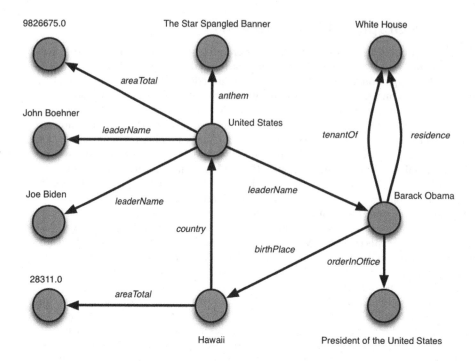

Figure 2-7. *A multigraph constructed from a set of triples from DBpedia*

Notice how all the triples that have Barack Obama as a subject result in an outgoing edge for the vertex representing Barack Obama. Also, notice how Barack Obama has two outgoing edges (tenantOf and residence) toward White House. The latter is an example of the multigraph property mentioned before.

Modeling Street Maps with Graphs and Weights

Street maps are good examples of graphs as well. Perhaps the first person who ever drew a map was also the first person who drew a graph. Modeling a street map with a graph is intuitive: cities, towns, and villages are modeled with vertices, and the roads and streets that connect them are modeled with edges. In general, any point where a different road can be taken is modeled with a vertex. Note that although streets cross, edges do not (they do when you draw them in two-dimensional space, but they do not conceptually). Hence, crossings must be explicitly modeled through vertices. (This will be clearer when you look at paths in chapter 4.) In a graph representing a street map, the edges can have the type (a highway or a road) or the number (a combination of both) as the label, and direction can be used to model the way the street can be followed (a one-way road). How can you model the length of the road? Clearly, the distance between two cities does not depend solely on the number of roads—and, hence, edges—that need to be followed, but also on the length of these roads.

You can use *weights* for this purpose. Weights are numerical properties of edges that can be used to represent quantitative properties of relationships. For example, they can represent the distance between two towns, the ranking of a movie by a user, the similarity between two users based on some profile data, or the strength of a social tie. Algorithms exploit weights to compute shortest paths, recommendations, and different metrics of the graph. You learn more about weights when you look at paths in chapter 4. Figure 2-8 shows a highway map of Italy and a portion of it modeled with a graph.

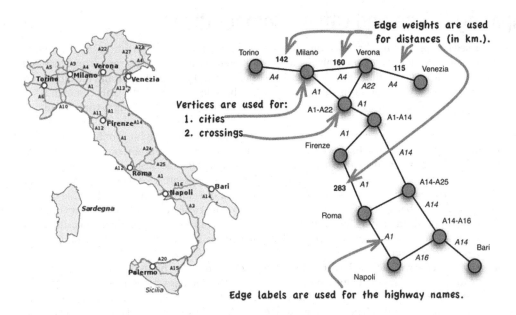

Figure 2-8. *A transportation network modeled through a graph with weights*

Graphs are flexible structures that are natural to think about, and their modeling strongly depends on their purpose. Software engineers use these structures in one form or another on a daily basis. Think of an ER diagram, an UML diagram, or a dependency tree; these are all graphs. Most of us naturally draw graphs when discussing a design or an architecture during brainstorming. That is why graphs are also said to be *whiteboard friendly*. Most data-structures are some kind of graph, from trees to linked lists. But data, data representations, and data structures are meaningful only when coupled with the algorithms that compute them. The following sections and chapter 4 look at what you can do with graphs. Table 2-1 summarizes the concepts presented in this section.

Table 2-1. *Core Concepts of Graphs*

Name	Description
Vertex	An entity in a graph
Identifier (ID)	The unique identifier of a vertex
Edge	A relationship between two entities
Edge label	A qualitative property of an edge (for example, a type)
Edge direction	Specifies the source and destination vertices of an edge
Edge weight	A quantitative (numerical) property of a relationship
Edge endpoints	The two vertices connected by an edge
Loop	An edge connecting a vertex to itself
Neighbors / Neighborhood	The vertices connected to a specific vertex
Directed graph	A graph comprising directed edges
Undirected graph	A graph comprising undirected edges
Multigraph	A graph allowing multiple edges between two vertices

Comparing Online and Offline Computations

Chapter 1 discussed Giraph with respect to graph databases and other tools for graph processing. You have seen Giraph positioned as a tool for graph analytics, and the differences between offline graph analytics and online interactive queries. This section digs deeper into these concepts. To better understand the differences between the two workloads, you explore an analogy with search in a filesystem with the Unix commands find and locate. After explaining the filesystem scenario, this section maps the concepts to Giraph and graph analytics.

One of the actions you probably perform often when using a computer is searching for files. For example, you may look for all the files that have a name starting with a particular prefix, or those created after a certain date. This does not mean a search based on the content of the file, which is performed differently. If you are a Unix users who uses the command line to perform searches, you must know the find command. find traverses a filesystem tree starting from a particular directory and returns all the file names that match a given pattern. To perform the pattern matching, find must visit each file and directory under the starting root and apply the matching function. This can be a costly operation, in particular because it may create extensive IO. Moreover, if you perform two or more consecutive searches, find must perform everything from scratch each time.

To avoid redoing all the work for each search, the locate command relies on its own database. It can be used like find to search for files matching a particular pattern. The database stores all the metadata information about the files and directories in the filesystem in a format that makes searches efficient. As files are created, deleted, and updated, periodically locate.updatedb is run in the background to scan the filesystem and update the locate database. There is clearly a sweet spot between how often to run the expensive indexing procedure in the background, the rate of file changes, and how often searches are performed. You do not want to re-index the database more often than the files are updated or searched, and you do not want the database to contain stale metadata. For this reason, if you work with few files in your home directory and modify them often, it makes more sense to rely on the find command and search only in the home directory. Figure 2-9 shows the differences between the two commands.

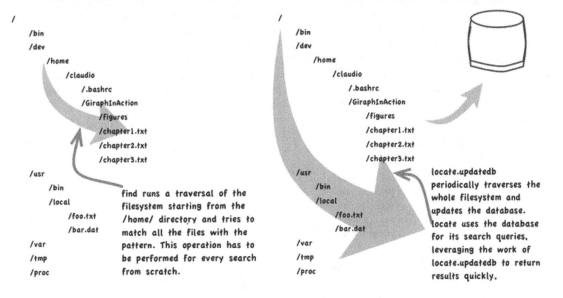

Figure 2-9. Searches in a filesystem with `find` *and* `locate`

The point is that for certain operations, you do not want to wait a long time before you get results. Such applications typically interact with the user or with other applications that need to make decisions in a short amount of time. As in the previous example, traversing an entire filesystem is an expensive operation because it requires processing large amounts of data and hence calls for periodic background computations aimed at making the interactive ones faster. These computations process indices and other data structures that allow the interactive operations to perform quickly. In the previous example, the `locate` command performs the interactive lookup in the database to return the list of matching files. It has also another command, `locate.updatedb`, that is run periodically and updates the database in the background. The side effect of this approach is that the application may not always have access to the latest version of the data, because the data may have changed after the databases was updated. For these cases, it is still necessary to compute the results for every query. However, to return results quickly, the query must be a fast operation that touches a small portion of the data.

In graph terms, a query to be performed quickly might be asking for the names of the friends of a particular user in a social network. It could also be something more complex that requires exploring a small portion of the graph that goes beyond the neighborhood of a vertex. Think of a dataset such as IMDb. You could ask the average age of actresses who starred in a movie set in France with Brad Pitt. Because the number of movies and actresses connected to Brad Pitt is quite small compared to the whole database of movies, this query should be performed quickly by a (graph) database. This is similar to using the `find` command in a specific small directory of the filesystem. However, if you want to ask the degree of separation between Brad Pitt and any other actor in the movie industry—supposing two actors are connected if they appeared in the same movie—this cannot be computed quickly with a database the size of IMDb. It requires exploring the entire graph and reaching every other actor in the graph, just like exploring the entire filesystem starting from the / directory. If you want to query the degree of separation between any two actors frequently, and you want the results quickly, you have to compute all the results in advance and store them in a database. Once they are computed, they can be looked up at query time. This process is similar to the functioning of the `locate` command and its background command, `locate.updatedb`.

Following this analogy, the ecosystem of data processing is divided into *online* and *offline* systems: online systems are designed to compute queries that are expected to conclude within seconds or milliseconds, and offline systems are designed to compute analyses that are expected to end, due to their size, in minutes, hours, or even days (see Figure 2-10). These systems have roles similar to the example using `find` and `locate`. Online systems are generally used to build interactive applications. Typical examples of these systems are databases like MySQL, Neo4J, and so on. They back applications that serve web applications, enterprise resource planning (ERP) systems, or any kind of software or service that can receive many short requests per second. Offline systems are generally used to compute analytics and batch computations. *Batch computations* are jobs that process batches of data without the need for human intervention. As mentioned earlier, Giraph is of this kind.

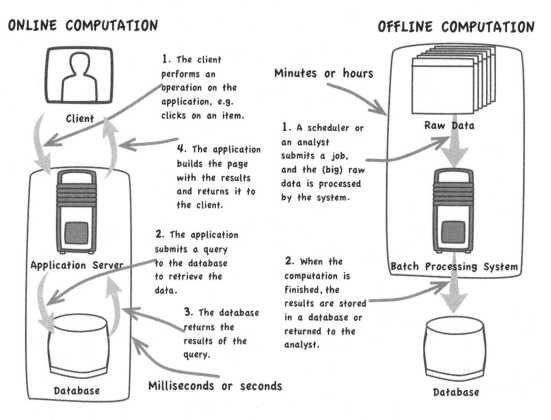

Figure 2-10. *Difference between online and offline workloads*

Now that you know the difference between online and offline systems, you are ready to see the architecture of an application that includes both and how each type of systems fits into this architecture. By looking at the architecture, you can learn how to position Giraph in the back end.

Fitting Giraph in an Application

As you saw in the previous section, online and offline systems can work together and are often both part of an application. This section presents a stereotypical architecture that is refined with more practical examples in the following sections. By the end of this section, you will learn how to position Giraph in your own architecture. Figure 2-11 shows a general architecture for an application composed of a number of different systems; let's zoom in in the components that are part of it.

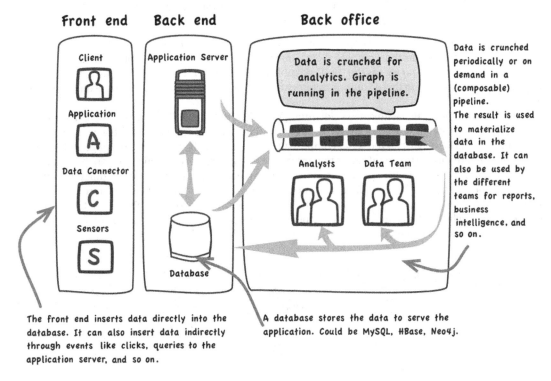

Figure 2-11. *System architecture for a stereotypical application*

The architecture is divided in three main components:

- *Front end*: The part of the system where the client interacts with the system. Depending on the client, the front end can insert data either directly into the database or indirectly by interacting with the application server, which inserts data in the database as a result. The application server inserts data into and reads data from the database to serve content to the front end. This is why a double arrow connects the database and the application server. In the front end, human clients are considered along with sensors, other applications, and so on.

- *Back end*: The part of the system where the application servers and database reside. The data and logic of the application and the online systems live here. The application servers compute the replies for client requests, depending on the type of request, the application logic, and the related data stored in the database. The back end can have multiple types of application servers and databases, depending on the number of applications running, the different architectures, and so on. The back end typically includes a plethora of different systems and technologies.

- *Back office*: The part of the system that is less interactive. It is also the component where human intervention is more involved. For example, in the back office, the teams responsible for data entry insert data into the database, and the data teams compute analytics to study the collected data and get insights about customer behavior. The offline systems reside here as well. Data collected from application logs in the form of raw logs, or stored in a structured format in the database, is crunched to compute analytics or to materialize new information to update the collections in the database. The materialized information is precisely the kind of data that is too expensive to compute online, and hence those results are computed periodically and cached in the databases in the back end.

Let's take a moment to look at an example and better understand how these three components of the architecture—and the systems that compose them—interact. The example uses a simple application for the sake of presentation; it is by no means the "best" possible architecture for this application, so bear with us. The definitions and borders in these kinds of architectures are also a bit fuzzy, so concentrate on the bigger picture.

Imagine a user interacting with a news site through her mobile phone. When the browser loads the homepage for the first time, it receives an HTML page, including a JavaScript script from the web server. The browser, the HTML page, and the script compose the front end. As the user scrolls the page and clicks news items, the script in the front end issues requests to the application server. Depending on the requests resulting from the user actions, the application server retrieves data from the database and different content stores: the content of the article and related multimedia. It then sends the data back to the browser, which updates the user interface so it displays the news items. The application servers, database and data stores, and, eventually, caches and message brokers compose the back end and run on the servers of the news site. Figure 2-12 shows the interaction between the front end and the back end.

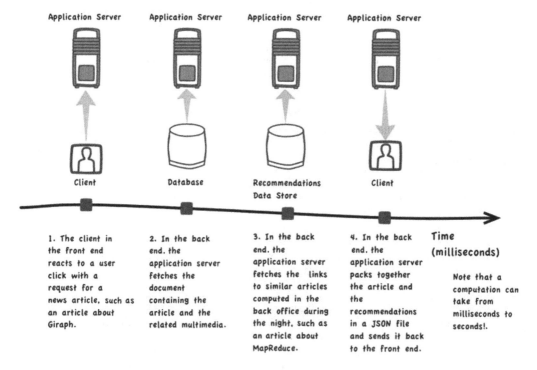

Figure 2-12. Low-latency interaction between front end and back end

When the browser requests the news items the client has clicked, the application server fetches and sends, along with the news items, recommendations for similar news articles. This feature helps the user find similar content and surf the news site. Finding similar news articles is excessively computationally expensive to be computed online at each client request, so recommendations are computed periodically offline (perhaps at night) and stored in a dedicated data store. The recommender system that computes the "similar articles" feature is in the back office, along with operations related to the providers of the news content, such as journalists and editors. To find similar articles, the system looks at the application server logs, using the "those who read this article also read those articles" paradigm (processing the clicks in the logs). Hence, the back office is connected to the back end to retrieve user-behavior data to compute the recommendations, and to store the recommendations to be served to the user. Figure 2-13 shows interaction between the back end and the back office. Note that interaction between the front end and the back end, and interaction between the back end and the back office, happen independently and at different time scales.

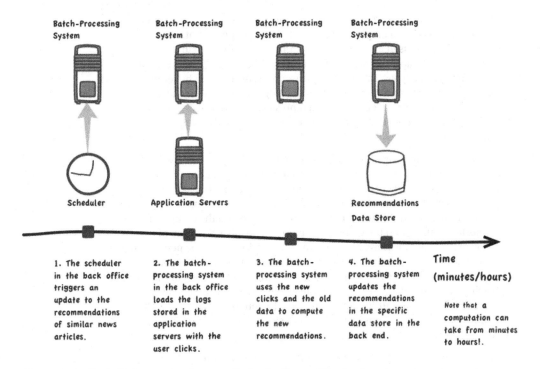

Figure 2-13. *Periodic interaction between the back office and the back end*

Giraph is positioned in the back office. It can be part of a number of pipelines, assuming the application is involved with graph data. A *pipeline* is nothing more than a sequence of jobs, where the output of the previous job acts as the input to the next one. Different pipelines can be computed in parallel or sequentially, depending on the workflows and the requests of the different teams. In fact, there are multiple pipelines of batch jobs executed in the back office, some to support certain application features, and others, for example, to support analytical queries from data scientists. A lot depends on the kind of results demanded at any given moment and, of course, on the applications. However, a general pattern—a common set of steps—can be identified. The following sections look at these steps in the context of real-world scenarios.

Figure 2-14 shows this pipeline where graph data is involved. The pipeline consists of the following steps.

1. *Input*: Data is collected from different sources. It is stored in different formats, sometimes following structured schemes, sometimes in simple formats such as raw text. This step of the pipeline can be implemented by storing data in a database or log files in a filesystem. In large architectures, it may require the deployment of a distributed system to collect logs streaming in real time from the application servers. In the "similar news articles" application, the inputs can be the articles themselves (for example, if articles are recommended because they have similar content), click logs coming from the application servers (if clicks are used to find similar articles), or additional metadata provided by the content provider (article categories, tags, and so on).

2. *Construction*: Data is collected in various formats and schemes, and it has not yet been processed to represent a graph. This data needs to be filtered, transformed, and aggregated until a graph is extracted. This step depends both on the type of data to be processed and on the algorithms to be computed in the following steps. Different graph algorithms often require different graph representations and different data types. In the news site example, articles can be represented by vertices, and they can be connected with an edge any time a user clicks from one article to another. The edge weight can represent the (normalized) number of times this has happened. A threshold can be applied to edge weights to avoid the graph becoming too dense.

3. *Computations*: Once the graph is constructed, it can be processed with Giraph. In this phase, one or multiple jobs are executed. The type and number of graph computations strongly depend on the mining application. Usually, graph algorithms are layered and built on top of each other. For example, some ranking algorithms use centrality measures, and those depend on the computation of shortest paths. Often the graph computed by one job is the input of the following one. For the news articles, two computations could be performed on the constructed graph. To compute the most popular articles, you could run PageRank on the clicks graph and use the rankings to sort the recommendations (such as most popular first). To compute which articles are related to another, you could run a community-detection algorithm on the clicks graph. Articles falling in the same community could be considered related and hence be recommended. You see both algorithms in chapter 4.

4. *Fusion*: The results of the computations performed in the previous step are transformed and joined or fusioned with other data, possibly coming from the database or from other pipelines. Keep in mind that graph computations are often part of a bigger application and hence may be solving only a sub-problem. Moreover, this step depends on the type of analytics being performed and on the type of materialization necessary in the databases. Fusion may actually require multiple steps, because the results of the computation may be reused for different reports and databases. For example, the rankings of the articles and their category computed through the community-detection algorithm could be joined along with the article metadata to prepare the injection into the recommendations data store.

5. *Output*: Once the data has been fusioned and packaged into a report or a materialized view, it is sent to the teams or injected into the respective databases. In the case of a report, the output can be a summarized view of the results; or, in the case of a materialized view, the output can be a data table to be injected into a database. In the latter case, the output phase can produce consistent load on the databases where the data is injected. For this reason, the injected data should be in a format that does not require further expensive manipulation by the database system. In the news article example, data can be injected into the recommendations data store following the schema of the store.

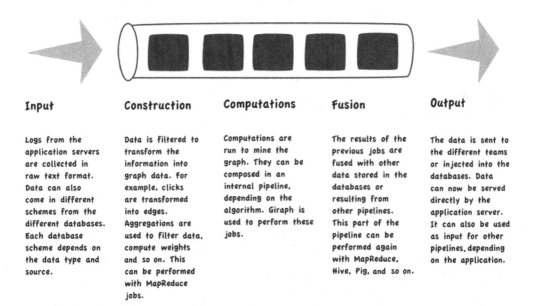

Input	Construction	Computations	Fusion	Output
Logs from the application servers are collected in raw text format. Data can also come in different schemes from the different databases. Each database scheme depends on the data type and source.	Data is filtered to transform the information into graph data. For example, clicks are transformed into edges. Aggregations are used to filter data, compute weights and so on. This can be performed with MapReduce jobs.	Computations are run to mine the graph. They can be composed in an internal pipeline, depending on the algorithm. Giraph is used to perform these jobs.	The results of the previous jobs are fused with other data stored in the databases or resulting from other pipelines. This part of the pipeline can be performed again with MapReduce, Hive, Pig, and so on.	The data is sent to the different teams or injected into the databases. Data can now be served directly by the application server. It can also be used as input for other pipelines, depending on the application.

Figure 2-14. *Data-processing pipeline in the back office*

Now that you have learned how Giraph fits in a system architecture, the following sections look at specialized architectures and pipelines for various real-world scenarios.

Giraph at a Web-Search Company

What is the core product of a company working in the web-search business? Its search engine. In the most general terms, providing results for web searches—searching the Web for specific content, as specified by the user through a query. However, when a user searches for pages through a query, the search engine does not surf the Web looking for pages that match that query. The Web is huge, and it would take days or weeks for such a search to return meaningful results. Instead, the search engine contains an index of the content of (a portion of) the Web, which is used to answer queries in a reasonable time (milliseconds). The problem is very similar to the one of the find and locate commands. In fact, it works analogously to locate. A search engine consists of a program that periodically updates a "database" (the index) of the content of the Web, and a program to search the database for content that matches a query, which is used by the *search page* (like the famous Google "white page" at www.google.com). The core of a search engine is composed of two main components: the *crawler* (also called a *spider*) and the *index*. The crawler is a program that constantly surfs the Web, jumping from page to page and following the links in the HTML of each page it visits. When the crawler reaches a new page or an updated version of a page it visited in the past, it sends the page to the indexer, which is responsible for keeping the index up to date with page content. Without getting into the details of how an index works, you can think of it as a map that links words to lists of URLs of pages that contain those words. Given a specific word, the map stores the URLs and the number of occurrences of that word in those pages. Figure 2-15 shows these components.

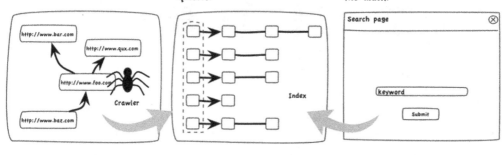

1. The crawler visits the pages on the Web, following the links in the HTML and collects the pages. This process is continuous.

2. The pages collected by the crawler are indexed by the indexer. The index can be used by the search engine to answer queries.

3. Users can submit queries through the search page. To answer the queries, the search engine looks up the keywords in the index.

Figure 2-15. *The components of a web search engine: a crawler, an index, and a search page*

To create the index, the indexer goes through each page passed to it by the crawler, divides the text in the page body into words, counts them, and updates the index accordingly. How does a search for a query in the index work—for example, a query expressed as a set of keywords? Suppose you submit the query Hello Giraph. The index searches for each specific single keyword and returns the URLs http://www.foo.com, http://www.bar.com, and http://www.baz.com, because they contain either both or only one of the keywords (see Figure 2-16). A page where the keywords appear more frequently than they do on another page is considered more relevant. The order in which search results are presented is also known as *ranking*. The search page returns the search results in this order because they are sorted by relevance (three for http://www.foo.com, two for http://www.bar.com, and one for http://www.baz.com). Keep in mind that the pure number of occurrences of the keywords on the pages is a simple metric of relevance. More sophisticated metrics consider, for example, document length and other parameters (search, for example, about TF-IDF to learn more). Indexers also use other techniques such as stemming, stop-words filtering, anchor text, and so on, but this example keeps things simple for the sake of clarity.

The index contains information about the three web pages that allows a search engine to look up keywords.

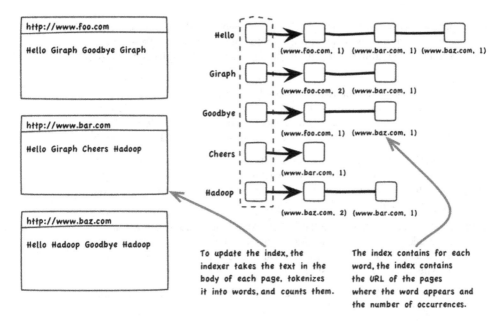

Figure 2-16. *An index generated from three web pages*

This is more or less how a very basic search engine worked before Google and its PageRank algorithm. You might have noticed that the structure of the Web is used by the crawler only to discover pages to visit; it is not used for anything else, such as computing rankings. In this search engine, rankings are computed based on the relevance of each web page. This is why search engine optimization (SEO) techniques before Google were based on polluting web pages with long lists of keywords hidden from the human user. SEO would "pump up" a page's rank for many different search queries. One of the things that made Google so successful in the beginning was its approach to ranking web pages. The PageRank algorithm uses the structure of the Web to rank pages. The intuition behind PageRank is very simple: the more pages link to a page, the higher rank the page should have. Moreover, the higher the rank of the pages that link to a page, the higher the rank of that page. It is a kind of meritocratic method to rank pages according to their popularity: if many popular pages link to a specific page, then that page must be popular and provide good content. All the necessary information lies in the link structure of the web pages. Although possible, it is more difficult to create pages with high popularity and add links from those pages to the page whose rank you want to raise.

Chapter 4 gets into the details of how PageRank works and how to implement it with Giraph. For now, consider that it is exactly the kind of graph algorithm that fits perfectly with an iterative computation with Giraph. The index can be built for an extremely large number of pages with a system like MapReduce, and the page rankings can be computed with Giraph. Both systems fit in the pipeline executed in the back office, along with the crawler. The only difference from the search engine architecture described so far would be to add a Giraph job to compute rankings. Once the rankings are computed, they can be used, together with the relevance of the pages, to decide the order in which the pages are presented to the user as search results. Today, many more metrics are added and considered by search engines like Google's to compute page rankings; for example, consider social media, click statistics, and user profiles. However, the PageRank algorithm shows how to look at data as a graph and at a solution as a graph algorithm.

Now, let's look at the pipeline in Figure 2-14 for the case of computing PageRank for a search engine with Giraph:

- *Input*: Input comes directly from the crawler, so it depends on the way the crawler stores the pages it has fetched so far. If you want to update the current PageRank instead of computing it from scratch, the current values for the graph can be used as input as well.

- *Construction*: The crawler can store the link graph separately from the pages as it parses it, or it may need to go through the pages and extract the links in the HTML to build the graph. URLs must be filtered out (if illegal), normalized, and de-duplicated. Often, URLs are transformed, with the domain converted to reverse-domain notation.

- *Computations*: Once the graph is transformed, it can be loaded into Giraph and the computation can start. During this phase, the PageRank is computed for the loaded graph. Other computations can also be performed on the graph, such as computing related pages through other specific graph algorithms.

- *Fusion*: The index was previously built through a MapReduce job. You may want to add the PageRank directly to the entries for each URL in the index. This would let you fetch URLs at query time, along with relevance and PageRank values, to compute the order in which to present the results.

- *Output*: The updated index data with the PageRank values is injected into the index store so it can be used directly by the search page.

This is just an example of a processing pipeline, and it depends on the features supported by the search engine. The page indexing, supposedly performed through MapReduce, can be considered part of this pipeline, executed right before the PageRank computation, or part of its own pipeline. Now that you have seen how Giraph can help a web-search company, let's move to another scenario.

Giraph at an E-Commerce Company

Users generally do two types of things on an e-commerce site: search for something specific they want, which probably brought them there in the first place; or click around and stumble onto something that triggers their attention and that they might decide to buy. Think about one of your browsing sessions on Amazon. To a certain extent, the infrastructure to support the first type of usage is similar to that of a web-search company. To support search, product data is indexed so that it can be found through queries. But because the data is already stored in the company's databases, it does not need to be discovered through a crawler. The moment the data is inserted into the site, it is processed and added to the index directly. The second type of usage is more similar to surfing the Web. From link to link, the user explores the available products on the site, looking for something they like. This means the links on a product page must be carefully selected to stimulate a user click and increase the likelihood of the user finding something interesting. But what makes a link a good link for a product page? What link would be effective on the page for *The Dark Side of the Moon*"? Other records by Pink Floyd, probably, or something that fits the taste of a fan of that record. And what is the taste of such a fan? It is complex to delineate, and it depends on different factors.

Recommender systems are designed to recommend items to users (and sometimes users to users), such as a product that might be interesting to them. In the current scenario, an effective recommender system generates links for a given product so a user can easily find something interesting without needing to search for it explicitly. Although this task can be performed by humans, such as expert clerks who place similar records close to each other on shelves in a store, it can be effective to let a computer program perform it: with millions of products, having humans organize recommendations does not scale; and a data-driven approach should produce more unbiased results.

What does all this have to do with Giraph? It turns out that e-commerce site data can be modeled through graphs. An e-commerce site has two types of entities: products and customers (or users). You can model both of them with a vertex. Technically, they can be represented by two types of vertices, one for each type of entity (for example, visualizing users with circles and items with squares). Edges connect vertices when customers buy products. Because edges always connect two vertices of different types, this type of graph is also known as a *bipartite graph*. Figure 2-17 shows an example. Sites can also support product ratings—for example, using stars. Ratings can be easily modeled with edge weights, where the weight is set to the rating given to a product by a customer. This graph data can be analyzed following the popular "people who bought this item also bought these other items" pattern. By performing this analysis, you can automatically discover the profiles of buyers of certain products, by using customers' purchase history. You can also answer the question about which items might be interesting to a customer of *The Dark Side of the Moon*. This profile data is often referred as a *latent* profile, because it is not directly observed but rather is inferred from user behavior. This approach is also known as *collaborative filtering*, because users collaborate (implicitly) in building filters that organize the item-recommendation database.

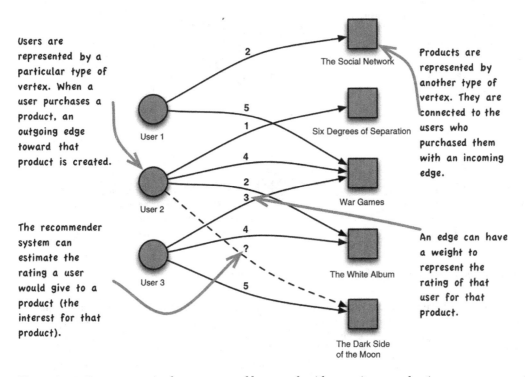

Figure 2-17. *E-commerce site data represented by a graph with users, items, and ratings*

Collaborative filtering is an effective technique to build recommender systems. However, it is not the only one. *Content-based* recommenders follow a different approach. Instead of looking (only) at purchase history, you can recommend products based on qualitative evidence. For example, imagine having metadata about music albums such as genre, year of release, place of recording, mood, style, tags, and so on. Based on this metadata, you could define a similarity metric to measure how similar two albums are— for example, whether they belong to the same genre and were recorded in the same period and country. Once the similarity between items has been established, you can build a graph in which similar products are connected by an edge and the weight of the edge encodes how similar those items are. Figure 2-18 shows an example of such a graph. With this information, if a customer buys *The Dark Side of the Moon*, the system can recommend other albums that have similar metadata (other psychedelic rock albums from the 1970s), maybe also considering other products the customer purchased in the past. This approach is similar to having the aforementioned expert clerk organizing the shelves of the shop for you, but in a more personalized and scalable way. Depending on the type of products, content-based similarity can be computed even without metadata. For years, scientists have been researching methods to recognize similarities between books by analyzing their text, songs by analyzing their audio signal, and images by analyzing shapes and colors. However, this goes beyond the scope of large-scale graph processing.

Categories

▶ Books
▼ Movies
    ▶ Comedy
    ▼ Documentaries
        ▶ The Beatles Anthology
        ▶ The Pink Floyd Live at Pompeii
    ▶ Drama
    ▶ Romance
    ▼ Sci-Fi
        ▶ 2001: A Space Odyssey
        ▶ World War Z
▼ Music
    ▶ Blues
    ▶ Classical
    ▶ Jazz
    ▶ Pop
    ▶ Reggae
        ▶ The Dub Side of the Moon
    ▼ Rock
        ▶ The Dark Side of the Moon
        ▶ The White Album

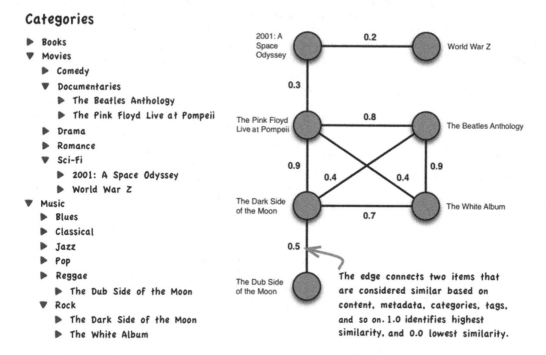

Figure 2-18. *E-commerce site data represented by a graph from metadata*

Both approaches have pros and cons. The advantage of collaborative filtering is that it does not require any type of additional data other than the purchase history. Moreover, it can compute recommendations across different genres of products. For example, it can recommend books based on purchase data about movies and music albums (maybe people who buy albums of metal rock music also buy black t-shirts). On the other hand, content-based recommendations do not suffer from the so-called "bootstrap problem," because items can be recommended from day one without the need for historical data. Often, the two approaches are combined, with content-based recommendations used in the early days and the system gradually migrated to collaborative filtering. This book is not dedicated to building recommender systems; other books cover that topic. The point is that it is possible (and often desirable) to model e-commerce data with graphs and recommendation algorithms as algorithms that explore them. These graphs can get very large, so you can use Giraph to compute graph algorithms to build a recommender system for an e-commerce site. Chapter 4 presents an algorithm that serves this purpose; you can use it as a building block for your particular data model and use-case scenario.

A recommender is only an example of how Giraph can be used to analyze data from an e-commerce site. Analyzing customer behavior can give you general information about the effectiveness of marketing campaigns or site organization, among other things.

In general, the processing pipeline for an e-commerce site may look like the following:

- *Input*: The input comes from historical purchase data, from raw web server logs (such as for the click graph), from a structured database in the case of content-based recommendations, and so on.

- *Construction*: The graph can be constructed by using products and customers as vertices and connecting them with purchase data or with click information. Usually, data is normalized to create weights—for example, the number of clicks on each page normalized on the total number of clicks. The similarity between items can also be computed according to different criteria—for example, through non-graphical models via MapReduce jobs or Apache Mahout.

- *Computations*: Once the graph is transformed, it can be loaded into Giraph and the algorithms can be computed. In the case of recommendation algorithms, the latent profile for each user is computed. These latent profiles can be used to compute predictions of ratings and recommendations in another step of the pipeline, perhaps through MapReduce.

- *Fusion*: The latent profiles and recommendations can be fused with structured data to produce new product pages with updated links to similar products. They can also be loaded into the database or personalized recommendations in each user's home page. If recommendations are computed online based on latent profiles computed offline, the profile data is prepared to be loaded into the online system.

- *Output*: Recommendations are sent to the recommender data store, and web pages are updated with new links.

You have now seen how Giraph can fit into an architecture to support the workload of an e-commerce site, in particular with an eye to building a recommender system. The next section focuses on a final scenario: an online social networking company.

Giraph at an Online Social Networking Company

One of the reasons for the success of social networking sites is that they allow us to connect with people and share information about what we do and what we like. The reason why we connect with people on these sites depends on each site. For example, on Facebook, users tend to connect with people they already know in real life, whereas on Twitter they connect with people they are interested in, and on Flickr they connect with photographers whose work they like. Regardless of the reason, the connections created with other individuals generate an interesting social fabric. Representing these networks as graphs and processing them through graph algorithms allows you to study the underlying social dynamics reflected in the graph.

Facebook published an image in December 2010.[1] Although the image resembles a satellite picture of the world at night (and that is what makes it so striking!), it is not. The picture presents a portion of the Facebook social graph. To draw this image, Paul Butler analyzed the friendship relationships of the users of Facebook and visualized the data according to the following criteria:

- For each pair of cities, Paul counted the number of friends between the two cities and connected the two cities with an edge.

- For each edge, he computed a weight as a function of the distance between the two cities and the number of friends between them.

---

[1]Paul Butler, "Visualizing Friendships," Facebook, `www.facebook.com/note.php?note_id=469716398919`.

- He placed the edges according to the geospatial information of the cities they connected. In other words, he placed each vertex (a city) according to its position on the map, but without visualizing it, hence visualizing only edges.

- To choose the color of the edges, he chose a shade from white to blue, with the highest weight mapped to the white color.

- Long edges, like those connecting cities in different continents, were drawn with arcs around the world, to minimize overlapping and increase readability.

This example clearly shows that individuals on Facebook are connected by geographical relationships, because people connect on Facebook mostly with people they hang out with in real life (or used to). The description of the criteria used to analyze the data and compute the edge weights for the visualization is in practice a graph algorithm.

One of the most interesting properties of social networks is their division into *communities*. In social network analysis (SNA), a community is a group of individuals who are tightly connected with each other more than they are connected with individuals outside of that community. Intuitively, you can think of a community as a *cluster*. The graph representing a social network has particular properties due to this division into communities, and graph algorithms can take advantage of these properties to detect communities. Figure 2-19 shows a portion of one of the authors' communities, extracted from his LinkedIn account and visualized through the LinkedIn Maps tool.[2] Here, each vertex represents a user on LinkedIn, and an edge connects two vertices if the two users they stand for are connected on LinkedIn. A community-detection algorithm was run on this graph, and the vertices were colored according to the community the algorithm assigned them to. Note that the algorithm did not use the information contained in each user's profile, such as past jobs, to detect communities. Instead, it used only the information contained in the graph about how users were connected. The algorithm successfully detected communities: each color is indeed mapped to one of the present or past communities the author is a member of, such as his previous and current co-workers, the Apache Giraph community, and so on. Chapter 4 presents an algorithm to detect communities with Giraph. Interestingly, the graph-layout algorithm used to draw the graph is also a graph algorithm. The graph-layout algorithm computes the position of the vertices in two-dimensional space such that connected vertices appear nearby and edge crossings are minimized (for readability). The graph-layout algorithm uses only the connections in the graph, and the fact that vertices with the same color (same community) appear together is another sign of the strong community structure in the graph.

---

[2]http://inmaps.linkedinlabs.com.

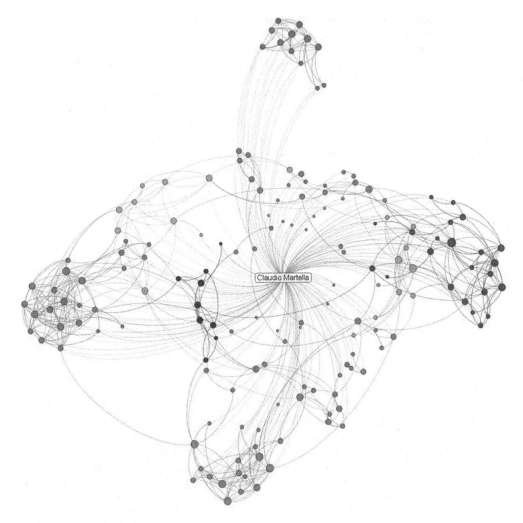

Figure 2-19. *Communities in the social network of a LinkedIn user*

The definition of *community* is not strict and not unique. Many real-world communities overlap, because each of us belongs to multiple communities at the same time. For example, on Facebook you may be connected to your school friends but also to your co-workers, your football teammates, and so on. Moreover, communities are often organized in *hierarchies*. You may be a member of your school community but also of the inner communities of your particular class, the school band, the basketball team you were a member of, and so on. This means a community-detection algorithm has to be tuned to the particular use case.

Why would a social networking site be interested in detecting communities? First, to study its users. By studying the communities of a social network, it is possible to understand how the users are organized and connected. This lets you target specific group behaviors and analyze the effectiveness of specific features. A site can also support features such as "people you may know." For example, Figure 2-20 presents a community of friends. By using the friendship relationships, the site can help users connect with old friends. But communities are not only useful for social networking sites. Think of the recommender system for the e-commerce site in the previous section. The algorithm presented in this book to compute recommendations generates profiles about customers and items that allow you to make predictions of ratings. It basically creates a function that, given a

39

customer and any item, predicts the rating the customer will give to that item based on past ratings. What is still required is the "matchmaking"—finding the items that are predicted to receive a high rating. The naive solution to this problem is to match the profile of a customer with all unrated items and keep the ones with the highest rating. However, this solution would mean matching every customer with every item, which would require a massive amount of computation and would be unfeasible. Another solution is to look into a customer's communities and match their profile only with items in those communities. Customer-item graphs, such as those presented in the previous section, tend to be organized into communities, usually delineating genres or tastes. It is very likely that people who like rock music will buy similar albums. This community structure tends to be even stronger if customers are allowed to connect on the site by creating an explicit social network. Using a community-detection algorithm on the customer-item graph is, in effect, a clustering computation.

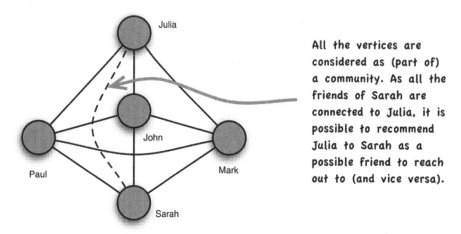

All the vertices are considered as (part of) a community. As all the friends of Sarah are connected to Julia, it is possible to recommend Julia to Sarah as a possible friend to reach out to (and vice versa).

Figure 2-20. *Recommendation of a friend based on a community structure*

Rankings can also be applied to social networks. Think of Twitter. On social media sites, users are usually ranked by their *influence*. Intuitively, a user with high influence is a user whose actions and behaviors reach deeply into the graph, effectively increasing the likelihood of influencing other users. Users with high influence are called *influencers*. On Twitter, influencers can be computed by looking at how their tweets spread across the network of followers, and by looking at actions such as retweets and mentions. You may want to give priority to influencers in the list of recommended individuals to follow and give their tweets a higher probability of being shown on a user timeline. Usually, algorithms to compute influencers use the same paradigm of PageRank (one of these algorithms is called TweetRank), but they also consider temporal aspects, such as how quickly tweets are spread.

Giraph fits in the back office of a social networking site in the pipeline that analyzes the social network. The different steps could be as follows:

- *Input*: The input comes from the different data stores where the social relationships and profiles (gender, age, country, and so on) are stored, and from raw web server logs (such as for the click graph). Which data is used depends on the type of application.

- *Construction*: Selecting a portion of the graph filtered depending on certain types of relationships and profile characteristics is often called a *projection* of the graph. For example, the social graph of male professionals living in the United States is a projection of the LinkedIn graph. The projection of the graph is constructed by filtering out unwanted data and normalizing edge weights. Often, different projections must be computed for different algorithms. Other times, different projects have to be merged in later steps of the pipeline.

- *Computations*: The graph is used to compute communities, rankings, influencers, and so on. The different computations are executed and pipelined according to the semantics of the analysis; for example, rankings can be used to compute influencers. At this point, for summarization and reports, data can be aggregated according to criteria such as membership in a community or profile data—for example, the average number of connections for male professionals in the US. Aggregations can be computed directly in Giraph, through tools like Hive, or directly through MapReduce.

- *Fusion*: The results of the computations are used to build the materialization data to be injected into the stores. Communities and rankings are used to build friend recommendations, and influencer rankings are used to recommend data items, depending on the sources.

- *Output*: Materialized data items are inserted into the databases or sent to the analytics teams.

Again, these are just some examples of features of an online social networking application. Through the examples provided so far, you have learned how to position Giraph in your architecture and how to integrate it with your existing data-processing pipeline.

Summary

Graphs are everywhere, and you can use them to describe many things in many different domains. By looking at data and problems through graphs, you can gain a better picture of your applications and products. Giraph helps you process this data at scale, as part of your application architecture.

In this chapter, you learned the following:

- Graphs are very simple structures, and they can be used to describe a number of different things.

- Depending on the type of data, you can use labels, directed edges, weights, and so on to represent specific features of your data.

- Computations that are computationally expensive but need to return results within milliseconds can be performed with a combination of online and offline computations.

- Giraph is an offline graph-processing engine that sits in the back office of an application architecture.

- You can use Giraph to compute rankings, recommendations, communities, and so on in cooperation with the other systems in your architecture.

Now that you have seen how to model your data with graphs and how to fit Giraph into your application, you are ready to look at how to program Giraph. The next chapter discusses the programming model and the API.

CHAPTER 3

■ ■ ■

The Giraph Programming Model

This chapter covers

- Giraph design goals for graph processing

- The vertex-centric API

- Using combiners to minimize communication

- Aggregations through aggregators

- The bulk synchronous parallel model

This chapter digs into the nature of graphs and graph algorithms, and how graph algorithms can be implemented and computed with Giraph. You learn how graph problems are inherently recursive and why graph algorithms therefore are usually solved iteratively. You see how Giraph is designed for iterative graph computations and explore the vertex-centric programming API and paradigm of Giraph. You then look at examples of simple algorithms to get acquainted with the model. The chapter concludes by opening the hood of Giraph to examine the underlying distributed engine that makes iterative computations so fast and simple.

Simplifying Large-Scale Graph Processing

Traditionally, graph algorithms have been designed following the model of sequential programming we are all accustomed to. The graph is represented in main memory with native data structures such as a matrix or lists. The algorithms assume a global view of the graph and a single thread of execution. Both the data structures and the execution logic are tailored to the solution. This approach has a number of drawbacks. First, a new graph problem brings a new graph algorithm, and probably with it a new approach and model redesigned from scratch. A tailored, ad hoc solution allows for fine-grained optimizations, but it requires practitioners to partially reinvent the wheel every time they implement a new algorithm in a system. Second, algorithms need to be specifically designed or modified to run in a parallel, distributed system. Again, this allows for fine-grained optimizations that exploit a particular platform, but it requires reinventing the wheel every time a new algorithm is implemented for a parallel, distributed system. Plus, it is nontrivial to parallelize a graph algorithm, because graph computations are unstructured and interleaved.

Giraph tackles both problems by providing a programming model that has been designed with graph algorithms in mind and that at the same time hides the complexity of programming a parallel, distributed system. Both characteristics minimize the effort of implementing a graph algorithm that works at large scale. The following two sections cover these characteristics.

Hiding the Complexity of Parallel, Distributed Computing

Giraph offers you much more than a library for executing graph computations; it offers a programming model and an API that let you focus on the semantics of the specific algorithm you are designing without having to worry about how the graph is stored in memory, the way the algorithm is executed, or how the computation is distributed across machines with tolerance to faults. Basically, the way the graph is represented in memory and the execution engine is general to any graph algorithm that can be expressed with Giraph. Practically, you have to write a user-defined function (UDF) that is executed iteratively on each vertex by the Giraph runtime across the processing units. This UDF is agnostic to the way data is shared across these units and the way code is executed concurrently. In other words, no locking or concurrent coordination is required on your side. As far as you know, your UDF is executed sequentially on a single unit. Figure 3-1 shows this conceptual stacked organization.

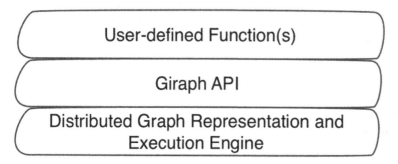

Figure 3-1. *Conceptual organization of an application in Giraph*

■ **Important**   According to the Giraph programming model, you have to develop a UDF that is executed iteratively on the vertices in the graph. You are agnostic of the way the graph is represented in memory, the way the function is executed in parallel across the distributed system, and how fault-tolerance is guaranteed.

The UDF defines how each vertex manages the messages it receives to update its value, and what messages it sends to what other vertices. Because vertices share data through messages, no locking is required. Also, because each vertex is executed at most once during each iteration, there is no need for explicit synchronization by the user. This means you have to focus only on how to express an algorithm from the perspective of being a vertex that can exchange messages with other vertices in a number of iterations. This is why the programming model is usually referred to as a *vertex-centric* paradigm. Although it is more restrictive, this model guides you toward developing algorithms that can reach massive parallelization. Chapter 4 develops this idea further.

Programming through a Graph-Specific Model Based on Iterations

The previous chapter presented graphs and how they can be used to model data in different domains. Now, let's have a quick look again at graphs, focusing on how they shape the programming model used to express graph algorithms. Graphs are characterized by a few concepts, making them very simple to understand. You have "just" a set of vertices with a set of edges connecting them, potentially with direction, label, and weight. There is nothing more to it. With a combination of these concepts, you can model pretty much anything you can think of. But simplicity comes at a price. The problem is that the information about each vertex is contained not only in its adjacent vertices and in the labels and weights assigned to the edges that connect them, but also in vertices that are farther away. Information about each vertex is often distributed all over the graph, making graphs complex and expensive to manage.

Let's look at an example of how you can gain more information about a vertex by looking further than its direct neighbors. In Figure 3-2, the leftmost circle represents the simplest graph of all (except, perhaps, an empty one). It consists of a single vertex. What does that graph tell you? As is, not much. But if you add a label to the vertex and connect it to a bunch of other vertices with labeled edges, you get a better picture. You realize that the graph contains information about Mark. He works at Banana Inc., he is in a relationship with Anna, and he lives in Berlin. Now look at graph on the right. It add a vertex representing Germany and connects it to the vertex representing Berlin, with an edge labeled "capital of".

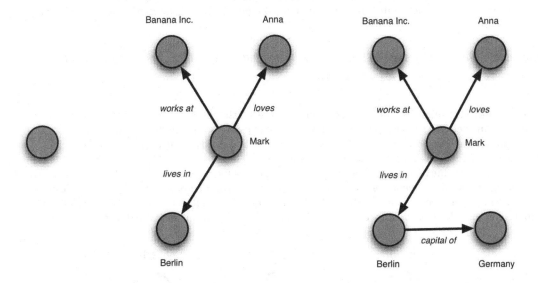

Figure 3-2. *The definition of a vertex through its neighbors*

The interesting thing here is that by adding a vertex to the graph and connecting it to a vertex that is *not* Mark, you find out more *about* Mark: that Mark lives in Germany. Moreover, assuming for this example that Mark is not telecommuting, you can infer that Banana Inc. has an office in Berlin. The more information is contained in the neighbors, either direct or multi-hops away, the more you can deduce about a vertex.

In other words, the information about a vertex depends on the information about the vertices in the neighborhood. Naturally, those vertices also depend on their neighborhood. This introduces a recursion in the information about each vertex, where each vertex depends on other vertices. That is why graphs are tough. *Each vertex depends on its neighbors.* Mark is defined by the vertices Banana Inc., Anna, and Berlin, but Berlin is defined by vertex Germany, making Mark also defined indirectly by vertex Germany.

Let's look at another example. It is commonly said that in order to judge a person, you have to look at her friends. But you may also say that to judge her friends, you have to look at *their* friends. As before, you end up having to look at the entire graph. Graph problems are often defined, in one form or another, through dependencies between vertices and their neighbors. Unfolding these dependencies is often what makes graph computations complex and expensive.

■ **Important**   The information about a vertex depends on the information about its neighbors. This makes graph problems *recursive*.

Fortunately, you know that these types of problems can be solved *iteratively*. Iteration after iteration, an iterative algorithm unfolds these dependencies one level at a time. For this reason, graph algorithms are often expressed as iterative algorithms, where some of the vertices are visited (once or multiple times) during each iteration. As the computation proceeds and intermediate states are computed, information about each vertex is updated in the face of updated intermediate state until the final results are computed. This is why the Giraph programming model is designed to express iterative algorithms and the Giraph execution engine is optimized for iterative computations. *You have to think in terms of iterative algorithms.*

■ **Important**   *Graph problems can be solved nicely through iterative algorithms. This is why Giraph is designed and optimized for iterative computations.*

Let's look at an example to clarify these concepts. Figure 3-3 shows a social graph with a number of people are connected by a friendship relationship. Imagine that they want to find out who is the oldest in the graph, but they can only communicate between friends. You can assign to each vertex a value that you initialize with the age of the person, and define the problem recursively by defining the value of each vertex as the largest value between its own value and the values of the neighbors. This definition is recursive because each vertex value depends on the value of the neighbors. This recursive definition works because the oldest person will affect the value of their friends, which in turn will affect the value of their friends, so that in the end the value of each individual depends on the age of the oldest person in the graph.

Figure 3-3. *A social network of four individuals connected by a friendship relationship*

But how do you solve this problem? You organize the computation in a series of iterations, where during each iteration each vertex sets its own value to the largest between its current value and the value of its neighbors. Iteration after iteration, the age of the oldest person flows through the graph until it reaches all the vertices. If at any iteration no vertex updates its value, the computation has reached its final iteration, and each vertex has in its value the age of the oldest person in the graph. Figure 3-4 shows the execution of this example. Note how the Carla's age reaches first Anna, then John, and ultimately Mark, iteration after iteration. Also, initially the Mark vertex updates its value based on John's and later updates it again, after the John vertex updates its value. The next section looks at how this maps to what is usually referred to as the *vertex-centric* approach of Giraph.

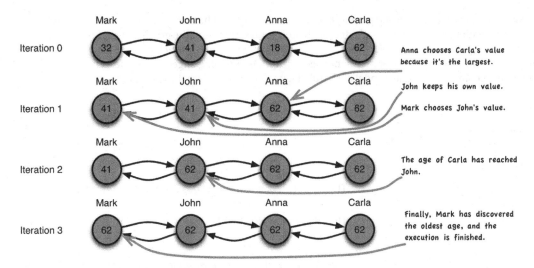

Figure 3-4. *Execution of the algorithm that finds the oldest age (maximum value) in the graph*

A Vertex-centric Perspective

According to the Giraph programming model and API, you have to put yourself in the shoes of being a vertex that has a value and that can exchange messages with other vertices in a number of iterations. This section presents how this works. To simplify the presentation and fit in as many examples as possible, we use pseudo-code instead of Java (in which Giraph is written and can be programmed). Starting in Chapter 5, after all the concepts are clear and you are familiar with the paradigm, the book's examples use Java code with the actual Giraph API. The two APIs (pseudo-code and Java) match perfectly.

THE USE OF PSEUDO-CODE IN THIS BOOK

This chapter and the next use a pseudo-code language to present the Giraph API and the implementation of the algorithms. This allows you to focus on the programming model without thinking about the particularities of the (Java) language. Also, because the pseudo-code is much less verbose than Java, these chapters can cover much more material.

The language is heavily inspired by Python, so if you know that language you won't have problems understanding the code. Most of the API uses the same naming as in the official Java API so you can easily move from the content learned here to the content present in Part 2 and Part 3, which present Java code.

The Giraph Data Model

This section presents the Giraph data model: the way a graph is represented. Look back at the example in Figure 3-4. Each person was represented with a vertex, and two vertices were connected if the people knew each other. This is how you build a social graph. You can use the name of a person as the identifier—for example, with a string. Each person also knows the friends they are connected to, so the vertex has outgoing edges. Keep in mind that with certain graphs, like street maps, you can have also values assigned to the edges. Vertices with their identifiers, and edges with their values, represent the graph. For the specific problem of finding the highest age, you needed to store an integer value with each person. This value was modified each time a vertex discovered a higher age. Vertex IDs, outgoing edges with values, and vertex values are the elements of the data model of Giraph. Figure 3-5 summarizes these elements.

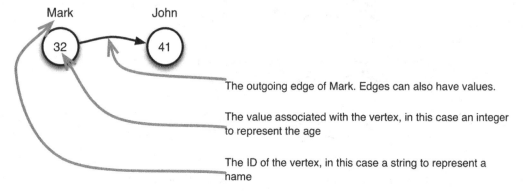

Figure 3-5. *The Giraph data model, with vertices, IDs, values, and edges*

A bit more rigorously, according to Giraph, a graph is a set of vertices and edges, and each vertex is defined as follows:

- It has a unique ID, defined by a type (an integer, a string, and so on)

- It has a value, also defined by a type (a double, an integer, and so on)

- It has a number of outgoing edges that point to other vertices.

- Each edge can have a value, also defined by a type (an integer, a double, and so on).

The first important thing to notice from this list is that the data model is a directed graph and edges are assigned to their source vertex. In principle, vertices are aware only of their outgoing edges, and if an algorithm needs to know the incoming ones, it must discover them as part of the algorithm. The section "Converting a Directed Graph to Undirected" presents an algorithm to do this. A second important thing to notice is that for each element (vertex ID and value, edge value), you have to define a type. This type can be either a primitive type, like an integer or a double, or a composite type, like a class.

Although vertex IDs depend on the graph—for example, a web graph has vertex IDs characterized by URLs (hence strings)—vertex values are often dependent on the algorithm. For example, shortest-distances algorithms define vertex values as integers for unweighted graphs and as doubles for weighted graphs, PageRank defines vertex values as doubles, and recommendation algorithms often define vertex values as vectors of floats. Chapter 4 presents all these algorithms and their implementation in Giraph. For each algorithm you write, you must decide which data type fits the vertex value.

Edge values, on the other hand, lie somewhere in between. For certain algorithms, you use no edge values; for others, you use values to model the label (if any) attached to the edge in the input graph; and for others, you use values to model the weight. For still other algorithms, you use a totally new edge value type that has nothing to do with the actual graph; the algorithm may use it to store intermediate results. Vertex values tend to change during the computation, because they are part of the intermediate results, but edge values tend to stay the same. But again, this is not a rule. If you look back at the example in Figure 3-4, the graph would be represented with string IDs for names, integer values for ages, and no edge values.

In Giraph, each vertex object is an instance of the Vertex class. The interface of the Vertex class, presented in Listing 3-1, lets you access the vertex value and the edges and their values and add and remove edges. For now, ignore the voteToHalt() method, which is presented in the next section. Giraph comes with a default implementation of vertices and edges, so you do not need to implement them yourself as part of your application (unless you want some specific behavior). You need to define the types of the vertex ID, the vertex value, and the edge value.

Listing 3-1. The Vertex Class

```
class Vertex:
    function getId() #1
    function getValue() #2
    function setValue(value) #3
    function getEdges() #4
    function getNumEdges() #5
    function getEdgeValue(targetId) #6
    function setEdgeValue(targetId, value) #7
    function getAllEdgeValues(targetId) #8
    function voteToHalt() #9
    function addEdge(edge) #10
    function removeEdges(targetId) #11
```

#1 Returns the ID of the vertex. The return type depends on the type of the ID.
#2 Returns the value of the vertex. The return type depends on the type of the vertex value.
#3 Sets the value of the vertex. The type of the parameter depends on the type of the value.
#4 Returns all the outgoing edges for the vertex in the form of an iterable of Edge objects.
#5 Returns the number of outgoing edges for the vertex
#6 Returns the value of the first edge connecting to the target vertex, if any. The return type depends on the type of the edge value.
#7 Sets the value of the first edge connecting to the target vertex, if any. The type of the parameter depends on the type of the value.
#8 Returns the values associated with all the edges connecting to a specific vertex. This methods is useful when managing multigraphs.
#9 Makes the vertex vote to halt
#10 Adds an edge to the vertex
#11 Removes all edges pointing to the target vertex

The Edge class is even simpler. It has only three methods: one to get the ID of the other endpoint, one to get the value, and one to set the value. The interface is presented in Listing 3-2.

Listing 3-2. The Edge Class

```
class Edge:
    function getTargetVertexId() #1
    function getValue() #2
    function setValue(value) #3
```

#1 Returns the ID of the target vertex. The return type depends on the type of the ID.
#2 Returns the value attached to the edge. The return type depends on the type of the edge value.
#3 Sets the value of the edge. The type of the parameter depends on the type of the edge value.

Now that you have seen how the graph is represented in Giraph and how you can access it programmatically, you are ready to look at how to express an algorithm.

A Computation Based on Messages and Supersteps

Once you have defined the way your graph looks through the type of vertex ID, vertex, and edge values, Giraph needs you to write a UDF called compute. As mentioned earlier, Giraph requires you to "think like a vertex." All this logic is put in the compute() method. But before you dig into the API, let's look at the way the computation is organized.

A Giraph computation is organized in a series of *supersteps*. They are called that because a superstep is composed of steps, as you see later in this chapter. Intuitively, you can think of a superstep as an iteration of an algorithm; this is not always the case but often is. At each superstep, a vertex can send messages to other vertices, access its vertex value and its edges, and vote to halt. Sent messages are delivered to the destination vertex at the beginning of the next superstep. Every vertex can be in either the active state or the inactive state. Only active vertices are computed during each superstep (once). At the beginning of the computation, every vertex is in the active state, and it can switch to the inactive state by voting to halt. A vertex votes to halt because it decides that from its local perspective, its work is done, and the computation may conclude. A vertex that is in the inactive state is not executed during a superstep unless it has received a message. The delivery of a message switches a vertex back from the inactive to the active state. A Giraph computation is over when all vertices have voted to halt and no message needs to be delivered. The diagram in Figure 3-6 illustrates the way a vertex can change state between active and inactive.

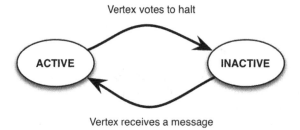

Vertex votes to halt

ACTIVE   INACTIVE

Vertex receives a message

Figure 3-6. *Diagram of transitions between vertex states*

A Giraph computation is said to be *synchronous*. A superstep is concluded when all active vertices have been computed and all messages have been delivered. Giraph will not compute the next superstep until all vertices have been computed. Because active vertices have to wait for the other vertices to be computed during the current superstep before their next superstep can be computed, the computation is referred to as *synchronous* and the waiting phase is called the *synchronization barrier*.

A Giraph computation is distributed and parallelized by spreading vertices, with their edges, across a number of processing units—for example, machines or CPU cores. During each iteration, each unit is responsible for executing the compute() method on the vertices assigned to it. Each unit is also responsible for delivering the sent messages to the units responsible for the vertices that should receive the messages. This means the more units involved in the computation, the more vertices can be executed in parallel. But the more units you have, the more communication is produced. Chapter 6 is dedicated to the architecture of Giraph and dives into the details. Figure 3-7 shows the computation of a vertex that receives three messages (5, 7, and 20), chooses the largest one (20), updates its value (from 2 to 20), and sends the value to its neighbors. Note that this chapter shows the synchronization barrier and illustrates the messages being sent with their own dashed arrows. To keep the figures leaner, later figures in the book do not include the synchronization barrier, and messages are drawn directly on the edges over which they are sent.

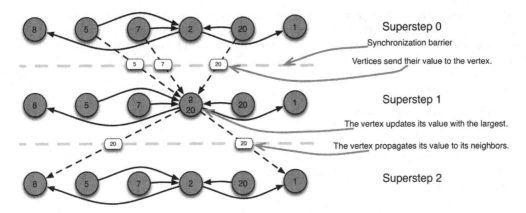

Figure 3-7. *Computation of a vertex that receives values through messages and propagates the largest*

In practice, to write an algorithm, you have to implement the compute() method of a class called
BasicComputation. At each superstep, Giraph calls the compute() method on all active vertices, delivering
the messages sent to that vertex (if any) during the previous superstep. The signature of the compute()
method is compute(vertex, messages), where the first parameter is the vertex that is being computed and
the second parameter is a container with the messages sent to that vertex (you can think of it for now as a
list of messages). The BasicComputation class must be defined with the same three types as the vertex it
operates on. However, in addition to these three types, you also need to define a fourth type: the message
type. Giraph needs this information to know how to store and deliver messages. Listing 3-3 presents the
most relevant methods of the BasicComputation class.

Listing 3-3. The BasicComputation Class

```
class BasicComputation:
    function compute(vertex, messages) #1
    function getSuperstep() #2
    function getTotalNumVertices() #3
    function getTotalNumEdges() #4
    function sendMessage(targetId, message) #5
    function sendMessagetoAllEdges(vertex, message) #6
    function addVertexRequest(vertexId) #7
    function removeVertexRequest(vertexId) #8
```

#1 The method to implement, which is called by the Giraph runtime
#2 Returns the current superstep
#3 Returns the total number of vertices in the graph
#4 Returns the total number of edges in the graph
#5 Sends a message to the target vertex
#6 Sends a message to the endpoints of all the outgoing edges of a vertex
#7 Requests the addition of a vertex to the graph
#8 Requests the removal of a vertex from the graph

This is all you need to know to program a basic graph algorithm in Giraph. The rest of this chapter presents the other parts of the basic API, but for now let's focus on how to implement the example in Figure 3-4 using what you have seen so far. Listing 3-4 presents the compute() method that implements the algorithm.

Listing 3-4. The MaxValue Algorithm

```
function compute(vertex, messages):
    maxValue = max(messages) #1
    if maxValue > vertex.getValue(): #2
        vertex.setValue(maxValue) #2
        sendMessageToAllEdges(vertex, maxValue) #2
    vertex.voteToHalt() #3
```

#1 Identify the largest value across those sent as a message.
#2 The value is larger than the value discovered so far by this vertex, so update and propagate.
#3 Vote to halt.

Figure 3-8 illustrates the execution of the algorithm on the graph from Figure 3-3. As you can see, the largest value propagates quickly through the graph. When vertices discover new, larger values, they are updated and propagated until all the vertices have discovered the largest value.

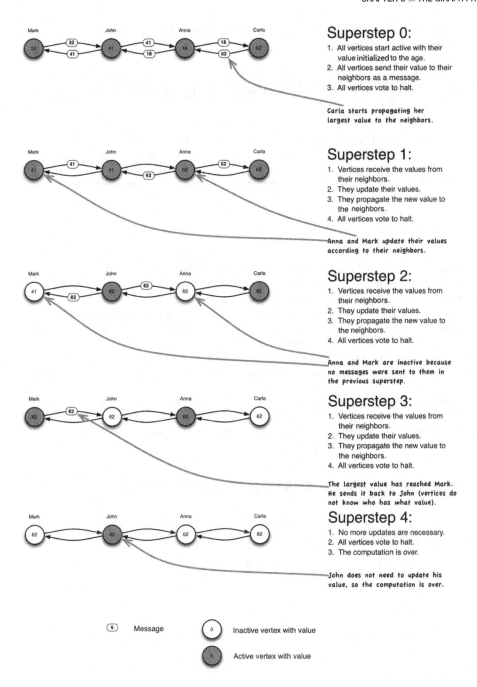

Superstep 0:

1. All vertices start active with their value initialized to the age.
2. All vertices send their value to their neighbors as a message.
3. All vertices vote to halt.

Carla starts propagating her largest value to the neighbors.

Superstep 1:

1. Vertices receive the values from their neighbors.
2. They update their values.
3. They propagate the new value to the neighbors.
4. All vertices vote to halt.

Anna and Mark update their values according to their neighbors.

Superstep 2:

1. Vertices receive the values from their neighbors.
2. They update their values.
3. They propagate the new value to the neighbors.
4. All vertices vote to halt.

Anna and Mark are inactive because no messages were sent to them in the previous superstep.

Superstep 3:

1. Vertices receive the values from their neighbors.
2. They update their values.
3. They propagate the new value to the neighbors.
4. All vertices vote to halt.

The largest value has reached Mark. He sends it back to John (vertices do not know who has what value).

Superstep 4:

1. No more updates are necessary.
2. All vertices vote to halt.
3. The computation is over.

John does not need to update his value, so the computation is over.

⬭ 6 Message

○ 3 Inactive vertex with value

● 6 Active vertex with value

Figure 3-8. Computation of the MaxValue algorithm in Giraph

53

Now what you have seen the simple Giraph API, the next section explores how you can make your algorithm more efficient through combiners.

THE BULK SYNCHRONOUS PARALLEL (BSP) MODEL

Some of the terminology in Giraph is borrowed from the bulk synchronous parallel (BSP) model, which inspired Pregel and Giraph. If you are interested in getting to know more about the BSP model, a section is dedicated to it at the end of this chapter. Keep in mind, however, that you do not need to know how the BSP model works to play with Giraph.

Reducing Messages with a Combiner

Messages play a very important role in Giraph, because they allow vertices to share information. Also, because Giraph uses messages instead of shared memory, graph computations can be parallelized without using (expensive) concurrency primitives. Still, exchanging messages has its cost and can impact on the total runtime of a computation. Can you reduce the number of messages?

A *combiner* is a function that combines messages sent to a vertex. Combining messages allows Giraph to send less data between processing units. A combiner is very simple and combines two messages into one. The messages it combines were sent to the same vertex during the current superstep. What is important is that there are no guarantees about how many times the combiner will be called or if it will be called at all. Basically, the only assumption you can make is that the messages passed to the combiner are all destined for the same vertex.

Because a combiner receives a partial collection of messages and can be called multiple times, it must apply a function that is commutative and associative. Listing 3-5 shows the interface of the MessageCombiner class. Clearly, because a combiner is defined on a specific type of message, an algorithm must use a combiner that matches the type of messages the vertices send.

COMMUTATIVE AND ASSOCIATIVE FUNCTIONS

A *commutative* function produces the same result regardless of the order in which it is applied on the elements. For example, sum is commutative:

$1 + 2 = 2 + 1$ or, more extensively, $1 + 2 + 3 + 4 = 3 + 2 + 4 + 1$, and so on.

An *associative* function can be applied to subgroups of the input elements and produce the same result. For example, sum is also associative:

$1 + 2 + 3 + 4 = (1 + 2) + (3 + 4) = 1 + (2 + 3) + 4$, and so on.

Listing 3-5. The MessageCombiner Class

```
class MessageCombiner:
    function combine(id, message1, message2) #1
```

#1 Returns the combination of the two messages

For the MaxValue algorithm, a viable combiner is one that returns the largest value of the two. Listing 3-6 shows the pseudo-code for this combiner.

Listing 3-6. The MaxValue Combiner

```
function combine(id, message1, message2):
    return max(message1, message2) #1
```

#1 Returns the largest value

Now, let's look at the effect of using the combiner on the example computation for the largest value (on a different graph). Figure 3-9 shows one superstep. On the left, messages are delivered as produced. On the right, however, Giraph applies the combiner two times, reducing the number of messages sent for that vertex to a third as many. Note also that the second combiner round might not have happened, because Giraph does not give any guarantees. In that case, the number of messages would be reduced only to two-thirds as many.

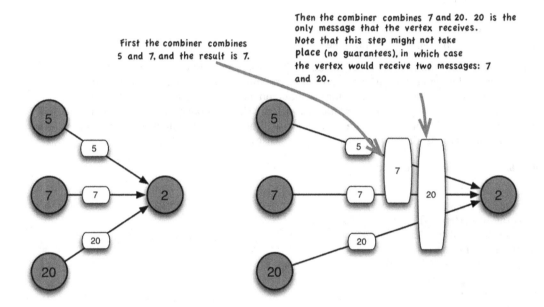

Figure 3-9. The effect of using the MaxValue combiner on the messages

The interesting thing about the combiner is that the compute() method cannot make any assumptions whether a combiner will be executed. For this reason, the combining logic of the combiner is often performed by the compute() method. Note how Listing 3-6 computes a max() function on the input messages in line maxValue = max(messages). This is the same function as the combiner. The combiner is executed before you enter the compute() method on some of the messages, but the result is the same.

Combiners are very useful for minimizing the use of resources by Giraph. Often you can apply a simple combiner, but not all algorithms can have one. Keep this in mind when you design your own algorithm.

Computing Global Functions with Aggregators

Computing the maximum value or other aggregations on the values associated with vertices can be expressed as a graph algorithm. But it would be easier if you could compute these aggregations without having to propagate messages across vertices in a number of supersteps. This is what *aggregators* were introduced for. Aggregators allow you to think of aggregations as global functions to which vertices can send values. During each superstep, these global functions aggregate the values, and the results are available to the vertices during the following superstep.

Aggregators compute global functions, but they are executed in parallel across processing units, and they are scalable. Aggregators, like combiners, require the function to be commutative and associative. The interface of an aggregator is presented in Listing 3-7.

Listing 3-7. The Aggregator Class

```
class Aggregator:
    function aggregate(value) #1
    function getAggregatedValue() #2
    function setAggregatedValue(value) #3
    function reset() #4
```

#1 Aggregates the value
#2 Returns the aggregated value
#3 Sets the aggregated value to the parameter
#4 Resets the aggregated value to a neutral value

Let's go back to the example. Instead of having vertices propagate their values, you can make vertices send their values to the aggregator during the first superstep and then vote to halt. This algorithm would execute only one superstep and finish the computation. Listing 3-8 presents the code for the MaxValueAggregator class. Note that for this kind of algorithm, Giraph probably is not the right tool for the job, because it does not exploit the structure of the graph. Still, algorithms can often use aggregators as part of their graph computations; Chapter 4 presents two such cases.

Listing 3-8. The MaxValueAggregator Class

```
maxValue = -Inf #1
function aggregate(value):
    maxValue = max(maxValue, value) #2
function getAggregatedValue():
    return maxValue #3
function setAggregatedValue(value):
    maxValue = value #4
function reset()
    maxValue = -Inf #5
```

#1 Local variable where the aggregated value is stored
#2 Updates the local value to the new value, if larger
#3 Returns the aggregated value
#4 Sets the value to the new value, used (for example) for initialization
#5 Resets the value to the neutral value –Inf (neutral to the max() function)

Listing 3-9 shows the pseudo-code for a trivial MaxValue algorithm that uses an aggregator. For simplicity, the code assumes you have the reference to an aggregator object called maxValueAggregator. In Giraph, however, you can use multiple aggregators at the same time—each with a string name to distinguish it—that need to be declared and initialized before they are used. Presenting this complete API would require introducing parts of the API that are not relevant to understanding how aggregators work and how to use them. Chapter 5 includes the complete API in its presentation of the Java API.

Listing 3-9. The MaxValue Algorithm with an Aggregator

```
function compute(vertex, messages):
    maxValueAggregator.aggregate(vertex.getValue()) #1
    vertex.voteToHalt()
```

#1 Aggregates the vertex value through the MaxValue aggregator

Figure 3-10 shows how the new computation is organized. Note how the aggregator is computed autonomously for each value, because the function is associative and commutative.

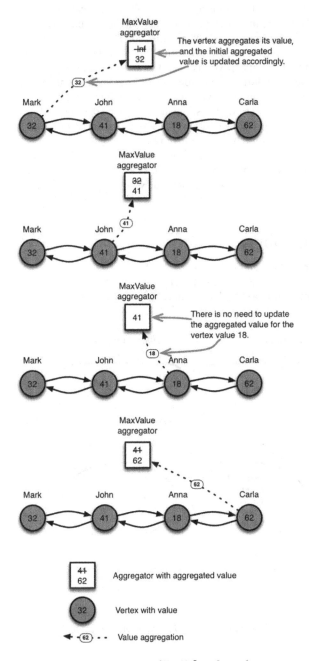

Figure 3-10. *Computation of* MaxValue *through an aggregator*

You need to know one last thing about aggregators before you move on. You might have noticed the reset() method in the Aggregator class. This method exists because there are two types of aggregators: regular and persistent. The value of a regular aggregator is reset to the initial value in each superstep, whereas the value of persistent aggregator lives through the application. Note that a call to getAggregatedValue() returns the value computed during the previous superstep, not the current one. The Giraph runtime uses the reset() method to reset the value of a regular aggregator. Hence, it should reinitialize the aggregator to a neutral value, like 0 for a sum aggregator or -Infinity for a MaxValue aggregator.

Imagine a simple application in which during each superstep, vertices send a value of 1 to an aggregator that sums all these values. If the aggregator is regular, then during each superstep (except the first one), the aggregator contains a number that is equal to the total number of vertices. Instead, if the aggregator is persistent the aggregated value will be the number of vertices times the superstep number. So, if you had four vertices in a computation of three supersteps, at the end of the computation a regular sum aggregator would have a value of 4, and a persistent sum aggregator would have a value of 12.

This chapter has presented the basic API and assumed that the graph was already loaded and initialized in memory and that the final superstep would be the last part of the computation. The next section looks at what happens before the first superstep and after the last superstep, to conclude your tour of a Giraph computation.

The Anatomy of a Giraph Computation

This chapter's presentation of the API has assumed that the graph was already loaded into memory, the values were initialized, and the last superstep would be the last phase of the computation. Let's look at a more complete overview. Figure 3-11 shows the anatomy of a Giraph computation.

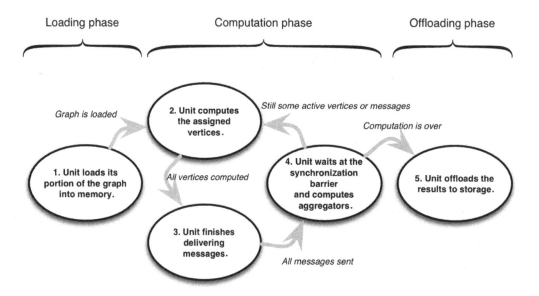

Figure 3-11. *The different phases and steps of a Giraph computation executed by each processing unit*

The computation starts with Giraph loading the graph into memory across the processing units. In Giraph, the loading phase is considered superstep -1. The graph is usually read from persistent storage, like a filesystem or a database (both most likely distributed). This phase, the loading phase, happens in parallel, when possible, and requires converting the data read from storage into the internal representation of a graph. As Chapter 7 explains, Giraph provides an API that allows you to define the format of the persistent data and how it should be parsed to load the graph. The loading phase loads all the vertices and their edges and initializes the vertex values and the edge values. Once the graph is loaded, the computation can start.

The computation phase consists of one or multiple supersteps. Each superstep is divided into three phases:

1. The processing units iterate over the active vertices and call the compute() method with the corresponding messages, if any. As the vertices produce messages, they are sent to the corresponding processing unit, depending on the destination vertex. At various points, combiners can be applied to messages, if defined, either before messages leave a source or when they reach the destination. The processing units finish computing the vertices and finish sending the remaining messages. Each processing unit waits at the synchronization barrier for the other units to finish.

2. The processing units conclude the computation of the aggregators, if any, and aggregate the local aggregations. Local computations that involve aggregators can be performed when computing the vertices, because aggregation functions are commutative and associative. This means the local aggregations are not necessarily computed after all vertices have been computed. When processing units are finished with aggregations, they can move to the next superstep, if the computation is not over.

3. In the offloading phase—the last phase of the computation—the processing units have in memory the vertices with their vertex values and the edges with their values. This data represents the results (often, if the algorithm does not change the graph, only the vertex values contain the end results). The processing units go through this data and write it back to persistent storage. The entire computation is now over, and the results can be used.

DIFFERENCE BETWEEN THE BSP MODEL AND GIRAPH

Traditionally, in the BSP model, a superstep is divided into three steps: the local computation, the communication, and the synchronization barrier. These steps are computed one after the other. When a units has finished computing its state locally, it starts sending data.

Giraph computes a superstep differently, and it overlaps local computation and communication. Processing units begin exchanging messages as soon as they are produced, instead of waiting for the computation of vertices to be finished. This helps dilute network usage over a longer period of time. This aspect is particularly helpful in Hadoop clusters that are running multiple jobs, of which Giraph is one, so that Giraph does not saturate the network.

You have seen the basic Giraph API and how the computation is organized. Part 2 and Part 3 of the book present the Giraph Java API. Before you move to Chapter 4, which presents Giraph implementations of algorithms for real-world graph analytics, let's spend the next two sections looking at some simple algorithms to help make you more familiar with what you have seen so far.

Computing In-Out-Degrees

To acquaint you with the Giraph programming model, this section presents a simple graph algorithm that could be considered the "Hello World" of graph computations (if you are familiar with Hadoop MapReduce, this example is like the "word counting" of Giraph).

As of December 31, 2012, Facebook was reported to have 1.06 billion monthly active users and 150 billion friend connections, which in graph terms means 1.06 billion vertices and 150 billion undirected edges. How many connections does a user have on Facebook or on Twitter? Figure 3-12 shows two social networks: one with symmetrical friend relationships (Facebook) and one with asymmetrical follower relationships (Twitter). In both, the graph has three vertices and three edges. In the first case, the average number of friends per user is two; and in the second case, the average number of followers per user is one. This is interesting because the ratio between the number of vertices and the number of edges is the same. However, because the Twitter graph is directed, and you are considering only the incoming edges (the followers), you do the counting differently. If you focus solely on the number of edges that "touch" a vertex, in both cases the average is two.

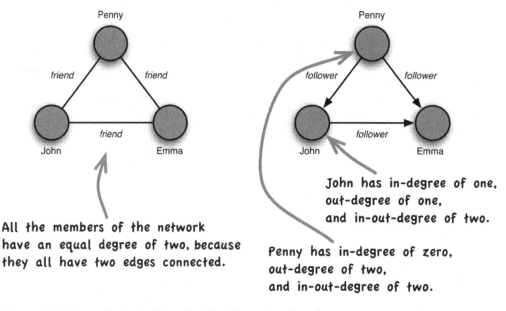

Figure 3-12. *Vertex degrees in directed and undirected graphs*

The number of edges that touch a vertex is called the *degree* of that vertex. For directed graphs, a vertex has an *in-degree* (the number of incoming edges), an *out-degree* (the number of outgoing edges), and an *in-out-degree* (the sum of both). If you consider the in-out-degree, then the degree of the vertices in a directed graph is computed the same way as for an undirected graph. If you were to convert the undirected Facebook graph to a directed one, where two directed relationships with opposite direction substitute for each undirected friend relationship, the graph would contain six edges.

The question is, how can you compute in-out-degrees with the Giraph vertex-centric programming model? The fact that each vertex only knows about its outgoing edges requires some exchange of information between each vertex and its neighbors. To discover its incoming edges, a vertex can take advantage of its messaging capabilities and inform its neighbors of its existence through the outgoing edges. Figure 3-13 shows a directed graph with seven vertices, for which you want to compute the in-out-degrees (you can think of it as a graph of Twitter followership).

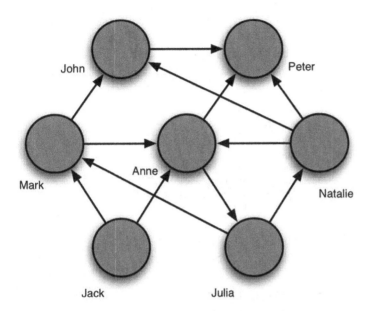

Figure 3-13. *A Twitter-like network of users and follower relationships*

The algorithm works as follows:

1. At the beginning, all vertices are active and have their value initialized to 0.

2. During superstep 0, each vertex computes the number of outgoing edges and sets its value to this number. If you wanted to compute the out-degree only, this would be enough, and the vertices would contain the out-degree in their value. However, the incoming edges are still unknown to each vertex, so each vertex sends a message through its outgoing edges and votes to halt. When all the vertices have finished computing their degree and have sent the messages, the following superstep begins.

3. Vertices with incoming edges are woken up by the messages sent to them during the previous superstep. Each vertex has received a number of messages equal to the number of vertices pointing to it with an incoming edge. Each vertex now counts the number of messages, effectively computing its in-degree, adds it to its out-degree, and votes to halt. At this point, all the vertices have their in-out-degree in their value, and the computation is over.

Figure 3-14 shows the flow of the computation for the graph presented in Figure 3-13.

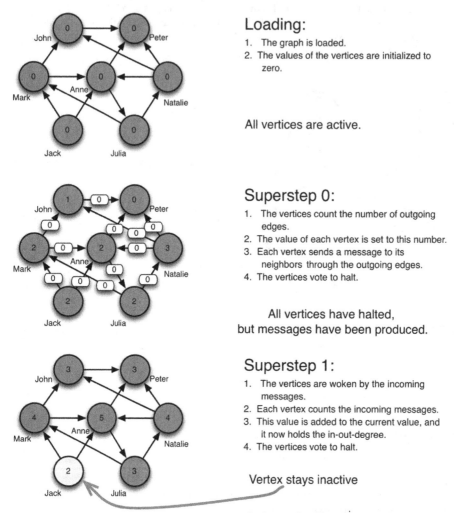

Loading:

1. The graph is loaded.
2. The values of the vertices are initialized to zero.

All vertices are active.

Superstep 0:

1. The vertices count the number of outgoing edges.
2. The value of each vertex is set to this number.
3. Each vertex sends a message to its neighbors through the outgoing edges.
4. The vertices vote to halt.

All vertices have halted,
but messages have been produced.

Superstep 1:

1. The vertices are woken by the incoming messages.
2. Each vertex counts the incoming messages.
3. This value is added to the current value, and it now holds the in-out-degree.
4. The vertices vote to halt.

Vertex stays inactive

At the end, all vertices are halted
and no messages are produced.
The computation is finished!

 Message

Active vertex

Figure 3-14. *Computing the in-out-degree in Giraph*

The computation of vertex degrees on the Facebook graph is trivial: it boils down to counting the number of edges. When an undirected graph is modeled through a directed one, as in Giraph, each incoming edge corresponds to an outgoing edge. Hence, a vertex can just count the number of edges it has, without having to depend on incoming messages. Listing 3-10 shows the pseudo-code for the algorithm.

Listing 3-10. The InOutDegree Algorithm

```
function compute(vertex, messages):
    if getSuperstep() == 0:
        vertex.setValue(vertex.getNumEdges()) #1
        sendMessageToAllEdges(vertex, 0) #1
    elif getSuperstep() == 1:
        inDegree = 0
        for msg in messages: #2
            inDegree += 1 #2
        outDegree = vertex.getValue()
        inOutDegree = outDegree + inDegree #3
        vertex.setValue(inOutDegree) #3
    vertex.voteToHalt()
```

#1 Initializes the vertex value to the out-degree, and propagates it
#2 Counts the incoming messages as the number of incoming edges
#3 Sums in-degree and out-degree to compute the in-out-degree, and sets the vertex value to it

In a few lines, you have developed an application that computes the in-out-degree of each vertex across hundreds or thousands of machines without having to worry about concurrency or parallel and distributed computing. With all this discussion about directed and undirected graphs, let's look at an algorithm to convert a directed graph to a (logically) undirected graph.

Converting a Directed Graph to Undirected

Certain algorithms require graphs to be undirected. Giraph supports only directed graphs, so this means a *logically* undirected graph. Here, *logically* means that each edge in the graph has a corresponding edge in the opposite direction. Because each edge has a corresponding edge in the opposite direction, direction is lost.

Take Figure 3-15 as an example. The original directed graph is converted to a logically undirected graph by adding an edge in the opposite direction if it is not already present. This simple conversion strategy may make you lose some edge information, but don't worry for now; you look into that in a moment. In the vertex-centric API, the vertex that will be the source vertex can add the edge.

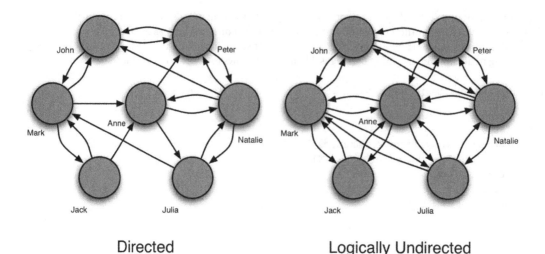

Directed   Logically Undirected

Figure 3-15. *The representation of a directed graph through a logically undirected graph*

The algorithm works in two supersteps, as follows. Note that it does not use vertex values, because it works only on the topology of the graph:

1. During superstep 0, each vertex sends its vertex ID to the neighbors.

2. During superstep 1, each vertex receives the IDs of the vertices that are endpoints of the incoming edges. That is, the vertex *discovers* its incoming edges. For each of these IDs, each vertex checks whether an edge in the opposite direction is already present. If this is not the case, the vertex creates an edge toward that vertex. In addition, active vertices vote to halt during both supersteps.

Listing 3-11 presents the pseudo-code for this algorithm.

Listing 3-11. The Graph Conversion Algorithm

```
function compute(vertex, messages):
    if getSuperstep() == 0:
        sendMessageToAllEdges(vertex, vertex.getId()) #1
    elif getSuperstep() == 1:
        for msg in messages:
            if vertex.getEdgeValue(msg) is None:
                vertex.addEdge(Edge(msg)) #2
    vertex.voteToHalt()
```

#1 Initially propagates the ID to all the neighbors
#2 Adds an edge if an edge toward the target does not exist already

In some cases, you may want to use the edge value to store information about whether the original graph contained only a single edge in one direction. For example, you might want to assign an edge value of 2 when the original graph had two corresponding edges, and a value of 1 in the case where only one edge was present. Figure 3-16, for example, converts the original graph in Figure 3-15 according to this new heuristic. The input graph is assumed to have a value of 1 initially assigned to each edge. The pseudo-code for this heuristic is presented in Listing 3-12.

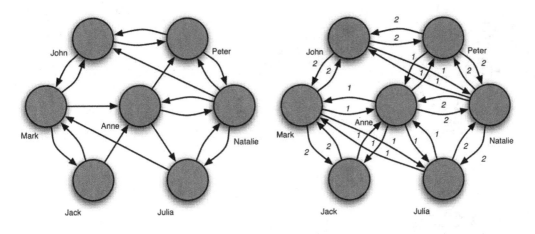

Directed   **Weighted Logically Undirected**

Figure 3-16. *The representation of a directed graph through a weighted logically undirected graph*

Listing 3-12. The Graph Conversion Algorithm with Weights

```
function compute(vertex, messages):
    if getSuperstep() == 0:
        sendMessageToAllEdges(vertex, vertex.getId()) #1
    elif getSuperstep() == 1:
        for msg in messages:
            value = vertex.getEdgeValue(msg):
            if value is None:
                vertex.addEdge(Edge(msg, 1)) #2
            else:
                vertex.setEdgeValue(msg, 2) #3
    vertex.voteToHalt()
```

#1 Initially propagates the ID to all the neighbors
#2 Adds an edge if an edge toward the target does not exist already
#3 Updates the weight to 2 if an edge toward the target already exists

Again, with a few lines, you can express an algorithm that can convert a large graph across many machines. Figure 3-17 shows the flow of the algorithm.

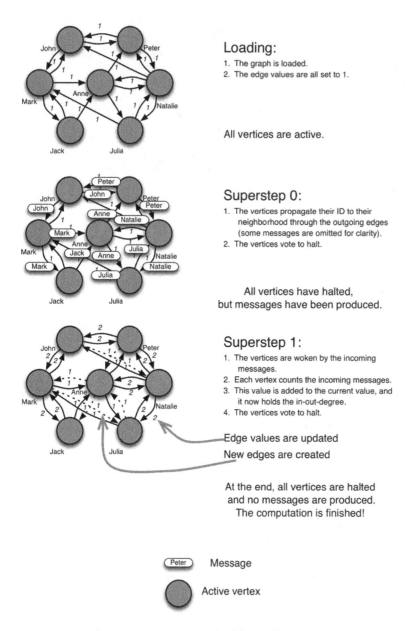

Loading:

1. The graph is loaded.
2. The edge values are all set to 1.

All vertices are active.

Superstep 0:

1. The vertices propagate their ID to their neighborhood through the outgoing edges (some messages are omitted for clarity).
2. The vertices vote to halt.

All vertices have halted,
but messages have been produced.

Superstep 1:

1. The vertices are woken by the incoming messages.
2. Each vertex counts the incoming messages.
3. This value is added to the current value, and it now holds the in-out-degree.
4. The vertices vote to halt.

Edge values are updated

New edges are created

At the end, all vertices are halted
and no messages are produced.
The computation is finished!

Message

Active vertex

Figure 3-17. *The computation in Giraph of the graph conversion to weighted logically undirected*

You have seen how to write iterative graph algorithms with the Giraph API and how the distributed engine executes the algorithm across a number of processing units. Before this chapter concludes, let's open the hood of Giraph and look at the computational model that inspired Giraph.

Understanding the Bulk Synchronous Parallel Model

This section looks at the model that inspired Giraph. As a Giraph user, you see many terms in the API that are borrowed from BSP model, such as *superstep* and *synchronization barrier*. As a Giraph developer, if you ever need to extend Giraph, its internals contain many references to the BSP model, and you need a better understanding of how the underlying computational model works. This section teaches you how Giraph works under the hood, but keep in mind that you don't need to think in BSP terms when you program Giraph. These concepts are borrowed by the Giraph model, or are hidden, so you can program Giraph using just the concepts you have learned up to now.

Imagine that you have a list of 20 positive integer numbers, and you want to find the largest value. On a sequential machine, this would require going through the entire list, saving the current largest value in a variable, and comparing it sequentially with each element in the list. At the end of the computation, this variable would contain the largest value in the list.

With ten machines, can you parallelize this computation? Yes, you can. You assign two numbers to each machine, and you let each machine find the largest value among those assigned to that machine. The problem you need to solve at this point is how to compare the 20 values assigned to the 10 machines to find the largest among them all (also in parallel, of course). Also, you want to avoid every machine sending its value to all the other machines. Toward this end, you can organize the computation in a tree:

1. Machine 1 sends its value to machine 0, which compares the two values and saves the largest. Machine 3 sends its value to machine 2, which compares the two values and saves the largest, and so on. Note that this step happens in parallel. You now have five machines with five values that need to be compared.

2. Again, you organize the process hierarchically. So, machine 2 sends its value to machine 0, which compares the two values and picks the largest, machine 6 sends its value to machine 4, which compares the results and picks the largest, and machine 8, which is alone, sends its value to machine 4 as well (which clearly has to compare three values this round).

3. You have one step to go: machine 4 sends its value to machine 0, which compares the two remaining values and picks the overall largest value.

Figure 3-18 presents this computation.

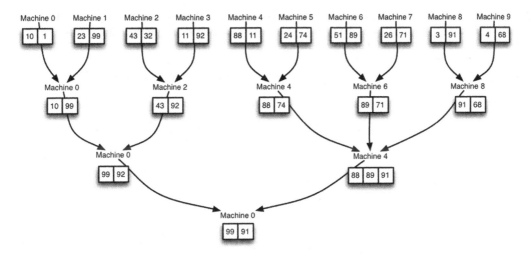

Figure 3-18. *The organization of a parallel MaxValue computation across ten machines*

This is one way of organizing a computation hierarchically to compute a problem in parallel. Not all problems need to be organized hierarchically, hence having all the machines busy only during the first step, but this example is simple enough to serve the purpose. Other algorithms would have each machine compute its part of the subproblem and communicate the result to some other machine. The BSP model generalizes these approaches in an abstract machine to compute parallel algorithms.

According to BSP, you have n processing units, which can communicate through a medium such as a network or a bus. You divide the input across the processing units, and you let each processing unit compute its intermediate solution to its subproblem locally. When the processing units have finished, they exchange the intermediate results according to the semantics of the algorithms. When a processing unit has finished computing its subproblem and sending its intermediate results, it waits for the others to finish as well. When all processing units have finished, they go on with the next iteration, computing their subproblem based on their previously computed state and the messages they have received. Each iteration is called a *superstep*, and the waiting phase is called the *synchronization barrier*—in Giraph, many concepts are borrowed from the BSP model. Figure 3-19 illustrates the conceptual organization of the BSP model.

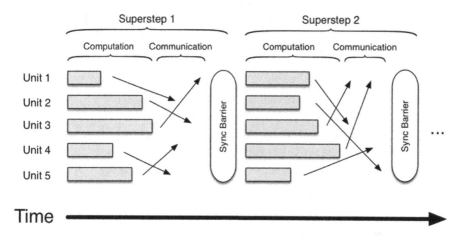

Figure 3-19. *The organization of a BSP computation across five processing units*

At this point, you should have noticed the matching between the BSP model and the Giraph model presented in this chapter. The graph is split across the processing units, and the intermediate results exchanged during the communication phase are the messages produced by the assigned vertices. Each processing unit keeps its local state in memory, represented by the assigned vertices with their values and the messages to be processed. Figure 3-20 shows the mapping between the BSP model and the Giraph model.

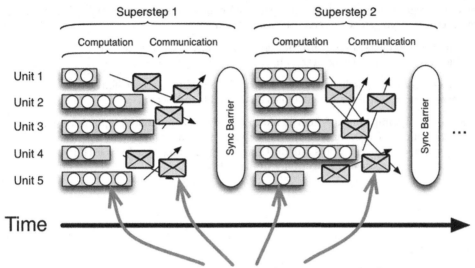

The local computation of a unit is the computation of the vertices. The communication phase is the delivery of the messages produced by the vertices.

Figure 3-20. *The mapping of the BSP model to Giraph*

As mentioned earlier, it is not necessary to think about a BSP machine when designing graph algorithms. Giraph builds on top of it so you can forget the underlying abstraction. Still, it can help you understand how Giraph was designed for iterative computations that are executed in a parallel, distributed system.

Summary

Designing and executing graph algorithms and a system to process them at scale is difficult. Giraph provides an intuitive programming paradigm that simplifies writing scalable graph algorithms, hiding the complexity of parallel, distributed systems. In this chapter, you learned the following:

- In graph algorithms, the computation of each vertex frequently depends on the computation of vertices nearby. For this reason, graph algorithms are often iterative.

- A platform for the processing of graph algorithms must support fast, iterative computations and possibly hide the complexity of distributed, concurrent programming.

- Giraph provides a simple vertex-centric API that requires you to "think like a vertex" that can exchange messages with other vertices.

- You can combine messages to minimize the number of messages transmitted between supersteps.

- You can use aggregators to compute aggregation functions across the values of the graph vertices. Aggregators are very scalable in Giraph.

- Computing vertex degrees and converting graphs from directed to undirected requires only a few lines of code in Giraph.

- The bulk synchronous parallel model defines parallel computations executed across a number of processing units. Giraph builds on top of this abstraction.

You have seen how to fit Giraph into a system architecture and how to program Giraph. You are now ready to look at more examples of algorithms and how to implement them in Giraph. The next chapter is dedicated to writing scalable graph algorithms with the Giraph programming model; toward this end, it presents the implementation of some of the algorithms from in Chapter 2.

CHAPTER 4

■ ■ ■

Giraph Algorithmic Building Blocks

This chapter covers

- Principles and patterns behind scalable graph algorithms

- Graph connectivity, paths, and connected components

- Ranking vertices with PageRank

- Predicting ratings for user-item recommendations

- Identifying communities with label propagation

- Graph types and how to characterize them

This chapter focuses on algorithmic building blocks for graph algorithms, with a particular emphasis on their scalability. Graph problems are commonly solved in Giraph using a number of patterns. Due to Giraph's vertex-centric paradigm based on message-passing, patterns use a type of value propagation. The chapter presents the general pattern and looks at a number of typical problems that can be solved with this pattern, such as finding paths, ranking vertices, identifying components and communities, and predicting ratings. You also look at different types of graphs and how they can be characterized by some of the algorithms described in this chapter.

Designing Graph Algorithms That Scale

In the previous chapter, you saw that Giraph provides a programming model that lets you express graph algorithms through a simple paradigm. Vertices have values and can exchange data through messages in a number of iterations. Under the hood, the system takes care of executing the algorithm in parallel in a distributed environment. Although more restrictive, the paradigm has been designed specifically to put you in a position to produce *scalable* algorithms. An algorithm expressed following this model is inherently *decentralized*. Each vertex makes independent decisions (such as whether and how to update its value, send messages, or vote to halt) based on *local information*, such as its current vertex value and the messages it has received. Because each vertex makes decisions during every iteration based on this set of its own data, the execution of the algorithm can be *massively parallelized*. In fact, the user-defined function can be executed independently on all vertices in parallel across the available processing units. A model based on message-passing avoids expensive concurrency primitives—in particular in a distributed environment—such as locks, semaphores, and atomic operations, which are required by a model based on shared memory.

Using a restrictive model allows you to focus on the semantics of the algorithm and ignore the execution model, which comes with the framework. This way, little has to be reinvented each time, and only problem-specific code needs to be developed. Although the process is simple, you still have to consider a few important decisions when designing a new algorithm. It takes some time to get acquainted with the

Giraph model, but once your brain clicks with the vertex-centric perspective, you'll begin looking at graph problems in a totally different way.

Chapter 3 used the example of a social network in which "a person is as good as her friends." That example made the point that the value of a vertex depends on the value of its neighbors; the values of the neighbors depend in turn on the values of their neighbors, and so on. This means to compute these values, information needs to be propagated through the structure of the graph, iteration after iteration. *Information propagation* is the basic and fundamental pattern behind the Giraph model. An algorithm in Giraph defines how information is propagated through the graph and how this information is used by each vertex to make independent decisions, such as whether and how to update its value. You'll see how this pattern is used throughout all the algorithms presented in this chapter.

■ **Note**   In Giraph, the global state of a computation is *distributed* across the vertex values, and it is *shared* through the *messages* that vertices send to each other.

According to this definition, if a vertex needs a certain piece of information owned by another vertex to compute its value, this piece of information must be delivered to the vertex through messages. To design a graph algorithm in Giraph and to build on the information-propagation pattern, you have to decide a number of things that define how the data associated with each vertex is *initialized, accessed, exchanged,* and *updated*—basically, how it is used.

In particular, you must define the following elements of the algorithm that are specific to a decentralized, vertex-centric approach:

- *Independent vertex decisions*: The decisions made by each vertex, such as whether it should send a message or vote to halt, should be based on information owned by the vertex itself (local), and only on that (independent/decentralized).

- *Initialization of vertex values*: Although it is obvious that vertex values need to be initialized correctly, value initialization is much more relevant for decentralized vertex-centric algorithms. Because vertices make decisions based on the current vertex value and the incoming messages, the path the computation takes depends on the initial values assigned to the vertices as much as on the graph structure.

- *Halting condition*: Vertices make decisions independently, so it is important to design a halting condition that is consistent, is well understood, and, most important, can be decided on collaboratively by the vertices.

- *Aggregators and combiners*: Sometimes global computations can simplify an algorithm or even be unnecessary. On these occasions, aggregators can prove very handy and do not undermine the scalability of an algorithm. When possible, an algorithm should use combiners to minimize the amount of resources—such as network and memory—used for messages.

The bottom line is that you should use a vertex-centric approach focusing on decentralized decisions based on local information, which are often more practical to parallelize.

The remainder of this chapter looks concretely at how existing algorithms define each of these points. The following sections have a two-fold function: they help you better understand the Giraph programming model and the concepts presented in this section, and they present solutions that often act as building blocks for more complex solutions. You may want to reuse some of the principles behind these solutions in your own applications.

Exploring Connectivity

A graph is nothing more than a bunch of vertices connected to other vertices, as you have learned. At this point, you consider two vertices to be connected if they are direct neighbors: if they are the two endpoints of a single edge. This is, however, a simplified view of the structure of a graph. Connectivity can be seen as a broader relationship between vertices. In fact, connectivity is a *transitive* relationship. In other words, if vertex A is connected to vertex B, and vertex B is connected to vertex C, then vertex A is connected to vertex C. The edge connecting A to B and the edge connecting B to C constitute *a path*. A path is a sequence of edges that connect vertices. If traversed, these edges "bring" from one vertex to another. Through paths, connectivity can be extended to vertices that are not neighbors; they are one of the fundamental tools to study the structure of a graph. This section looks at how you can compute shortest paths and use them to identify components in the graph that are subgraphs.

Computing Shortest Paths

Imagine that you have a social graph, and you want to find out whom you should ask to introduce you to tennis superstar Roger Federer. If none of your acquaintances knows Roger, chances are they may know somebody who knows him. Or they may know somebody who knows somebody who knows Roger, and so on. Basically, you are looking for a path of "friend of a friend" relationships that allows you to reach Roger. You are probably familiar with the theory of *six degrees of separation*. According to this theory, in our social relationships we are all separated on average by six steps. For example, each of us is on average six steps away from Roger Federer (or anybody else on Earth), in a long chain of "friend of a friend" relationships.

At the end of the 1960s, Stanley Milgram executed a simple experiment. He sent a letter to a number of randomly selected individuals in the United States; these people were called *sources*, and each letter contained the full name of a *target* individual (in Milgram's experiment, living in Boston) and a roster. The recipient was asked whether they knew the target on a first-name basis. If that was the case, the recipient was asked to forward the letter directly to the target. If the recipient did not know the target, they were asked to forward the letter to a friend who they thought was more likely to know the target. Each recipient of the letter put their name on the roster, to keep track of the chain. When the letter was received by the target, the roster contained the entire chain of "friends of friends" connecting the initial randomly chosen source and the target individual. This procedure was performed for more than 600 individuals, and the result was that, for those letters that actually reached the target, the rosters contained on average between five and six names. The list of names contained in each roster was a path. Figure 4-1 illustrates the possible path of one of the letters; for this example, the target is Maria, and the randomly chosen source is Sam.

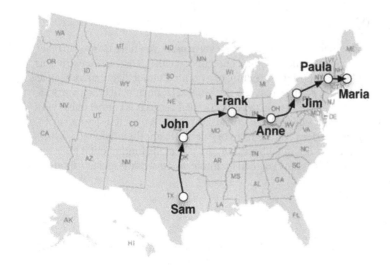

Figure 4-1. *A fictitious path for one of the letters from Milgram's experiment*

In practice, what Milgram did with his experiment was to compute an approximation of the average path length of the social graph in the United States—and this number turned out to be about six. Often, multiple paths exist between two vertices, and you usually care about the shortest one. In an unweighted graph, the *shortest path* between two vertices is nothing more than the path between two vertices with the smallest number of edges. There are many ways to compute the distance between two vertices in a graph and the path with such length. Most of them can be thought of as variations of the technique used by Milgram in his experiment.

An algorithm to search for the shortest path(s) between two vertices—or the shortest path between a vertex and any other (reachable) vertex—is called a *single-source shortest paths (SSSP) algorithm*. A very intuitive and simple algorithm that can be used to this end is the so-called *breadth-first search* (BFS).[1] Starting from a source vertex, the algorithm visits all its neighbors. These neighbors are considered to be at distance 1, because they are separated from the source vertex by one edge. The next vertices to be visited are the neighbors of these neighbors, except those that have been already visited. These vertices are considered to be at distance 2 from the source, or one hop more distant than the vertices they were visited from. The algorithm continues like this until all vertices have been visited and all distances have been computed. Intuitively, the way BFS visits the vertices in a graph starting from a source vertex follows a wave-like pattern, just like the waves produced on a flat surface by a stone thrown into water. The paths are explored in breadth, all of them one hop at a time.

In the case of a weighted graph, edge weights represent distances between neighbors. Hence, the length of a path is computed as the sum of the weights of the edges that constitute the path. To support weights, the algorithm has to be slightly modified. Figure 4-2 shows the execution of SSSP on a weighted graph starting from the leftmost vertex, Mark. Any time vertices are visited, instead of adding one hop to the distance of the vertices they were visited from, the weight of the traversed edge is added. Look, for example, at how the distance of John is defined as the weight of the edge that connects him to Mark, and how the distance of Peter is this value plus the distance from John. If a vertex is visited multiple times, either in the same iteration or in a future one, its distance is updated only if the distance has improved. For example, notice

---

[1]BFS is a strategy to traverse a graph that can be used for computing shortest paths but also for other operations. There are also other algorithms that can be used to compute shortest paths.

how Maria decides her distance (6) based on the path that goes through Julia, because it is shorter than the distance that goes through Peter (8). Also, notice how Sophia updates her distance in two steps, as soon as the shorter distance is determined.

Vertices are visited in a wave-like pattern, starting from the source with an initial value of 0.

Vertices set their distance as the sum of the distance of the source vertex (0) and the weight of the respective edge.

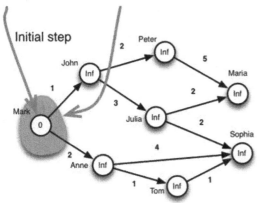
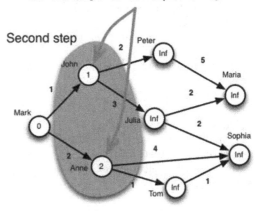

Maria chooses the shorter path passing through Julia rather than the longer one passing through Peter.

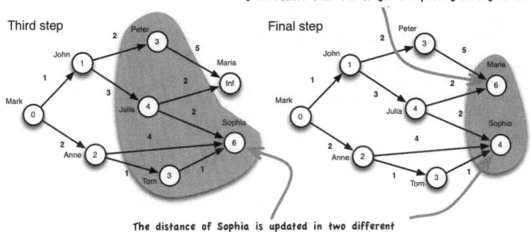

The distance of Sophia is updated in two different steps as soon as a new shortest distance is found.

Figure 4-2. *Example of execution of SSSP on a weighted graph*

You can consider SSSP on an unweighted graph to be like SSSP on a weighted graph where all the edge weights are set to 1. In Figure 4-2, the current distance from the source is written inside each vertex, and the currently visited vertices are contained in the bag-like shape. Distances are initialized to infinity, except for the source vertex, which is initialized to 0. Initializing the distances to infinity guarantees that vertices use a new distance as soon as it is discovered (any distance is smaller than Infinity).

Implementing SSSP becomes very natural in the vertex-centric programming model provided by Giraph. First, all vertices store their distance from the source vertex in their vertex value. Edge values can store the edge weights. Messages can be used to propagate distances from neighbors to neighbors. Hence, *distance* is the information propagated through the graph. Listing 4-1 presents the pseudocode for this algorithm, and Figure 4-3 shows the flow of the computation in Giraph.

Listing 4-1. Weighted Single Source Shortest Paths (SSSP)

```
function compute(vertex, messages):
  if getSuperstep() == 0:                #1
    if isSource(vertex.getId()) is True: #1
      vertex.setValue(0)                 #1
      propagateDistance(vertex)          #1
    else:                                #1
      vertex.setValue(Inf)               #1
  else:
    minDistance = min(messages)          #2
    if minDistance < vertex.getValue():  #2
      vertex.setValue(minValue)          #2
      propagateDistance(vertex)          #2
  vertex.voteToHalt()

function propagateDistance(vertex):
  minDistance = vertex.getValue()
  for edge in vertex.getEdges():
    endpoint = edge.getTargetVertexId()
    weight = edge.getValue()
    sendMessage(endpoint, minDistance + weight) #3
```

#1 Vertex distances are initialized.
#2 Vertices compute the shortest distances.
#3 Vertices propagate the distances, considering edge weights.

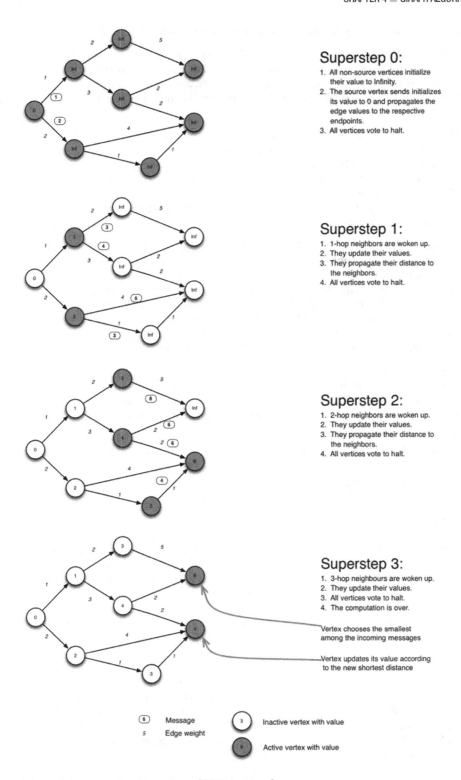

Superstep 0:

1. All non-source vertices initialize their value to Infinity.
2. The source vertex sends initializes its value to 0 and propagates the edge values to the respective endpoints.
3. All vertices vote to halt.

Superstep 1:

1. 1-hop neighbors are woken up.
2. They update their values.
3. They propagate their distance to the neighbors.
4. All vertices vote to halt.

Superstep 2:

1. 2-hop neighbors are woken up.
2. They update their values.
3. They propagate their distance to the neighbors.
4. All vertices vote to halt.

Superstep 3:

1. 3-hop neighbours are woken up.
2. They update their values.
3. All vertices vote to halt.
4. The computation is over.

Vertex chooses the smallest among the incoming messages

Vertex updates its value according to the new shortest distance

Figure 4-3. *Example of execution of SSSP in Giraph*

According to the points outlined earlier, for the SSSP algorithm you make the following decisions:

- *Independent vertex decisions*: Each vertex computes its distance based on its current value and the incoming messages. It selects the smallest distance contained in the messages and compares that distance to the shortest distance discovered so far. The vertex also decides whether to send new messages depending on the outcome of this comparison only, hence independently.

- *Initialization of vertex values*: Each vertex makes decisions based on the comparison between incoming distances and its currently stored shortest distance, so you initialize the value of each destination vertex (thus excluding the source vertex) to infinity. This guarantees that as soon as the first shortest distance is discovered, the vertex will store this distance as its vertex value. On the other hand, the source vertex initializes its vertex value to 0. In this sense, the vertex-centric algorithm behaves like the general BFS algorithm outlined earlier.

- *Halting condition*: The algorithm halts when all shortest distances have been discovered. Messages are sent only when a vertex discovers a new shortest distance, so no messages are produced when a distance is not improved. For this reason, it is sufficient for each vertex to vote to halt every time after it has evaluated a set of messages. If new candidate distances need to be evaluated, the vertex is woken up. Otherwise, the vertex remains inactive while storing the shortest distance discovered so far. Eventually, all vertices discover their shortest distance and stop sending messages. This heuristic allows a fully decentralized halting condition.

- *Aggregators and combiners*: Aggregators are not necessary to compute shortest distances, unless general statistics need to be computed, such as the maximum or minimum shortest distance from the source. Instead, a combiner can play an important role in BFS, minimizing the number of messages that need to be transmitted.

Listing 4-2 shows the pseudocode for the combiner.

Listing 4-2. MinValue combiner for Single Source Shortest Paths

```
function ssspCombiner(messages):
  return min(messages)
```

The semantics of the combiner are simple. It produces a message that contains the minimum value contained in the messages it combines. Basically, it corresponds to line minDistance = min(messages) in Listing 4-1. This correspondence allows you to use (or not use) combiners transparently, without breaking the semantics of the algorithm. The effect of using the combiner is that each vertex may need to evaluate only a subset of the original messages sent to it (from the spectrum of all the original messages down to a single message). In other words, line 9 will not change its functioning as a result of a combiner being used. The only result is that unnecessary messages are discarded early by the combiner, and you avoid transmitting them.

Paths are very important in the study of graph connectivity. They measure the *reachability* of vertices. In a social network scenario like Facebook, reachability helps clarify the extent to which information can propagate in the network, or, in other words, to what extent content can go viral via the *share* functionality. A graph is said to be *connected* if all the vertices in the graph are connected: if for each pair of vertices, at least one path connects them. Intuitively, a social graph is connected if for each person you can find a chain of "friend of a friend" relationships to all the others. In principle, in a connected graph, the content shared by a user can potentially reach any other user.

This is no different from a more traditional question applied to computer networks: to what extent can computers transmit data to the other computers in a network? Computers are connected through routers and other networking infrastructure that allows them to communicate even if they are not directly connected to each other. But does the topology of the network allow any computer to exchange data with any other computer in the network? You can easily see that this is exactly the same as the dissemination of information in a social network. If you think of a computer network as a graph, the traceroute program does nothing more than compute a path between two computers.

Computing Connected Components

So far, you have looked mostly at connected graphs. Another related concept is that of a *connected component*. A connected component, in an undirected graph, is a connected subgraph of the graph. A graph can be composed of multiple subgraphs. Think of a social network composed of two groups of friends, where none of the individuals in the first group know any of the second (and vice versa, of course). Each of these groups of friends is a different component, because the members of each group are disconnected from the members of the other. You can think of connected components as graphs composing other graphs (hence the term *subgraph*).

For an undirected graph, the definition of a connected component is pretty straightforward. For directed graphs, the definition is trickier. The nature of directed graphs imposes that, for example, in a Twitter-like network, the existence of a path from user Mark to user Julia does not imply that the inverse also exists (paths depend on follower edge directions). Lady Gaga has millions of followers on Twitter, and if she shares content, this content may reach users very far away in the network. However, regardless of her millions of followers, it is less likely that content shared by the "average John" reaches Lady Gaga, because she probably does not follow him back.

For this reason, a directed graph can either be *weakly* connected or *strongly* connected. In the former case, if you substitute each directed edge with an undirected edge, you obtain a connected graph: there is a path that connects each pair of vertices in both directions. Note that certain vertices may be connected only because of the conversion to undirected; this is why the graph is considered *weakly* connected. In the latter case, a graph is already connected without the need to convert the graph. Figure 4-4 shows an example of a weakly connected directed graph and a strongly connected graph. To obtain the second graph, we added to the first a bunch of edges that help connect unreachable pairs. Weakly and strongly connected components are respectively weakly and strongly connected subgraphs of a graph.

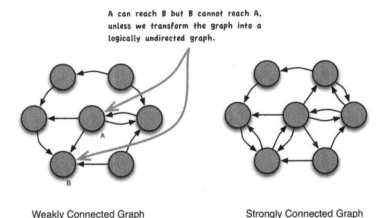

Weakly Connected Graph   Strongly Connected Graph

Figure 4-4. *Example of a weakly connected graph and a strongly connected graph*

What makes connected components interesting is that they allow you to study the initial problem about dissemination of information in a social network. You can identify disconnected islands where information can propagate; but propagation eventually encounters a barrier that does not allow information to reach the rest of the graph because such a path does not exist. Many real-world graphs are characterized by multiple connected components, with one component comprising most of the vertices. This component is also called the *largest connected component*. This component is usually the focus of further analysis, and the remaining components are ignored. This filtering has two main causes. First, the smaller connected components may be considered noise, such as people who try Facebook once and never connect to anybody. Second, some algorithms assume a connected graph, and hence the largest connected component is usually selected as the most representative for the graph.

Computing weakly and strongly connected components in Giraph is straightforward. Let's look at a approach that computes connected components in undirected graphs and weakly connected components in directed ones. The algorithm builds on the idea that in a connected component, information can propagate to all the vertices. In Giraph terms, this means if a vertex sends a message to its neighbors, and each neighbor forwards this message to its neighbors, the message eventually reaches all the vertices. In particular, if vertices have comparable IDs such that you can find the maximum or minimum one (for example, integers or strings), each connected component can be characterized by the maximum (or minimum) ID of a vertex that belongs to that component. If each vertex initially uses its ID as its value and sends that to its neighbors, and later the vertex assigns and propagates another ID only if it is larger than the currently assigned ID, eventually all vertices receive the largest ID in the component. Because the components are disconnected from each other, this information stays within the boundaries of each component, and at the end of the computation vertices belonging to different components are assigned a different value (the largest ID of a vertex in that component).

Listing 4-3 presents the pseudocode that implements this algorithm. For directed graphs, the graph must be first converted to a logically undirected graph—for example, using the algorithm presented in Chapter 3. Figure 4-5 shows the flow of the algorithm in Giraph.

Listing 4-3. Weakly Connected Components (WCC)

```
function compute(vertex, messages):
  if getSuperstep() == 0:                         #1
    vertex.setValue(vertex.getId())               #1
    sendMessageToAllEdges(vertex, vertex.getId()) #1
  else:
    maxID = max(messages)                         #2
    if maxID > vertex.getValue():                 #2
      vertex.setValue(maxID)                      #2
      sendMessageToAllEdges(vertex, maxID)        #2
  vertex.voteToHalt()
```

#1 Vertices initialize their value to their ID and propagate it.
#2 Vertices update their value, if necessary, and propagate it.

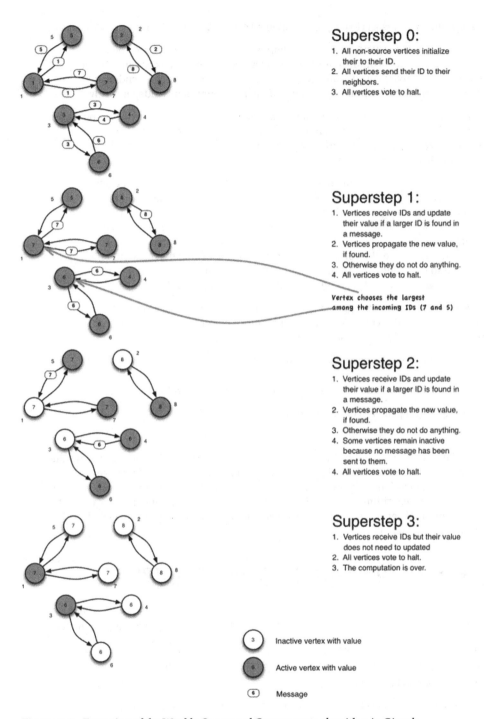

Superstep 0:

1. All non-source vertices initialize their to their ID.
2. All vertices send their ID to their neighbors.
3. All vertices vote to halt.

Superstep 1:

1. Vertices receive IDs and update their value if a larger ID is found in a message.
2. Vertices propagate the new value, if found.
3. Otherwise they do not do anything.
4. All vertices vote to halt.

Vertex chooses the largest among the incoming IDs (7 and 5)

Superstep 2:

1. Vertices receive IDs and update their value if a larger ID is found in a message.
2. Vertices propagate the new value, if found.
3. Otherwise they do not do anything.
4. Some vertices remain inactive because no message has been sent to them.
4. All vertices vote to halt.

Superstep 3:

1. Vertices receive IDs but their value does not need to updated
2. All vertices vote to halt.
3. The computation is over.

⬜ Inactive vertex with value

⚫ Active vertex with value

⬭ Message

Figure 4-5. *Execution of the Weakly Connected Components algorithm in Giraph*

81

Analyzing the algorithm with respect to the points presented earlier, you can outline the following decisions:

- *Independent vertex decisions*: Each vertex computes its component ID based on its current value and the incoming messages. It selects the maximum ID contained in the messages and compares that to the current ID. It also decides whether to send new messages depending on the outcome of this comparison only, hence independently.

- *Initialization of vertex values*: Each vertex makes decisions based on the comparison between incoming IDs and its currently stored ID, so you initialize the value of a vertex to the vertex ID. This guarantees that as soon as a larger ID is discovered, the vertex stores it as its vertex value.

- *Halting condition*: The algorithm halts when the maximum ID has been propagated to all the vertices. Messages are sent only when a vertex discovers a new maximum ID, so no messages are produced when a new maximum ID is not discovered. For this reason, it is sufficient for each vertex to vote to halt every time after it has evaluated a set of messages. If new candidate IDs need to be evaluated, the vertex is woken up. Otherwise, the vertex remains inactive while storing the maximum ID discovered so far. Eventually, all vertices discover the maximum ID and stop sending messages. This heuristic allows a fully decentralized halting condition.

- *Aggregators and combiners*: Aggregators are not necessary to compute connected components. Instead, a combiner can play an important role in this algorithm, minimizing the number of messages that need to be transmitted.

Listing 4-4 presents the pseudocode for the combiner.

Listing 4-4. MaxValue combiner for the Weakly Connected Component Algorithm

```
function weaklyConnectedComponentsCombiner(messages):
  return max(messages)
```

The semantics of the combiner are simple. It produces a message that contains the maximum value contained in the messages it combines. Basically, it corresponds to line maxID = max(messages) in Listing 4-2. Again, this correspondence allows you to use (or not use) combiners transparently, without breaking the semantics of the algorithm. As with BFS, the effect of using the combiner is that each vertex may need to evaluate only a subset of the original messages sent to it (from the spectrum of all the original messages down to a single message).

POPULAR DISTANCE-BASED GRAPH METRICS

SSSP and WCC are just two examples of algorithms that can be used to explore graph connectivity. Many metrics that are used to study graphs are based on paths and distances. Here are some of these metrics:

- *Eccentricity*: The distance between that vertex and the furthest vertex. In other words, the largest distance between that vertex and any other vertex. Eccentricity is a metric used by the following metrics.

- *Diameter*: The largest eccentricity of any vertex in the graph. The diameter is the longest shortest path in the graph. It can also define the maximum number of iterations needed for an algorithm to complete. For example, computing SSSP will not take more iterations than the length of the diameter of the graph (because it is the longest of the shortest paths to be discovered).

- *Radius*: The smallest eccentricity of any vertex in the graph. It represents the shortest, among the longest shortest paths for any vertex. The radius gives a minimum number of iterations used by SSSP to complete. This the case when you compute SSSP starting from the center of the graph.

- *Center*: Any vertex whose eccentricity equals the radius of the graph. The center of the graph are vertices that can reach any other vertex with few steps. These are central vertices to the graph because they can connect other vertices and shorten their paths. The information starting from these vertices can quickly reach any other vertex.

- *Peripheral*: Any vertex whose eccentricity equals the diameter of the graph. As opposed to the center of the graph, peripheral vertices need more steps to spread their information. To reach the furthest of the reachable vertices, a peripheral vertex requires a number of steps equal to the diameter.

Centrality measures are also interesting ways to study a graph. Like the metrics just listed, they are based on paths and distances between vertices. Unfortunately, a thorough presentation of all the metrics goes beyond what this book can cover; you can learn more at the related Wikipedia pages, such as `http://en.wikipedia.org/wiki/Centrality` and `http://en.wikipedia.org/wiki/Distance_(graph_theory)`.

You now understand how to compute paths and use them to measure different characteristics of a graph. Let's move on to see how to rank vertices according to their importance in the graph.

Ranking Important Vertices with PageRank

A graph can have many vertices, but are they are equally important? Are all web pages equally important? Are all Twitter users equally important? Can you define a notion of importance in a graph? *Important* in this context is by no means a correct term, but it can often be useful to define metrics to rank vertices. Some of these metrics are domain-dependent, capturing the definition that is relevant for a particular problem—for example, by counting the number of Olympic medals as a way to rank athletes. Others look solely at the structure of the graph, trying to capture a more general picture. An example of such an algorithm to rank vertices in a graph is PageRank. As you saw in Chapter 2, PageRank was designed by Brin and Page at Google to identify important web pages and rank them in search results. You have also seen that PageRank is a graph algorithm that looks at the Web as a graph, with pages being vertices and hyperlinks being edges.

This section looks at how PageRank works and how to implement it in Giraph. There is some math involved, but we break it down and show you how it is translated into actual code, so you don't need to remember it after you understand the underlying idea. First you look at the general model of travelling through a graph. Then you use this model to define the ranking of vertices through PageRank. After the basics are set, you learn how to implement PageRank in Giraph.

Ranking Web Pages

Although it was initially applied to the Web, PageRank can be applied to many other scenarios. To see how it works, let's look at the Web as an example. Imagine that you were to open your browser and point it to a random URL (assuming you could pick a valid URL at random). You would then click a random link on the page and jump to that page. You would follow this procedure forever: an infinitely long sequence of random clicks from page to page. The question is, would you end up on certain pages more often than on others? After all, if a page is important on the Web, there should be many links pointing to that page. More links means more probability of clicking such links and more probability of ending up on certain pages. Moreover, it should be safe to assume that important pages often contain links to other important pages. This increases the likelihood of clicking a link to an important page. Figure 4-6 shows an example of a web graph where vertices have been ranked with PageRank; the vertex size reflects the PageRank.

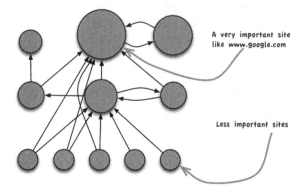

Figure 4-6. *A web graph where a page is represented by a vertex whose size depends on its PageRank*

This is the model of the so-called *random surfer*. The model can be extended with *teleportation*: the random surfer sometimes jumps to a new random page, without following a link. This is a realistic scenario, if you think about it—you sometimes open a new tab and start surfing the Web from another page, either because you are interested in something new, or because the last page contained no links. Intuitively, performing random surfing on the Web for a long time, and counting the number of times you end up on each page, should give you a pretty good idea of how important pages are and let you rank them accordingly.

PageRank

The idea of infinitely surfing the Web is the idea behind PageRank. This concept is expressed formally like this:

$$PageRank(page_i) = \frac{1-d}{\#Pages} + d \cdot \sum_{page_j \in N(page_i)} \frac{PageRank(page_j)}{\#Links(page_j)}$$

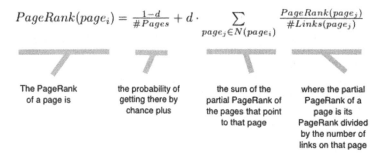

| The PageRank of a page is | the probability of getting there by chance plus | the sum of the partial PageRank of the pages that point to that page | where the partial PageRank of a page is its PageRank divided by the number of links on that page |

Don't be scared by the math, and bear with us for a moment. What this formula says is as simple as the following. The PageRank of a page is the sum of two components: the probability of landing on that page due to chance, and the probably of arriving there from another page. The *dumping factor* d allows you to control, as a weight, how much emphasis you want to give the first and the second components. The first component is obtained by dividing 1 by the total number of pages in the graph and is hence a uniform distribution. The second component is obtained by summing the partial PageRanks coming from the pages that have a link to the page for which you are computing the PageRank. The partial PageRank of a page is its PageRank divided by the number of links appearing on that page. It's as simple as that!

One thing to notice in the definition of the PageRank is that it is recursive. The PageRank of each page depends on the PageRank of the incoming neighbors. How can you then compute the PageRank values? Every vertex PageRank depends on that of its neighbors. You need a starting point. Well, you can assign an initial value to each page and then iteratively compute the new PageRank value for each page with the formula. Although each vertex starts with the same initial value—typically 1 divided by the number of vertices in the graph—to mimic the initial random choice of page to start the random surf, iteration after iteration certain pages collect a higher PageRank due to the underlying structure of the graph. After some iterations, the PageRank values converge to stability, meaning the values change only slightly between the previous value and the next one. The pseudocode for PageRank is presented in Listing 4-5. Figure 4-7 shows the flow of the algorithm in Giraph when using a dumping factor of 0.85.

Listing 4-5. PageRank

```
function compute(vertex, messages):
  if getSuperstep() == 0:
    vertex.setValue(1 / getTotalNumVertices())        #1
  else:
    prSum = sum(messages)                             #2
    pr = (1 - D) / getTotalNumVertices() + D * prSum  #3
    vertex.setValue(pr)
  msg = vertex.getValue() / vertex.getNumEdges()      #4
  sendMessageToAllEdges(vertex, msg)
  if getSuperstep() == NUMBER_OF_ITERATIONS:
    vertex.voteToHalt()
```

#1 Initialization of the vertex value
#2 The sum of the partial PageRank values
#3 Computation of PageRank following the formula
#4 Current partial PageRank value to be sent

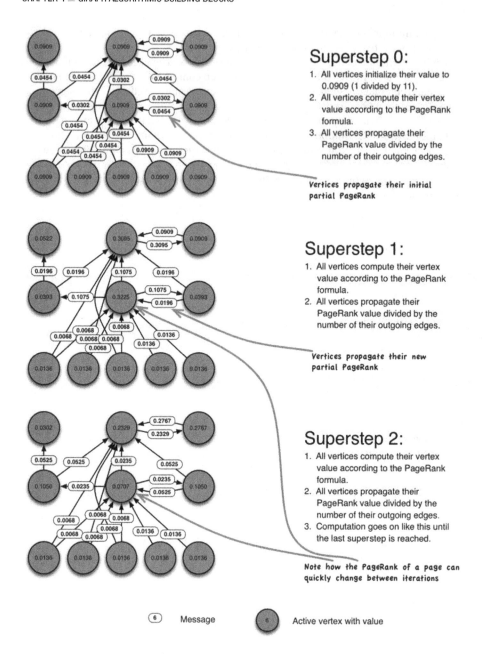

Superstep 0:

1. All vertices initialize their value to 0.0909 (1 divided by 11).
2. All vertices compute their vertex value according to the PageRank formula.
3. All vertices propagate their PageRank value divided by the number of their outgoing edges.

Vertices propagate their initial partial PageRank

Superstep 1:

1. All vertices compute their vertex value according to the PageRank formula.
2. All vertices propagate their PageRank value divided by the number of their outgoing edges.

Vertices propagate their new partial PageRank

Superstep 2:

1. All vertices compute their vertex value according to the PageRank formula.
2. All vertices propagate their PageRank value divided by the number of their outgoing edges.
3. Computation goes on like this until the last superstep is reached.

Note how the PageRank of a page can quickly change between iterations

⑥ Message   ⑥ Active vertex with value

Figure 4-7. *Three iterations of PageRank in Giraph*

Analyzing the algorithm with respect to the points presented at the beginning of the chapter, you can outline the following decisions:

- *Independent vertex decisions*: Each vertex computes its PageRank based on the incoming messages and the number of vertices (and the constant d). The computation is thus completely independent for each vertex.

- *Initialization of vertex values*: Each vertex is initialized based on the number of vertices in the graph. This value is used to compute the first iteration.

- *Halting condition*: Ideallys the algorithm would complete at convergence. Because convergence is reached only asymptotically, PageRank is usually computed for a fixed number of iterations (often fewer than 10 to 15 iterations). The halting condition is based on this number. An alternative approach requires you to compute how much the PageRank values of the vertices have changed between two iterations and use a threshold to decide to halt the computation.

- *Aggregators and combiners*: An aggregator could be used to implement the second halting condition just mentioned. This would require having an aggregator collect the PageRank values between two iterations to perform the thresholding. A combiner can also play an important role in this algorithm, minimizing the number of messages that need to be transmitted.

Listing 4-6 shows the pseudocode for the combiner.

Listing 4-6. SumValues PageRank Combiner

```
function pageRankCombiner(messages):
  return sum(messages)
```

The combiner sums the values sent to a particular vertex; its semantics correspond to the line prSum = sum(messages) from Listing 4-5. Being the sum of an associative operation and a commutative operation, the combiner works transparently to the normal execution of the algorithm.

PageRank, with its model of the random surfer (or random walker), has been used for a number of applications in addition to ranking web pages. For this reason, understanding this algorithm and how to implement it in Giraph may prove useful to you; you may find an application or adaptation that can help you with a problem. This example is more than just the implementation of the PageRank algorithm; it shows a pattern in the design of an algorithm in Giraph. In particular, the algorithm implements a (stationary) iterative method that can be used to compute an approximate solution to a system by iteratively improving the solution until convergence. In this case, the vertex values represent the variables of the system represented by the graph and the PageRank formula. The algorithm discussed in the next section is similar in this respect.

Predicting Ratings to Compute Recommendations

Chapter 2 presented a use-case scenario that used a graph to model the relationships between users and items rated by those users. You saw a high-level way to build a recommender system on top of such a graph. This section presents an algorithm that follows the *collaborative filtering (CF)* approach to predict the ratings a user will give unrated items, based on the user's past ratings. You first learn how to model users, items, and ratings as a graph, and then you see the design of an algorithm in Giraph to predict ratings. This section contains some math, but it's broken down for you, and you see how to implement it in the code. You do not have to remember the precise mathematical formulation—just the idea behind it.

Modeling Ratings with Graphs and Latent Vectors

Think of all the sites where users can rate items like movies, books, and so on using, for example, stars on a scale from 1 to 5. An interesting question is whether you can predict, based on past ratings only, the ratings a user will give unrated items. If you can, then you can recommend to each user items with the highest predicted ratings. Many techniques try to achieve this goal, and discussing their differences goes beyond the scope of this book. This section examines the *stochastic gradient descent (SGD)* algorithm and how to implement it in Giraph. You learn how this algorithm requires minimal modification to implement another popular algorithm in this class: the *alternating least squares (ALS)* algorithm.

Figure 4-8 shows a conceptual version of a ratings database. Each column represents a user, and each row represents an item. Hence, each cell contains a user's rating for an item. A question mark appears in each cell where no rating was provided; these are the ratings you want to predict. The basic assumption is that a user has assigned ratings based on a profile that describes the user's taste. The other assumption is that the same kind of profile can be used to describe the items. Conceptually, you can think of this profiling data as a vector of variables, where each element of the vector represents a dimension of a profile space. You can think of each dimension as something like a genre or other feature describing the user/item space, but in practice this space is abstract and cannot be mapped to a human-friendly representation of particular genres or features. The elements of the vectors represent variables that are not directly observed by the system about the user's taste or the item. For this reason, they are usually called *latent variables*, and the vector is called the *latent vector* of a user and of an item. Figure 4-9 shows such a project space of two dimensions, where items and users that are close together have similar latent vectors.

	Mark	Anna	John	Sophia
The white album	?	3	?	?
War games	?	4	5	?
The social network	3	?	1	?
Kid A	2	?	?	3

Figure 4-8. *An example of a database of ratings. A question mark (?) indicates ratings that you want to predict*

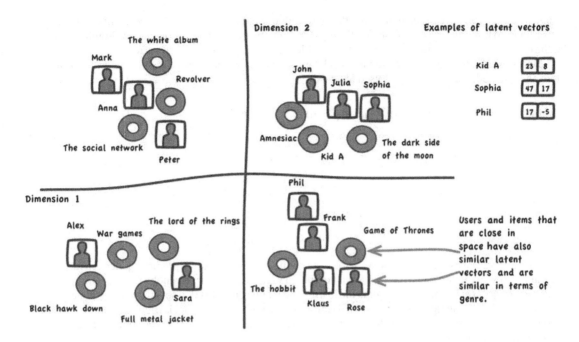

Figure 4-9. *An example of a two-dimensional space of features of latent vectors. Users and items that are close together should have a respectively high predicted rating*

■ **Definition**   The dot product between vector [1, 2, 3] and vector [4, 5, 6] is the sum of the products of the corresponding elements: $(1 * 4) + (2 * 5) + (3 * 6) = 32$.

The interesting thing about these latent vectors is that if you compute a dot product between the latent vector of a user and the latent vector of an item, you can interpret the result as a prediction of a rating that the user would give to that particular item. The extent to which this prediction will be correct depends on the values that compose the vectors. With your algorithm, you want to find for each user and each item the latent vectors that will produce the most accurate predictions of ratings. But how can you do that? You can use the ratings you know already. To start, you initialize the latent variables of each user and item at random, and you compute the predicted rating for each user-item pair for which you know the correct rating. You can then compare the predictions with the known ratings, and you can try to update the latent variables to minimize the error. In principle, if you can get your set of variables to predict correctly the past ratings, they should be able to predict the future ones correctly. In other words, you want to train your system, and you want to use the known ratings as a training set.

Minimizing Prediction Error

The question now is, how do you modify the latent variables to minimize the prediction error? In the answer to this question lies the difference between SGD and ALS (and other optimization methods of this kind). First, the procedure is the same from the perspective of both a user and an item. The principle is that for a given user and their latent vector, you take the latent vectors of the items that user has rated, and you compute the predicted ratings for those items through a dot product. Conversely, for a given item and its latent vector, you take the latent vectors of the users who have rated that item, and you compute the predicted ratings of those users to that item through a dot product. You then modify the latent variables accordingly to minimize this error. Before digging in to how you perform this in parallel in Giraph, let's look at the update function that modifies the latent variables to minimize the prediction error.

According to SGD, the update function of a latent vector for a user, given another latent vector of an item and the prediction error, can be used like this:

$$update(user_i) = user_i - \alpha(\lambda \cdot user_i - error \cdot item_j)$$

| The new latent vector of a user is | the old latent vector minus alpha times | lambda times the old latent vector of the user minus | the prediction error times the latent vector of the item |

This formula gives you a new latent vector that produces a more accurate prediction. Keep in mind that the same formula can be used to compute the new latent vector for an item, given the latent vector of a user and the prediction error. Alpha represents the learning rate and can be used to speed up convergence at the cost of suboptimal convergence, and lambda is the regularization constant that is used to avoid over-fitting to the training data and improve prediction accuracy. As mentioned, the error can be computed via the dot product between the two vectors.

To this point, the update to a user latent vector or item latent vector has been a one-time operation. Chapter 2 introduced the concept of a *bipartite graph*: a graph that has two types of vertices, with the vertices of the first type connected only to the vertices of the second type, and vice versa. In this case, one type is users and the other type is items. You connect two vertices of different types with an edge if a rating was issued. The edge weight is used to store the rating. You keep the edge logically undirected by having two specular edges, one for each direction. Now comes the tricky part: you organize the computation so that at a given iteration, only one class of vertices is updated. During an iteration, you let only the vertices of one type execute their update function; during the following iteration, you let only the second type of vertices execute it. You continue like this, alternating which vertex type is updated during each iteration. For example, during an iteration, you let only users update their latent vectors based on the latent vectors of the items, and during the following iteration you let only the items update their latent vectors based on the latent vectors of the users (computed during the previous iteration). You divide the iterations according to their number and let one type of vertex execute during odd iterations and the other during even iterations.

This mechanism allows you to update the latent vectors of the users based on the freshly computed latent vectors of the items, and vice versa. This creates a "ping pong" computation, where a group of vertices updates their values based on the values of the other group. This computation continues for a number of iterations or until the latent vectors converge to a stable state: that is, they do not change more than a certain threshold between two iterations. Letting both types of vertices update their latent vectors during the same iteration would yield inaccurate results.

It is important to understand what is happening in the graph during the computation. You can see from the formula that the latent vectors are recursively interleaved. The latent vector of a user is updated based on the latent vectors of the items the user has rated, and these items' latent vectors are updated as well, based on the other users who have rated them. And these users may have rated different items whose latent vectors have been influenced by yet another set of users. The latent vectors are influenced not only by direct neighbors but also throughout the entire graph, iteration after iteration. In practice, the user vertices act as bridges between the item vertices (because users are not directly connected), and vice versa. The computation of the latent vectors of each vertex in the graph is in the end influenced by the latent vectors of all the other vertices.

In the Giraph model, SGD is implemented following this pattern. At each superstep, only one of the two types of vertices is active and able to update the latent vectors; the other type is inactive. The latent vectors from the neighbors used to update a single latent vector are transmitted from each vertex to its neighbors through messages. Edge values are used to store the ratings.

Listing 4-7 presents the code for SGD, and Figure 4-10 shows the computation of the algorithm in Giraph. In particular, note in the figure how only one type of vertex is active at each iteration.

Listing 4-7. Stochastic Gradient Descent (SGD)

```
function compute(vertex, messages):
  if getSuperstep() == 0:
    value = generateRandomLatentVector()                    #1
    vertex.setValue(value)                                  #1
    if vertexType(vertex.getId()) == getSuperstep() % 2: #2
      propagateValue(vertex, value)
  elif getSuperstep() < NUMBER_OF_ITERATIONS:
    errorRMSE = 0
    for message in messages:
      rating = vertex.getEdgeValue(message.getSourceId())
      msgValue = message.getValue()
      oldValue = vertex.getValue()
      newValue = computeValue(oldValue, msgValue, rating)  #3
      error = computeError(newValue, msgValue, rating)     #4
      vertex.setValue(newValue)
      errorRMSE += pow(error, 2)
    aggregate(errorRMSE)                                    #5
    propagateValue(vertex, vertex.getValue())
  vertex.voteToHalt()

function computeError(value, msgValue, rating):
  prediction = value · msgValue                             #6
  return max(min(prediction, 5), 1) - rating

function computeValue(value, msgValue, rating):
  error = computeError(value, msgValue, rating)
  return value - ALPHA * (LAMBDA * value - error * msgValue) #7

function propagateValue(vertex, value):
  message = Message(vertex.getId(), value)
  sendMessageToAllEdges(vertex, message)
```

#1 Latent vectors are initialized at random.
#2 You divide the vertices according to their type.
#3 Compute the new latent vector.
#4 Compute the error with the new latent vector.
#5 Aggregate the error.
#6 Use a dot product to compute the prediction error.
#7 This is the actual SGD formula in the code.

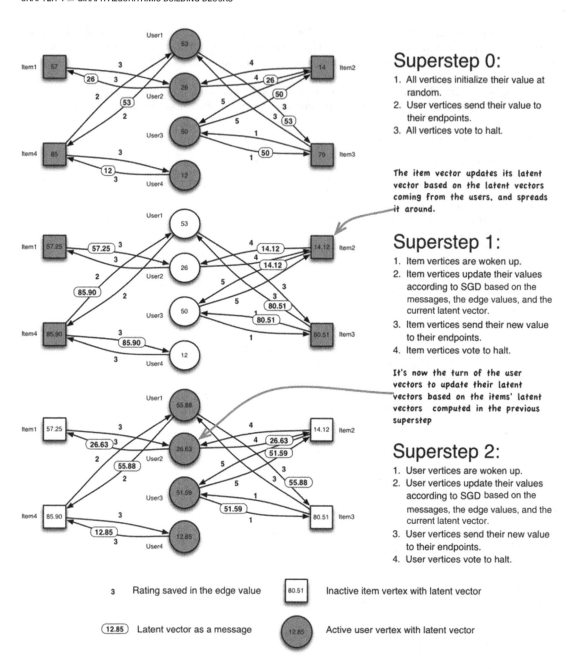

Superstep 0:

1. All vertices initialize their value at random.
2. User vertices send their value to their endpoints.
3. All vertices vote to halt.

The item vector updates its latent vector based on the latent vectors coming from the users, and spreads it around.

Superstep 1:

1. Item vertices are woken up.
2. Item vertices update their values according to SGD based on the messages, the edge values, and the current latent vector.
3. Item vertices send their new value to their endpoints.
4. Item vertices vote to halt.

It's now the turn of the user vectors to update their latent vectors based on the items' latent vectors computed in the previous superstep

Superstep 2:

1. User vertices are woken up.
2. User vertices update their values according to SGD based on the messages, the edge values, and the current latent vector.
3. User vertices send their new value to their endpoints.
4. User vertices vote to halt.

3	Rating saved in the edge value
(12.85)	Latent vector as a message
80.51	Inactive item vertex with latent vector
12.85	Active user vertex with latent vector

Figure 4-10. *The flow of the computation of SGD in Giraph*

Note that the operations on the vectors are actually vector operations, so you have to implement the dot and scalar products yourself. Also, this code assumes that given a vertex ID, you can identify the type of vertex (user or item)—for example, via a prefix in the ID. This is what vertexType() does: it returns a value of 0 or 1 depending on the type of vertex. The code assumes a class called Message that is a wrapper for the normal message and also stores the ID of the sender vertex, because you need to identify in which edge to store the respective latent vector.

Analyzing the algorithm with respect to the points presented at the beginning of the chapter, you can outline the following decisions:

- *Independent vertex decisions*: Each vertex updates its latent vector based on the incoming latent vectors and its current latent vector. This way, each vertex can independently compute its latent vector in parallel.

- *Initialization of vertex values*: Each latent vector is initialized randomly during the first superstep so that during the following ones, the compute() method can update them according to the SGD formula.

- *Halting condition*: Ideally, the algorithm would complete at convergence. Because a convergence is reached only asymptotically, usually SGD is computed for a number of iterations. The halting condition is based on this number. An alternative approach requires computing how much the overall mean prediction error has been improved between two iterations. When this error is not improved overall a certain threshold, the computation can halt.

- *Aggregators and combiners*: An aggregator could be used to implement the second halting condition mentioned. This would require having an aggregator collect the RMSE values between two iterations to perform the thresholding. Combiners have no use in this application, because you need to keep the latent vectors distinct to compute the SGD formula.

Listing 4-8 shows the pseudocode for the aggregate() method for the aggregator used in SGD.

Listing 4-8. SGDAggregator

```
function aggregate(value):
  setAggregatedValue(getAggregatedValue() + value)
```

The aggregator is initialized to 0 at the beginning of each superstep, and it sums the values sent by each vertex. At the end of the computation, the aggregator will contain the sum of all the errors at the last superstep. By dividing this value by the number of ratings, you can compute the average error. This aggregator is just summing values, so there is nothing specific to SGD; it is usually called SumAggregator, which can be used as is in different algorithms.

Keep in mind that this algorithm only computes the latent vectors. The latent vectors are then used by another application to compute recommendations. The naive approach would be to perform a dot product between each user and all the items that user has not rated yet, and select the top-K. With millions of users and items, this approach most likely is not feasible. Instead, you can use the topology of the graph and compute predictions only between a user and the items that have been rated by other users who have some ratings in common with that user.

Another interesting aspect of this algorithm is that with a little modification, you can also implement another recommendation algorithm such as ALS. The only part you need to modify is the logic behind computeValue(), which is responsible for the method-specific math.

Now that you have learned how to rank vertices in a graph, you are ready to move to the next problem: how to identify communities in a social network. The next section presents a very popular algorithm called *label propagation*.

Identifying Communities with Label Propagation

Social networks often share properties in their graphs. Look, for example, at Figure 4-11, which shows a social network of individuals. It is a simplified example, and the structure is overemphasized, but it should make the point. What do you see in this figure? The graph is divided in *communities*, or *clusters*. Communities are groups of vertices that tend to be connected to each other more than they are connected to vertices outside the community. Two good friends tend to have many friends in common—not all of them, but many. And these friends they have in common also tend to be friends with each other—again, not all of them, but this is the characteristic of groups of friends (and other groups, such as colleagues). In graph terms, vertices that are embedded in a community tend to have a high *clustering coefficient*, which measures the extent to which the neighbors of a vertex are also connected with each other.

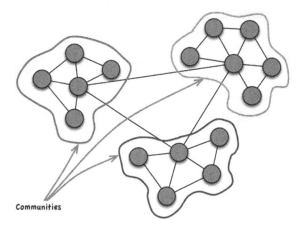

Communities

Figure 4-11. *Example of a community structure in a graph*

Identifying communities is an interesting problem that may yield useful information. For instance, you could recommend other users to connect to. Another option is to use the identified community structure to study other dimensions, perhaps for analytics. Does information exit the boundaries of communities (such as shared pictures or posts)? Do people who belong to the same community share characteristics such as age, hobbies, or type of work?

The basic idea of identifying communities is to assign labels to vertices, where each label represents a community. Typically, community-detection algorithms do not require specifying the *number* of communities in advance, as clustering algorithms tend to require (for example, k-means). The idea is that you should find those that are in the graph (you do not know how many). Keep in mind that in the real world, individuals belong to multiple communities at the same time: they have different groups of friends, co-workers, family, and so on. Algorithms that assign multiple labels (multiple communities) to vertices are also said to identify *overlapping* communities. This section, however, focuses on a simple algorithm called the *label propagation algorithm (LPA)* that does not detect overlapping communities, but that can be (and has been) adapted to accomplish that goal.

How do you assign labels to vertices? Well, the idea is simple and the intuitions behind it are as follows. First, at the end of the computation, each vertex in the same community should have the same label. If two vertices belong to the same community, then this should be reflected by the fact that they have the same label. Second, a vertex should have the same label as its neighbors, or at least most of them (remember that a user might have friends who belong to a different community, whose members are closer friends with that user).

The algorithm works as follows. Each vertex has a label, initially its own ID, that it sends to its direct neighbors through messages. The simple heuristic that each vertex applies is to acquire the label that is occurring most frequently among its neighbors, breaking ties randomly (that is, if two labels occur most frequently across neighbors and have the same frequency, choose randomly). When a vertex acquires a new label, it sends the label to its neighbors. This is why the algorithm is called the label propagation algorithm. For example, Lady Gaga initially propagates the string "Lady Gaga", which is her ID. At each iteration, each vertex acquires a new label if it finds a label that occurs more frequently across its neighbors than its current one. In the case of Lady Gaga, after some iterations, the label appears more frequently in the neighborhood of her fans should be "Lady Gaga". In that case, the vertex propagates the label to its neighbors through a message. The algorithm halts when no vertex changes label. At the end of the computation, each vertex holds the label representing its community. This label corresponds to the ID of a vertex in its community—a vertex that is more "central" to the community it represents (in this example, "Lady Gaga").

The algorithm is as simple as that. At the first iteration, there are as many labels as vertices. When each vertex receives IDs from its neighbors (with each label occurring exactly once), it must choose randomly among them. IDs of vertices with many outgoing edges are more likely to be chosen by the random tie-breaking heuristic. In the following iteration, more vertices in the neighborhoods of these vertices have that same label, hence increasing the likelihood of that label being assigned to more vertices. Iteration after iteration, labels spread in the neighborhoods and compete with each other, until a point of equilibrium is reached and no more label changes occur. How and when this point of equilibrium is reached depends on the topology of the graph and on its community structure.

■ **Warning**   Running LPA in a synchronous system like Giraph requires some ad hoc arrangements, because it can result in unstable states in which the algorithm oscillates between two or more solutions and never halts.

To run LPA in Giraph, you must modify the algorithm slightly. The reason is that LPA does not play well with synchronous systems like Giraph. Running LPA in synchronous systems can cause unstable states where some vertices keep oscillating between two or more labels, avoiding convergence. Consider this simple example. Imagine a graph with two vertices A and B, connected only to each other. At the first superstep, vertex A receives label B and vertex B receives label A. At this point, each vertex decides to acquire the received label, because it is the one occurring most frequently across their neighbors, and each vertex sends its label again through a message. At the next superstep, the same situation occurs, but with inverse labels. The problem is that the algorithm continues, with the two vertices trying to "catch" each other forever. This happens not only with the trivial case of two vertices connected by one edge, but also with more complex topologies. The good news is that this can be fixed. First you need to ensure that a vertex has a deterministic way to break ties—for example, by acquiring the smallest label between the (equally) most frequently occurring labels. Second, a vertex needs to avoid re-acquiring the label it had before the current label. This guarantees that vertices will reach a non-oscillating state.

Listing 4-9 presents the pseudocode of LPA. LPA works best on undirected graphs, so you assume the graph has been converted to a logically undirected graph. Figure 4-12 shows the flow of LPA in Giraph. Note that for the sake of presentation, the figure shows the flow of the algorithm without the ad hoc modifications to work in Giraph. The code uses an LPAValue object as a vertex value that contains both current and previous values, to allow you to implement the technique mentioned to guarantee convergence. Also, note that you store labels in edge values. This is because vertices send new labels only when they change. When a vertex computes label frequencies, it needs the labels of all the neighbors, including labels sent a few supersteps earlier. This is why you store labels in the corresponding edge. This comes in handy because labels are easily serialized for checkpointing by Giraph, and you don't need additional data structures for labels.

Listing 4-9. Label Propagation Algorithm (LPA)

```
function compute(vertex, messages):
  if getSuperstep() == 0:                                        #1
    vertex.setValue(LPAValue(vertex.getId(), vertex.getId()))    #1
    propagateValue(vertex, vertex.getId())                       #1
  else:
    for message in messages:                                     #2
      label = message.getValue()                                 #2
      source = message.getSourceId()                             #2
      vertex.setEdgeValue(source, label)                         #2
    labels = DefaultDictionary(0)
    for edge in vertex.getEdges():                               #3
      labels[edge.getValue()]++                                  #3
    mfLabel = computeMostFrequentLabel(labels)
    cLabel = vertex.getValue().getCurrent()
    lLabel = vertex.getValue().getLast()
    if labels[cLabel] < labels[mfLabel] and mfLabel != lLabel:   #4
      vertex.setValue(LPAValue(cLabel, mfLabel))                 #4
      propagateValue(vertex, mfLabel)                            #4
  vertex.voteToHalt()

function computeMostFrequentValue(labels):
  maxFreq, mfLabel = -INFINITY, None
  for label, freq in labels:
    if freq > maxFreq or (freq == maxFreq and label < mfLabel):  #5
      maxFreq, mfLabel = freq, label
  return mfLabel

function propagateValue(vertex, value):
  message = Message(vertex.getId(), value)
  sendMessageToAllEdges(vertex, message)

class LPAValue:
  function LPAValue(lastLabel, currentLabel):
    this.lastLabel = lastLabel
    this.currentLabel = currentLabel
  function getCurrent():
    return this.currentLabel
  function getLast():
    return this.lastLabel
```

#1 Vertices use their own ID as the initial label.
#2 Vertices store the new labels as edge values.
#3 Occurrences of labels are computed.
#4 The current label is updated if necessary.
#5 Find the most frequently occurring label with the smallest ID.

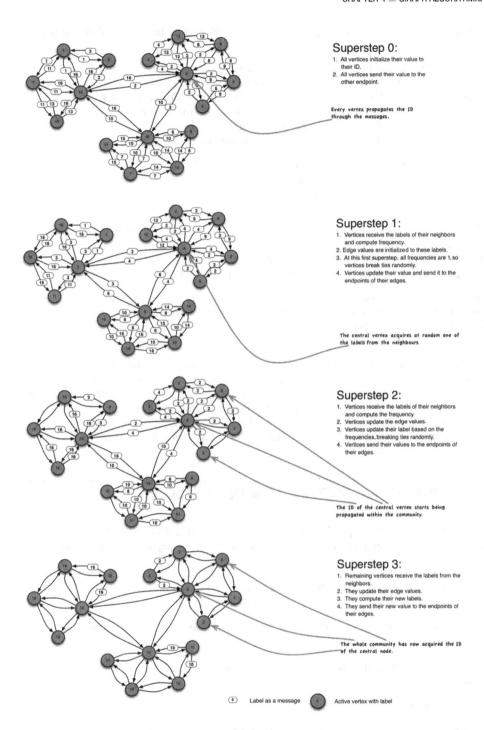

Superstep 0:
1. All vertices initialize their value to their ID.
2. All vertices send their value to the other endpoint.

Every vertex propagates the ID through the messages.

Superstep 1:
1. Vertices receive the labels of their neighbors and compute frequency.
2. Edge values are initialized to these labels.
3. At this first superstep, all frequencies are 1, so vertices break ties randomly.
4. Vertices update their value and send it to the endpoints of their edges.

The central vertex acquires at random one of the labels from the neighbours

Superstep 2:
1. Vertices receive the labels of their neighbors and compute the frequency.
2. Vertices update the edge values.
3. Vertices update their label based on the frequencies, breaking ties randomly.
4. Vertices send their values to the endpoints of their edges.

The ID of the central vertex starts being propagated within the community.

Superstep 3:
1. Remaining vertices receive the labels from the neighbors.
2. They update their edge values.
3. They compute their new labels.
4. They send their new value to the endpoints of their edges.

The whole community has now acquired the ID of the central node.

⬡ Label as a message   ⬤ Active vertex with label

Figure 4-12. *The flow of the computation of LPA in Giraph*

Analyzing the algorithm with respect to the points presented at the beginning of the chapter, you can outline the following decisions:

- *Independent vertex decisions*: Each vertex chooses its label according to the label of the neighbors and its current label. Because the neighbors' values also depend on their respective neighborhood, a vertex can only consider its direct neighbors for its own label. Hence it has all the necessary information in these elements. The tie-breaking strategy is also important, because due to its deterministic nature (it is based on an ordering of labels), it is consistent across different vertices.

- *Initialization of vertex values*: Each vertex uses its own ID as the initial value. This value is by definition unique, which makes it possible to uniquely identify communities at the end of the computation.

- *Halting condition*: Each vertex propagates its value only when the vertex changes the value. The vertex votes to halt at the end of each superstep. Hence, when no vertex changes its label, no message is produced, and all the vertices are inactive. Because of the synchronous model implemented by Giraph, you have to introduce the tie-breaking heuristic described earlier to guarantee that the job will halt.

- *Aggregators and combiners*: Combiners are not usable given the way the algorithm is implemented. You want to keep the messages uncombined because you need to keep track of senders. An aggregator could be used to keep track of the number of communities found and their size.

You have seen how to detect communities with an algorithm that propagates membership to communities following the Giraph information-propagation pattern. The algorithm is very scalable and simple to implement. The algorithms presented so far should give you a taste of how to design an algorithm for Giraph and what to pay attention to. Hopefully you'll be inspired to get your hands dirty with the Giraph API.

By now you should understand what is relevant and what to avoid when designing an algorithm that scales. For convenience, the following checklist summarizes the items you should consider when designing an algorithm for Giraph.

GIRAPH ALGORITHM DESIGN CHECKLIST

Keep he following things in mind when you design an algorithm for Giraph. Don't consider this a definitive and complete selection of items that will guarantee scalability; but this list should bootstrap your analysis in the right direction.

- The decisions a vertex makes are based on local information, such as its current vertex and edge values; the incoming messages; and the current superstep.

- A vertex sends messages mostly to known vertices, such as its neighbors, and it does not have to look up vertex IDs.

- The graph and the algorithm do not imply patterns in which vertices receive messages from all other vertices.

- The halting condition is clear and consistent.

- Vertex values have a well-defined initial value that is coherent with the heuristics applied by the vertices.

- If possible, messages are combined.

- A vertex value's size is not proportional to the size of the graph.

- The algorithm considers side cases like a vertex with no edges or only outgoing or incoming edges, and so on.

- The size of messages does not increase too much over time (for example, exponentially), such as when incoming messages are concatenated and propagated to all neighbors iteration after iteration.

- Although the API supports adding and removing vertices and edges, the algorithm should not assume that an entire graph is constructed iteration after iteration from scratch.

- Aggregators are associative and commutative functions, and they do not occupy space proportional to the size of the graph (such as a "fake" aggregator that stores all the incoming values in memory).

- When a received message needs to be reused, if possible store it locally instead of requiring it to be sent again, to save network I/O.

Characterizing Types of Graphs and Networks

Graph algorithms can be useful to build applications and run analytics on your graph. Depending on your business scenario, if you are modeling your data as a graph, you can find a bunch of graph algorithms that will prove useful to you. Some of them will probably belong to one of the classes of algorithms presented so far in this chapter. However, running graph algorithms on your graph can also help you better understand the processes or phenomena underlying your data. For example, you could study the type of social interactions that result in certain relationships in a social network, or the hyperlink structure of web pages.

Although all graphs consist of a number of vertices connected by edges, modeling various aspects of the world, the topologies of graphs modeling different data often have characteristics in common. There are many *types* of graphs, each with specific properties, many of which are common in real-world graphs—in particular the graphs described so far, such as social networks, the Internet, and the Web. This section presents a selection of the most common graph types and explains how to recognize whether a graph is of each type. It is very important for you to get to know your graph, and some of these algorithms are tools that allow you to do so.

Three characteristics are at the core of this analysis:

- *Average path length*: The average number of hops between two nodes chosen at random. This number is usually compared to the number of vertices in the graph.

- *Degree distribution*: The way vertex degrees are distributed across the vertices. Do vertices have approximately the same number of edges, or do a few have many edges while most of the others have a small number of edges?

- *Clustering coefficient*: The extent to which vertices are embedded in tightly connected clusters.

These are only some of the graph characteristics you may want to look at to identify a graph type, but they are usually sufficient to give you a good picture of the graph. Note that these characteristics relate only to the topology of the graph and do not depend on what the graph actually represents (web pages, individuals, users and items, and so on).

For example, consider a graph that represents the United States' numbered highway system. As you saw in Chapter 2, the road network can be modeled with a graph by using a vertex for each point where multiple roads can be chosen—such as a crossing, town, or city—and edges to represent the roads. If you look at Figure 4-13 and consider the road network as a graph, you can see that the graph has the following properties:

- All vertices tend to have a low degree. Most have five or six edges, some more, some fewer. This is natural, because crossings and cities are connected directly only to the closest neighbors.

- The clustering coefficient is very low, and neighbor vertices tend to share very few neighbors, if any.

- The graph has a very large average path length. If two cities are far from each other, such as New York and Los Angeles, the path that connects them in the graph is long.

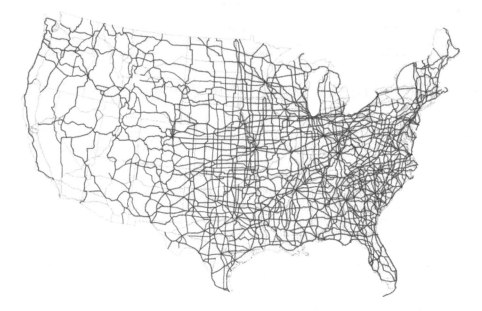

Figure 4-13. *The United States numbered highway system*

These properties are expected, because a road network resembles a *grid* (or lattice, or mesh).

As the name hints, a grid is a graph that, when drawn, has regular tiles, like squares. Figure 4-14 shows such a graph with 15 vertices. This is a very regular graph with a large average path length, which increases quickly as the graph grows in the number of vertices and edges. The degree distribution is very regular: the vast majority of the vertices have exactly four neighbors, except the vertices at the borders of the graph. The average clustering coefficient is zero, because vertices do not share any neighbors with their neighbors. This is a boring graph, and few real-world graphs have such a strict topology (although the road network in Figure 4-13 resembles this topology in a more relaxed way). However, as a toy graph, it serves the purpose of illustrating the discussion. Things get more interesting in a second.

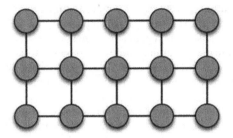

Figure 4-14. *Example of a grid graph*

Now look at the routes of domestic airlines in the United States, shown in Figure 4-15. This is a very different graph, although it still connects cities in the United States. First, notice how certain (few) cities have a very high degree, whereas most have a low degree. Those with high degree are called *hub airports*, because they can be used to connect many cities with connecting flights. In graph terms, vertices with this kind of role in a graph are also called *hubs*. This degree distribution resembles a power-law distribution. This kind of connectivity pattern allows an important drop in the average path length. Graphs with these type of connectivity are also called *scale-free networks*, because when you add vertices to the graph, the average shortest path length tends to remain low. Every city can reach any other city by passing through hubs, without the need to connect every city with every other city. The clustering coefficient of this graph is still pretty low, because airlines try to remove redundant flights that can be replaced by two connected routes through hubs. Of course, the flight routes do not allow all cities to be reached—only those that have a (large enough) airport.

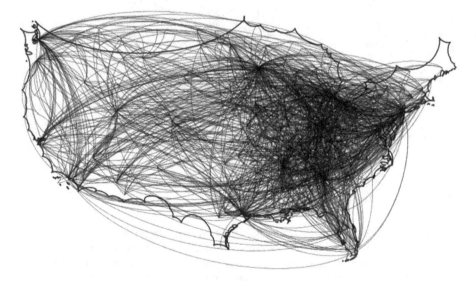

Figure 4-15. *United States domestic flight routes*

Suppose you merged the two graphs together, connecting cities through roads and airlines routes. What would their effect be on each other?

Because you would be connecting cities far from each other through flight routes, you would still have a graph with a smaller average path length than the pure road map. Still, the average path length would be larger than the average path length of the airline map, because certain paths would need to be traversed through the road map. Also, the degree distribution of the merged graph would have a power-law distribution, due to the hubs. The clustering coefficient would remain low, because both graphs have a low clustering coefficient and you would conveniently only add long-range connections to the road map. In graph terms, merging the two graphs is analogous to *rewiring*.

Let's go back to the lattice in Figure 4-14. Take a number of edges from this graph, and rewire them at random. This means you take a number of the edges (say, 18%) and reconnect one endpoint to a vertex chosen at random in the graph. How does this rewiring affect the three characteristics of the graph? Well, most important, it decreases the average path length. In the grid topology, the average path length tends to be large because there can be vertices that are far from each other, such as those at far left and those at far right in Figure 4-14. Figure 4-16 presents such a rewired graph. Rewiring vertices in the grid increases the chance that areas that are far from each other in the graph are connected through the rewired edge. Basically, these rewired edges act like bridges. Shortest paths can now traverse these bridges, with an impact on their length. As far as the degree distribution is concerned, because you choose endpoints at random, you may end up adding more edges to certain vertices than the edges you remove from those same vertices. Hence, their degree may change, but on average edges should still be uniformly distributed. This means you have a slightly different degree distribution than the deterministic distribution of the grid. In other words, the average vertex degree should be very similar, but this time the deviation should be larger than in the grid topology. For the same reason, the average clustering coefficient should be different than 0 (although still very small), because there is a certain likelihood that a rewired edge introduces a triangle.

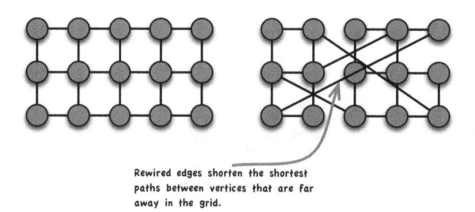

Rewired edges shorten the shortest
paths between vertices that are far
away in the grid.

Figure 4-16. *The grid and the rewired grid with 18% of edges rewired at random*

To take this example to an extreme, you can generate a fully random graph, by taking the 15 vertices and connecting them completely at random with the same number of edges as before (continuing to rewire at random until you obtain a connected graph). For the same reasons discussed, this random graph, while having the same number of vertices and edges, has an even shorter average path length, a degree distribution with a similar mean but higher deviation, and a slightly higher clustering coefficient (although still small). Figure 4-17 shows such a graph. Grids and random graphs are not very realistic representations of real-world networks, but they serve to introduce the three characteristics that are commonly used to describe graph types.

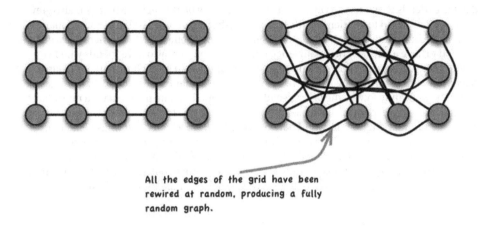

All the edges of the grid have been
rewired at random, producing a fully
random graph.

Figure 4-17. A random graph can be obtained by rewiring all the edges at random

Before moving on to a realistic graph, let's go back to the rewiring of the grid. If you used another rewiring strategy when rewiring a percentage of the edges, you could obtain a graph that resembles the graph that merges airlines map and the road map: it would have a power-law degree distribution. To obtain such a graph, when you choose the new endpoint of a rewired edge, instead of choosing randomly, you choose a vertex with a probability that is proportional to its degree. Vertices with high degree have a higher probability and are chosen more often. The more a vertex is chosen as the new endpoint for the rewired edge, the more likely it is that the next rewired edges will choose that vertex as an endpoint (due to its increasing degree). This process tends to make big vertices bigger. At the end of the process, you obtain a power-law distribution with few hubs and a long tail of vertices with low degree (this strategy is also called *preferential attachment*)—exactly like the merged graph. Figure 4-18 shows such a graph.

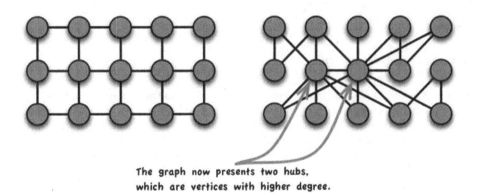

The graph now presents two hubs,
which are vertices with higher degree.

Figure 4-18. A grid with 60% of the edges rewired following preferential attachment

These examples have shown the relationships between graph topology and the degree distribution, the clustering coefficient, and the average path length. You have also learned about hubs and the way they influence these metrics. Let's now look at a final important type of graph. In the last decade, the study of real-world networks has shown that many of these networks have *small-world* properties (that in turn makes them *small-world networks* or graphs). Social graphs tend to have such properties. Think of your network of friends: you tend to have friends in the region where you live (which may be the region where you also grew up), and many of these friends tend to be friends with each other. After all, people tend to be part of many communities. This means social networks tend to have a high clustering coefficient. Such a network with a local nature would tend to have a large average path length. However, you may also have friends who live far from you and who belong to very different communities. For example, you might have a friend in Japan, which puts you only a few hops from many Japanese people. These particular friendships act like bridges in the rewired grid, which significantly lowers the average path length. Finally, some people have a much larger network of friends than the average, perhaps because they travel a lot, or due to their job. This means social networks also have hubs. Summarizing, small-world networks tend to have a high clustering coefficient, small average path length, and a power-law degree distribution. Small-world properties are the reason Milgram discovered such a small average path length in a large social network like the population of the United States. The same sort of experiment was recently executed on the Facebook social graph, revealing a small average path length of four to seven! Figure 4-19 shows a graph with small-world properties.

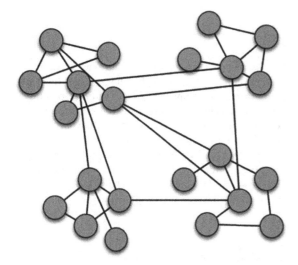

Figure 4-19. *A graph with small-world properties*

You need a better understanding of your graph in order to design, build, and master graph applications. If you were to build a social network such as Twitter, you would be interested in the way information (tweets) propagates through the followers network. In a social network with a low average path length, you know that only a few retweets would be necessary to reach distant areas of the graph. Moreover, a graph with a high clustering coefficient would increase the likelihood that information would reach a large portion of a community from which it started.

Graph topology has many more aspects, such as how resilient a graph is to attacks or failures, and so on; but network science constitutes a larger body of work than this book can present. However, with the tools provided in this chapter, you should be ready to start your journey toward a better understanding and management of your graph data.

Summary

Modeling data as a graph is the first step toward using Giraph. The next step is to design an algorithm that can process it:

- In Giraph, algorithms build on the principle of information propagation, where the state of a computation is distributed across the vertex values, and vertices share it through messages.

- Paths are sequences of edges that connect vertices. SSSP is an algorithm to compute the distance between a source vertex and all other reachable vertices, and it fits the Giraph model naturally.

- Paths can be used to define connected components, which are subgraphs where each vertex can reach any other vertex. Connected components can be used to study how information propagates in the graph.

- Certain vertices are more important than others, and it is useful to identify them. PageRank is an algorithm that lets you rank vertices according to their importance in the graph.

- If you model users and items and their ratings as a graph, you can compute recommendations for users about new items through the stochastic gradient descent (SGD) algorithm.

- Graphs, and in particular social networks, can be characterized by communities (or clusters) where vertices are tightly connected to each other. Label propagation allows you to identify these communities and label each vertex with the community it belongs to.

- Different types of graphs, regardless of the data they represent, are characterized by particular topological properties. Using a few metrics and some graph algorithms to compute them, you can find out the type of a graph.

Giraph provides a programming API that helps developers design scalable graph algorithms. You are now ready to dig into the Giraph Java API, to implement and run algorithms on a graph with a Hadoop cluster.

PART II

Giraph Overview

CHAPTER 5

■ ■ ■

Working with Giraph

Previous chapters introduced the generic Giraph programming model and talked about a few common use cases that lend themselves easily to being modeled as graph-processing applications. This chapter covers practical aspects of developing Giraph applications and focuses on running Giraph on top of Hadoop.

Writing a Giraph application boils down to plugging custom graph-processing logic into the Giraph framework. This is done by providing custom implementations of the interfaces and abstract classes that are later orchestrated by the Giraph BSP machinery. In this chapter, you learn how to quickly build graph-processing applications by extending a basic abstract class BasicComputation and providing a custom implementation of the compute function. This is the absolute minimum code that needs to be written in order to have a graph-processing application, and quite a few practical Giraph applications do just that; more complex ones, however, opt to use various pieces of Giraph's framework that this chapter reviews in detail.

You then look into what it takes to execute a Giraph application by submitting it to Hadoop as a MapReduce job. Because the Giraph framework is implemented on top of Hadoop, graph-processing algorithms don't care whether they are running on a single node or on a cluster of tens of thousands of nodes. The process of executing a Giraph application remains the same.

The rest of this chapter focuses on a single node execution via Hadoop's local job runner. Because the local job runner executes jobs in a single Java virtual machine (JVM), it has the advantage of being extremely easy to use with common IDEs and debugging tools (the only disadvantage being that you can't use a full dataset—you have to run it on a small sample of your graph data). Once you get comfortable with the implementation running locally, you can either go all the way to a fully distributed Hadoop cluster or take an intermediate step of running against a pseudo-distributed Hadoop cluster. The pseudo-distributed cluster gives users a chance to experience all the bits and pieces of Hadoop machinery while still running on a single host (they are running in different JVMs communicating over the loopback networking interface). Either way, the good news is that you don't have to change anything in your implementation—you just need to use a different set of configuration files. If you want to find out more about all the options for using Hadoop, head straight to Appendix A and then come back to this chapter.

Unlike traditional graph databases, Giraph excels at turning unstructured data into graph relationships. This chapter looks at how to achieve this flexibility via input and output formats and how you can either use built-in implementations as is or extend them to let Giraph slurp data from any source and deposit it to any kind of sink.

Finally, you see some of the more advanced use cases for the Giraph API, such as combiners for messages, master compute, and aggregators. This is quite a bit of infrastructure to review in a single chapter—buckle up for the ride!

"Hello World" in Giraph

Following the grand old tradition established by Brian Kernighan and his fellow C hackers at Bell Labs, your first Giraph application is "Hello World": as simple an implementation as possible, which outputs a few text messages. One notable difference between the classic "Hello World" and a graph-processing "Hello World" Giraph application is the need for an input. Your Giraph application can't function without a graph definition given to it; Giraph, after all, is a graph-processing framework.

For the rest of this chapter, you use the graph of Twitter followership that you saw in Chapter 3 (alongside the pseudo-code example of computing in-out-degrees). In this section, though, you start with much simpler code that works on the graph. All it does is print a "Hello World" message for every vertex (in this case, person) and output the neighbors the vertex is connected to (those the person is following). To make the code as simple as possible, it uses numbers instead of people's names. The output will eventually appear as follows:

```
Hello world from the 5 who is following: 1 2 4
```

All the output that you'll see is based on the graph Figure 5-1. It's a Twitter followership graph illustrating who is following whom on Twitter. Look at the node for #5. The arrows pointing toward nodes 1, 2, and 4 indicate that user #5 is following those other three.

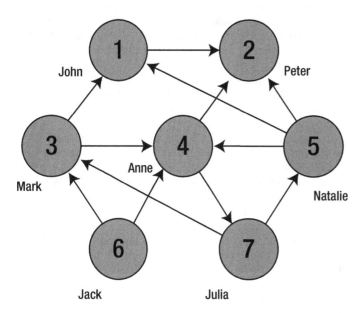

Figure 5-1. *Twitter followership directed graph*

Defining the Twitter Followership Graph

The simplest textual representation of the graph in Figure 5-1 that your Giraph application can recognize is the ASCII file shown in Listing 5-1. It encodes the adjacency lists for each vertex, with each line in the input file formatted as follows: vertex neighbor1 neighbor2....

Listing 5-1. Content of src/main/resources/1/graph.txt

```
1 2
2
3 1 4
4 2 7
5 1 2 4
6 3 4
7 3 5
```

If you're familiar with how input data is provided to Hadoop MapReduce jobs, you see that Giraph follows the same approach: Giraph applications process all the files in a given subdirectory (in this case, src/main/resources/1) and consider them to be part of the same graph definition. You could put every line (defining a given vertex's topology) into a separate file or group them in a number of files and achieve the same result. For example, the input shown in Listings 5-2 and 5-3 that consists of two files is considered identical to the input in Listing 5-1.

Listing 5-2. Content of src/main/resources/1.1/1.txt

```
1 2
7 3 5
```

Listing 5-3. Content of src/main/resources/1.1/2.txt

```
2
3 1 4
4 2 7
5 1 2 4
6 3 4
```

Creating Your First Graph Application

With the input data in place, let's now turn to the Java implementation of your first "Hello World" Giraph application (GiraphHelloWorld.java, shown in Listing 5-4). Remember that you are trying to print as many lines as there are vertices in a graph, with each line reading "Hello world from the X who is following: Y Z...."

Listing 5-4. GiraphHelloWorld.java

```java
import org.apache.giraph.edge.Edge;
import org.apache.giraph.GiraphRunner;
import org.apache.giraph.graph.BasicComputation;
import org.apache.giraph.graph.Vertex;
import org.apache.hadoop.io.IntWritable;
import org.apache.hadoop.io.NullWritable;
import org.apache.hadoop.util.ToolRunner;


// Giraph applications are custom classes that typically use
// BasicComputation class for all their defaults... except for
// the compute method that has to be defined
```

```
public class GiraphHelloWorld extends
   BasicComputation<IntWritable, IntWritable,
                    NullWritable, NullWritable> {
  @Override
  public void compute(Vertex<IntWritable,
                      IntWritable, NullWritable> vertex,
                 Iterable<NullWritable> messages) {
    System.out.print("Hello world from the: " +
      vertex.getId().toString() + " who is following:");

    // iterating over vertex's neighbors
    for (Edge<IntWritable, NullWritable> e : vertex.getEdges()) {
      System.out.print(" " + e.getTargetVertexId());
    }
    System.out.println("");

    // signaling the end of the current BSP computation for the current vertex
    vertex.voteToHalt();
  }

  public static void main(String[] args) throws Exception {
    System.exit(ToolRunner.run(new GiraphRunner(), args));
}
```

Even in this trivial example, there's quite a bit going on. You start by extending a basic abstract class for performing computations in Giraph's implementation of the BSP model: BasicComputation. Immediately you have to specify the four type parameters that tell Giraph how to model the graph. Recall that the Giraph framework models every graph via a distributed data structure that is parameterized by the type of each vertex (VertexID), the type of label associated with each vertex (VertexData), the type of label associated with each edge (EdgeData), and the type of messages the BSP framework will communicate over the network (MessageData). Even in this simple example, you have to make choices about all four of those based on the representation of your input graph (integer IDs for vertices with no edge labels). Thus the type parameters you need to pick are IntWritable for VertexID and VertexData and NullWritable for EdgeData and MessageData. Once again, the choices are predetermined by the simple input format for the graph data you've settled on; they must also be compatible with an actual implementation of the vertex input format given to the Giraph command-line executor when a computation job is submitted to Hadoop (shown later in Listing 5-8). Finally, these choices mean the rest of the GiraphHelloWorld implementation expects to have access to the topology of the graph, with every vertex having an Integer ID and a list of edges connecting to other Integer IDs; but it doesn't expect any labels to be associated with the edges and, most important, doesn't send messages between the vertices of the graph (NullWritable signals the lack of data).

THE MIGHTY FOUR OF GIRAPH PROGRAMMING

Throughout Giraph, you see the following quadruple of type variables used for various generic types:

- VertexID is a type that can be used for referencing each vertex.

- VertexData is a type of a label attached to each vertex.

- EdgeData is a type of a label attached to each edge.

- MessageData is a type of message that could be sent between two vertices.

All four types are required to extend a basic building block of the Hadoop serialization framework called a `Writable` interface. On top of that, because objects representing vertices have to be partitioned between different nodes in the cluster, `VertexID` has a stricter guarantee of extending a `WritableComparable` interface. Reusing Hadoop's approach to object serialization means Giraph can use a lot of common code. Also, given that message passing between vertices residing on different nodes in the cluster is one of the key features of the BSP framework, the types you use must be extremely efficient when it comes to serialization and deserialization. In other words, they need to be optimized for network traffic. Fortunately, Hadoop has solved the same problem for its MapReduce framework by creating its own approach to serialization and deserialization built around the `Writable` interface. In any Giraph application, you can use a wide collection of Hadoop-provided classes that implement the `Writable` interface and map to all the usual data types a Java application might need (`boolean`, `int`, `Array`, `Map`, and so on).

Giraph expects you to implement all the computation in a method called `compute()` that is called for each vertex (at least once initially and then for as long as there are messages for it). Because you are extending an abstract class, you must provide a definition for `compute()`. This example prints a message with a vertex ID followed by a list of all the IDs of this vertex's neighbors. The neighbors are determined by iterating over all the edges and looking up an ID of a vertex that happens to be a target of each edge. Finally, you call `voteToHalt()`, thus signaling the end of the computation for a given vertex. When all vertices call this method, that signals the end of the current iteration of the BSP computation (and, unless there are messages left, the end of the entire Giraph application run).

Finally, in case you are wondering why a `main()` method is defined, it is not required but is convenient for cases in which you want to manually execute the example without having to call the `giraph` command-line executor. One such example is given in Chapter 12, when you run Giraph applications on Amazon's cloud.

With the Java implementation of `GiraphHelloWorld` in place, the only other bit of housekeeping is to create a project (either Maven or Ant) that helps with pulling in the right dependencies and compiling the example. If you decide to go with Maven, Listing 5-5 shows how your Maven `pom.xml` file for the project should look. Note that the only required dependencies are `giraph-core` and `hadoop-core`.

Listing 5-5. Maven Project Object Model Definition: `pom.xml`

```
<?xml version="1.0" encoding="UTF-8"?>
<project>
  <modelVersion>4.0.0</modelVersion>

  <groupId>giraph</groupId>
  <artifactId>book-examples</artifactId>
  <version>1.0.0</version>

  <dependencies>
    <dependency>
      <groupId>org.apache.giraph</groupId>
      <artifactId>giraph-core</artifactId>
      <version>1.1.0</version>
    </dependency>

    <dependency>
      <groupId>org.apache.hadoop</groupId>
      <artifactId>hadoop-core</artifactId>
```

```
        <version>1.2.1</version>
      </dependency>
    </dependencies>

  <build>
  </build>
</project>
```

At this point, make sure pom.xml is at the top of the source tree and the Java implementation under src/main/java. This will let you easily build the project by issuing the command shown in Listing 5-6.

Listing 5-6. Building the HelloWorld Project

```
$ mvn package
```

Once Maven is done with the build, you should see the resulting jar file appear under the target subdirectory. The next section explains how to use this jar to execute the GiraphHelloWorld application.

Launching Your Application

The only remaining two things you need in order to run the example are standalone installations of Hadoop and Giraph (Appendix A has detailed instructions for installing both). Once you have those in place, update your shell environment. If bash is your shell of choice, Listing 5-7 will do the job (if you are using a different shell, make sure to consult its documentation regarding how to export environment variables).

Listing 5-7. Defining the Shell Environment for Giraph Execution

```
$ export HADOOP_HOME=<path to a root of Hadoop installation tree>
$ export GIRAPH_HOME=<path to a root of Giraph installation tree>
$ export HADOOP_CONF_DIR=$GIRAPH_HOME/conf
$ PATH=$HADOOP_HOME/bin:$GIRAPH_HOME/bin:$PATH
```

As long as you keep the previous environment part of your shell session, you can run the example Giraph application using Hadoop's local execution mode, as shown in Listing 5-8. Note that executing Giraph applications results in a fairly long single command line. To fit everything on the book page, the single command is broken with backslash (\) characters. When typing this command line into your terminal window, you can either do the same (make sure the backslash is the last character on each line) or type it as a single line.

Listing 5-8. Running the Giraph Application

```
$ giraph target/*.jar GiraphHelloWorld -vip src/main/resources/1          \
  -vif org.apache.giraph.io.formats.IntIntNullTextInputFormat             \
  -w 1 -ca giraph.SplitMasterWorker=false,giraph.logLevel=error
...
Hello world from 2 who is following:
Hello world from 1 who is following: 2
Hello world from 3 who is following: 1 4
Hello world from 4 who is following: 2 7
Hello world from 5 who is following: 1 2 4
Hello world from 6 who is following: 3 4
Hello world from 7 who is following: 3 5
```

Congratulations! You've developed and executed your first Giraph application. Granted, it is pretty simple, and the output could be made better by printing the names of people instead of their numeric IDs (the next section deals with this), but this is a self-contained Giraph application in just under a couple of dozen lines of code. By the way, don't be alarmed if you see the output lines in a slightly different order than they appear in Listing 5-8: the order of vertex processing in Giraph is non-deterministic, although, of course, all vertices are guaranteed to be processed eventually.

Let's take one more look at Listing 5-8 and go through all the components of the command line. The first argument (target/*.jar) specifies the location of the jar file containing the implementation of your graph-processing "Hello World" application. The second argument (GiraphHelloWorld) specifies the name of the class with the compute method to be applied to all the vertices in the graph. The third argument (-vip) specifies the input path to the graph data. The fourth argument (-vif org.apache.giraph.io.formats. IntIntNullTextInputFormat) tells the Giraph implementation what input format to use for parsing the Graph data. The fifth argument (-w 1) makes Giraph use only one worker to process all vertices in the example graph (because you are not running on a cluster, you can only use one worker). Finally, the last command-line argument (-ca giraph.SplitMasterWorker=false,giraph.logLevel=error) tweaks a few internal knobs of the Giraph implementation to allow execution with a single worker and reduce the verbosity level to errors only.

Although it is useful to configure Giraph application execution on the fly using the –ca command-line option, a more practical approach is to put any reusable configuration into an XML configuration file. Anybody familiar with the Hadoop Configuration API will find the Giraph configuration very similar (internally, Giraph simply reuses Hadoop's implementation of the Configuration API). For example, to create a permanent Giraph configuration for the local Hadoop execution environment, you can put the configuration shown in Listing 5-9 into the giraph-site.xml configuration file under $GIRAPH_HOME/conf folder (see Appendix A for more information on how to install and configure Giraph).

Listing 5-9. Giraph Configuration

```
<configuration>
  <property>
    <name>giraph.SplitMasterWorker</name>
    <value>false</value>
  </property>

  <property>
    <name>giraph.logLevel</name>
    <value>error</value>
  </property>
</configuration>
```

With this configuration in place, you no longer have to supply the last line of the command-line options when executing Giraph applications.

■ **Note** Keep in mind that everything you put in the static configuration file applies to all your giraph runs, so it's important to keep track of the properties defined there.

Using a More Natural Definition of a Twitter Followership Graph

As mentioned before, one tiny problem with the example application is that the messages it outputs are not particularly descriptive. Wouldn't it better if instead of numerical identifiers, you had the original names of the people, as shown next to each vertex in Figure 5-2?

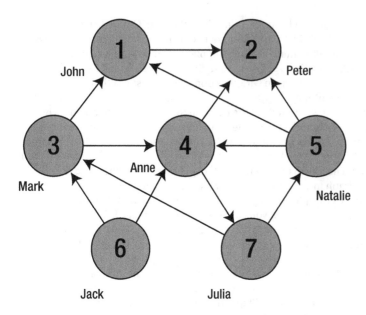

Figure 5-2. *Twitter followership directed graph*

Even the definition of the graph itself looks more natural specified as shown in Listing 5-10 (the definition uses dummy 0s as the values for the vertex and edge labels because you are not interested in them for now).

Listing 5-10. A More Natural Definition of the Twitter Followership Graph stored in `src/main/resources/2/g.txt`

```
John 0 Peter 0
Peter 0
Mark 0 John 0 Anne 0
Anne 0 Peter 0 Julia 0
Natalie 0 John 0 Peter 0 Anne 0
Jack 0 Mark 0 Anne 0
Julia 0 Mark 0 Natalie 0
```

The definition of the graph looks more complete now, but the format still has a few annoying restrictions. Everything must be delimited by a single tab (\t) character (whitespaces won't work), and dummy 0s designating vertex and label edge values need to be present even though they are not used. A more flexible format would be nice; but believe it or not, Giraph doesn't support such a format out of the box. You have to wait until Chapter 8 to see how you can extend the built-in implementation of the I/O formats to tweak them to your own liking. For now, the annoying 0s will stay because `TextDoubleDoubleAdjacencyListVertexInputFormat` expects this format.

Although it is natural to assume that at this point you can simply feed this new graph definition to your existing Giraph application, the fact that you changed the vertex representation from integers to strings requires you to make a change to the type variables used in defining the `compute()` method and its enclosing class. The changes to `GiraphHelloWorld` are trivial; the new implementation is shown in Listing 5-11. Note that you use Hadoop's `Text` type for string representations. This is consistent with the earlier advice to use Hadoop's types to manage data in an I/O friendly way.

Listing 5-11. GiraphHelloWorld2.java

```java
import org.apache.giraph.edge.Edge;
import org.apache.giraph.GiraphRunner;
import org.apache.giraph.graph.BasicComputation;
import org.apache.giraph.graph.Vertex;
import org.apache.hadoop.io.DoubleWritable;
import org.apache.hadoop.io.Text;
import org.apache.hadoop.io.NullWritable;
import org.apache.hadoop.util.ToolRunner;

public class GiraphHelloWorld2 extends
  // note a change in type variables to match new input
  BasicComputation<Text, DoubleWritable, DoubleWritable, NullWritable> {
  @Override
  public void compute(Vertex<Text,DoubleWritable,DoubleWritable> vertex,
                      Iterable<NullWritable> messages) {
    System.out.print("Hello world from the: " +
      vertex.getId().toString() + " who is following:");

    for (Edge<Text, DoubleWritable> e : vertex.getEdges()) {
      System.out.print(" " + e.getTargetVertexId());
    }
    System.out.println("");

    vertex.voteToHalt();
  }
}
```

Once you rebuild the jar using the previous source code, don't forget to also specify the new InputFormat on the command line, as shown in Listing 5-12.

Listing 5-12. Running the Giraph Application with a Different InputFormat

```
$ giraph target/*.jar GiraphHelloWorld2  -vip src/main/resources/2 –vif        \
   org.apache.giraph.io.formats.TextDoubleDoubleAdjacencyListVertexInputFormat  \
  -w 1 -ca giraph.SplitMasterWorker=false,giraph.logLevel=error
Hello world from Julia who is following: Mark Natalie
Hello world from Natalie who is following: John Peter Anne
Hello world from Jack who is following: Mark Anne
Hello world from Anne who is following: Peter Julia
Hello world from Peter who is following:
Hello world from Mark who is following: John Anne
Hello world from John who is following: Peter
```

As you can see, switching between input formats was pretty easy. All you had to do was update the type variables in the Java implementation and specify a different input format (org.apache.giraph.io.formats. TextDoubleDoubleAdjacencyListVertexInputFormat) in a fourth argument on the command line. The rest of the command line is exactly the same as Listing 5-8.

Now that you have seen the basics of Giraph applications, let's move on to how more complex graph-processing algorithms are expected to use the Giraph APIs.

Counting the Number of Twitter Connections

At this point, you should be comfortable with the basics of the Giraph framework. It is time to kick it up a notch and implement a few more applications. In this section, you turn pseudo-code examples from Chapter 3 into actual Java code that can be executed with Giraph the same way you executed the "Hello World" Giraph application. The first example, counting followership on Twitter, is shown in Listing 5-13.

Listing 5-13. TwitterConnections.java

```java
import org.apache.giraph.GiraphRunner;
import org.apache.giraph.graph.BasicComputation;
import org.apache.giraph.graph.Vertex;
import org.apache.hadoop.io.DoubleWritable;
import org.apache.hadoop.util.ToolRunner;
import org.apache.hadoop.io.Text;


public class TwitterConnections extends
// Note that the last type variable is no longer NullWritable.
// This is because we are going to send Text messages to vertex's
// neighbors as a way to notify target vertices that we are
// connected to them.
   BasicComputation<Text, DoubleWritable, DoubleWritable, Text> {
@Override
  public void compute(Vertex<Text, DoubleWritable, DoubleWritable> vertex,
                      Iterable<Text> messages) {

    if (getSuperstep() == 0) {
    // this indicates we're in the very first superstep of BSP
    // and it is time for us to send messages that will be processed
      // in the next superstep (else clause)
      vertex.setValue(new DoubleWritable(vertex.getNumEdges()));
      sendMessageToAllEdges(vertex, new Text());
    } else {
      int inDegree = 0;
      for (Text m : messages) {
        inDegree++;
      }
      vertex.setValue(new DoubleWritable(vertex.getValue().get() +
                                    inDegree));

    }

    vertex.voteToHalt();
  }
}
```

An overall idea for the previous algorithm is to use empty messages as a way of notifying vertices receiving them that there's a connection (recall that Giraph doesn't store the set of all the vertices that are connected to a give vertex). You send messages during the first superstep of the BSP computation (getSuperstep() returning 0). Given that there are messages to be processed, the compute() method is called again for the next superstep. You process those messages (sent in the previous superstep) by counting them and assigning the sum to each vertex's label. This is how you arrive at a value of indegree for each vertex.

A natural question to ask after reading the code is, how in the world are you supposed to find out the final indegrees of each vertex if the code doesn't produce any output? The answer lies in the Giraph facility that is complementary to that of the input format: an output format. An output format is just that: a way for Giraph to dump the topology of a graph and all the labels once it is finished running. Using output formats is as simple as specifying two command-line options: -vof to request a particular implementation and –op to specify an output subdirectory. Listing 5-14 shows the command line required to specify an output format that is exactly the same as the input format used in Listing 5-12. Note that the first thing you do is removing the output subdirectory; this is because Giraph expects that the requested subdirectory does not exist. Also note that the last command outputs the content of all the files into which Giraph's output format could have split the resulting graph. For big graphs, this could be quite a few individual files named part-XXX. Finally, although running Hadoop in local mode results in output in the local filesystem, in general the last command would be based on calling Hadoop's filesystem utilities, as you see in the next section.

Listing 5-14. Running the TwitterConnections Example with an OutputFormat

```
$ rm -rf output
$ giraph target/*.jar  TwitterConnections  -op output    \
-vof org.apache.giraph.io.formats.AdjacencyListTextVertexOutputFormat \
-vip src/main/resources/2     \
-vif \
org.apache.giraph.io.formats.TextDoubleDoubleAdjacencyListVertexInputFormat \
-w 1 -ca giraph.SplitMasterWorker=false,giraph.logLevel=error
$ cat output/part*
Julia    3.0    Mark    0.0    Natalie  0.0
Natalie  4.0    John    0.0    Peter    0.0    Anne    0.0
Jack     2.0    Mark    0.0    Anne     0.0
Anne     5.0    Peter   0.0    Julia    0.0
Peter    3.0
Mark     4.0    John    0.0    Anne     0.0
John     3.0    Peter   0.0
```

Let's recap what you have learned so far. Every Giraph application starts with reading the graph definition from the set of input files. This data describes the initial Graph topology and consists of Vertex objects connected by Edge objects. The Giraph application runs in a distributed fashion, iterating over the set of Vertex objects and calling a compute() method provided in a user-defined computation class. Implementing the Computation interface is the only requirement for the custom computation class. In practice, though, most applications subclass one of the two abstract classes: AbstractComputation or BasicComputation. The latter is a slightly more restricted form of the former. Compute() methods for all the required vertices are executed in series of supersteps. A superstep finishes when all the vertices call voteToHalt(). The supersteps are numbered starting from 0, and the number of the current superstep can always be requested by calling getSuperstep(). The "Hello World" example consisted of exactly one superstep: superstep 0. The current example runs in two supersteps; most real-world Giraph applications iterate through a variable number of supersteps as part of the convergence of the algorithm. Finally, compute() can elect to call voteToHalt() on its Vertex object, signaling that, unless new messages are delivered to it in the next superstep, it wishes not to be called again. The entire Giraph application finishes when all of its vertices are halted and there is not a single pending message. At this point, if a user requests that the entire state of the graph be flushed to storage, an output format takes over and makes sure the data is serialized in files under the desired location.

Keeping input and output formats compatible makes it possible to chain Giraph applications in arbitrary pipelines, each operating on the output of the previous one(s). Giraph algorithms are close to the Unix tooling philosophy where one tool is meant to "do one thing and do it well." The expressiveness comes not from the complexity of each tool or algorithm but rather from the flexibility of building arbitrary pipelines from multiple stages, all operating on common datasets.

With this theoretical background, let's turn to the second example from Chapter 3: converting directed graphs to undirected graphs.

Turning Twitter into Facebook

From a graph theory perspective, the biggest difference between Facebook and Twitter is *directionality*. The Twitter social graph is directional (if a user follows another one it doesn't mean the reverse is true), and the Facebook graph is not (your friends consider you to be a friend). So far, in the examples, you've been using a Twitter followership graph. This has the additional benefit of mapping one to one on Giraph's graph model of directional graphs. If you want to model a Facebook graph with Giraph, you have to make sure that vertexes are connected in both directions.

Let's use Giraph to turn your existing Twitter connection graph into a Facebook one by running the code shown in Listing 5-15.

Listing 5-15. Twitter2Facebook.java

```java
import org.apache.giraph.GiraphRunner;
import org.apache.giraph.graph.BasicComputation;
import org.apache.giraph.graph.Vertex;
import org.apache.giraph.edge.EdgeFactory;
import org.apache.hadoop.io.DoubleWritable;
import org.apache.hadoop.util.ToolRunner;
import org.apache.hadoop.io.Text;

public class Twitter2Facebook extends
  BasicComputation<Text, DoubleWritable, DoubleWritable, Text> {

  // the following two constants will represent existing edges and the ones we
  // created to model Facebook connections
  static final DoubleWritable ORIGINAL_EDGE = new DoubleWritable(1);
  static final DoubleWritable SYNTHETIC_EDGE = new DoubleWritable(2);

  @Override
  public void compute(Vertex<Text, DoubleWritable, DoubleWritable> vertex,
                      Iterable<Text> messages) {

    if (getSuperstep() == 0) {
      sendMessageToAllEdges(vertex, vertex.getId());
    } else {
      for (Text m : messages) {
        DoubleWritable edgeValue = vertex.getEdgeValue(m);
        if (edgeValue == null) {
          vertex.addEdge(EdgeFactory.create(m, SYNTHETIC_EDGE));
        } else {
          // for existing edges just label them as such
          vertex.setEdgeValue(m, ORIGINAL_EDGE);
        }
      }
    }

    vertex.voteToHalt();
  }
}
```

At this point, you could run the code exactly the same way you ran your previous examples. That, however, would produce Giraph-friendly, but not necessarily human-friendly, graph output. Instead, you can use a different output format that can be used as an input to a variety of graph visualization tools. A picture, after all, is worth a thousands words. Listing 5-16 executes Giraph almost identically to the previous example, aside from using a different output format (org.apache.giraph.io.formats. GraphvizOutputFormat).

Listing 5-16. Running Twitter2Facebook with a Visualization-Friendly OutputFormat

```
$ rm -rf output
$ giraph target/*.jar Twitter2Facebook    -op output             \
-vof org.apache.giraph.io.formats.GraphvizOutputFormat        \
-vip src/main/resources/2                                     \
-vif \
org.apache.giraph.io.formats.TextDoubleDoubleAdjacencyListVertexInputFormat \
-w 1 -ca giraph.SplitMasterWorker=false,giraph.logLevel=error
```

To visualize the output generated from running this command, you first ask Hadoop to merge different sections of the output into a single file. Then you add a header and a footer required by the Graphviz format, and finally you run a Graphviz visualization command-line utility to generate a file that can be viewed on screen. These four commands are shown in Listing 5-17, and Figure 5-3 shows the resulting picture.

Listing 5-17. Visualizing the Output of Giraph's Graphviz Format

```
$ hadoop fs -getmerge output data.txt
$ echo "digraph Giraph {" > giraph.dot
$ cat data.txt              >> giraph.dot
$ echo "}"                  >> giraph.dot
$ circo giraph.dot -Tpng -ogiraph.png
```

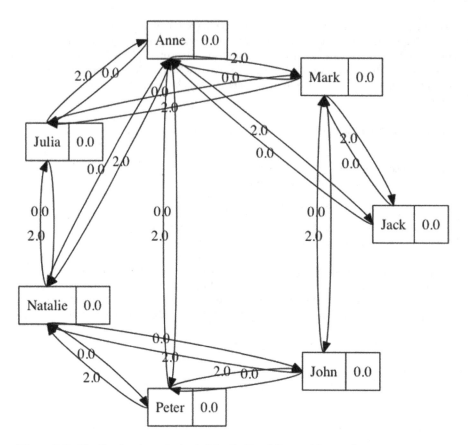

Figure 5-3. *The Facebook equivalent of the Twitter followership graph*

GRAPHVIZ (DOT): A LINGUA FRANCA FOR GRAPH VISUALIZATION

Graphviz is open source graph-visualization software. Graphviz and its input language, DOT, were developed at AT&T Labs Research for quickly visualizing arbitrary graph data and have taken on a life of their own. A reference implementation is available on most Unix platforms, but there are also a number of cleanroom visualizer implementations for the DOT format. If the DOT visualization utility (circo) is not available on your platform, you can try visualizing the Facebook graph by pasting the content of the giraph.dot file into one of the online Graphviz tools, such as the one at www.webgraphviz.com.

The graph visualization lets you see the vertex value associated with each vertex and the edge value associated with each edge. Because you are not using vertex values in the example, in Figure 5-3 all of them appear as 0.0 in a box next to the person's name. Edges, on the other hand, are labeled according to whether the edge was added as part of your conversion (2) or whether it was always present in the graph (1).

This was the first example of mutating a graph's topology by adding new edges. The next section covers additional ways to change the graph structure on the fly.

Changing the Graph Structure

Adding new edges to a graph was easy enough, but what about adding new vertices? What about building entire graph segments directly in memory without reading the data from external sources? An extreme example of that would be generating a graph from scratch based on certain criteria. If you think of Giraph jobs forming a pipeline, generating a graph could be the first step in that pipeline. This is a typical approach for simulation and system testing of complex graph algorithms. The next example, shown in Listing 5-18, generates your old friend the Twitter followership graph starting from nothing but an empty graph.

Listing 5-18. GenerateTwitter.java

```java
import org.apache.giraph.GiraphRunner;
import org.apache.giraph.graph.BasicComputation;
import org.apache.giraph.graph.Vertex;
import org.apache.giraph.edge.EdgeFactory;
import org.apache.hadoop.io.DoubleWritable;
import org.apache.hadoop.util.ToolRunner;
import org.apache.hadoop.io.Text;
import java.io.IOException;

public class GenerateTwitter extends
  BasicComputation<Text, DoubleWritable, DoubleWritable, Text> {

  // two static arrays defining the same graph as input data from Listing 5.10
  private static final String[] twitterMembers = { "seed",
    "John", "Peter", "Mark", "Anne", "Natalie", "Jack", "Julia" };
  private static final byte[][] twitterFollowership = {{0},
    {2},    {},       {1, 4}, {2, 7}, {1, 2, 4}, {3, 4}, {3, 5}};

  @Override
  public void compute(Vertex<Text, DoubleWritable, DoubleWritable> vertex,
                      Iterable<Text> messages) throws IOException {

    if (getSuperstep() == 0) {
      for (int i = 1; i < twitterFollowership.length; i++) {
        Text destVertexID = new Text(twitterMembers[i]);
        addVertexRequest(destVertexID, new DoubleWritable(0));

        for (byte neighbour : twitterFollowership[i]) {
          addEdgeRequest(destVertexID, EdgeFactory.create(
                new Text(twitterMembers[neighbour]),
                new DoubleWritable(0)));
        }
      }

      removeVertexRequest(new Text("seed"));
    } else {
      vertex.voteToHalt();
    }
  }
}
```

When you run this example, a very logical assumption would be that because the entire point of this Giraph application is to generate the graph from scratch, you shouldn't give Giraph any input files. A more advanced use of Giraph would allow you to do that, but for this simple example you face a small challenge. The code that generates a graph is part of the compute() method, and because compute() methods are called for vertices of an input graph, the easiest way to do what you want is to give Giraph the smallest graph possible: a graph with a single vertex and no edges, as shown in Listing 5-19.

Listing 5-19. Smallest Graph Definition with a Single Vertex and No Edges (src/main/resources/3/g.txt)

```
seed 0
```

With this input in place, you can run Giraph exactly as you did for all the previous examples. Let's run it with the Graphviz output format, as shown in Listing 5-20, so you can quickly check the visualized graph and make sure the code generated the original graph from Figure 5-2.

Listing 5-20. Running GenerateTwitter with the Graphviz Output Format

```
$ rm -rf output
$ giraph target/*.jar GenerateTwitter    -op output              \
-vof org.apache.giraph.io.formats.GraphvizOutputFormat           \
-vip src/main/resources/3                                        \
-vif \
org.apache.giraph.io.formats.TextDoubleDoubleAdjacencyListVertexInputFormat \
-w 1 -ca giraph.SplitMasterWorker=false,giraph.logLevel=error
```

Two of the Giraph API calls used in this example to modify the topology of the graph (addVertexRequest and addEdgeRequest) are part of a broader API set known as *graph mutation APIs*. Most of these API calls modify the overall graph structure in one way or another. The changes can be available to all members of the computational process either immediately or in the next superstep (that's why in the example you can't exit immediately but must wait for the next superstep). Here's a summary of the most popular mutation API methods used in real-world applications:

- setValue(): Modifies its own value. The result of this action is immediately visible (in the same superstep) to the entire computation.

- setEdgeValue(): Modifies the edge value of the outgoing edges. Even though the result is also immediately visible to the entire computation, the application of this method is limited because typically only a source vertex has access to the outgoing edges.

- addEdge(), removeEdges(): Modifies the local topology of the graph by adding or removing edges, with the source being the same vertex.

- addVertexRequest(), removeVertexRequest(), addEdgeRequest(), removeEdgesRequest(): Asks to modify the global topology of the graph starting from the next superstep.

By using these methods, you can create graphs with a very different structure and topology compared to the graph available to Giraph at the start of execution. Interestingly enough, in addition to the straightforward methods of mutating a graph presented here, a less obvious one has to do with that jack of all trades in Giraph: messages. This is the subject of the next section.

Sending and Combining Multiple Messages

As mentioned, the Giraph messaging topology is completely independent of graph topology. In other words, every vertex can send a message to every other vertex. Sending messages to a nonexistent vertex creates the vertex. (See? Messages can also be considered part of the graph mutation APIs!)

This is the trick the next example exploits to build a graph topology in parallel. Instead of a single loop running on a single host and iterating over all the vertices and their edges, you use a two-stage approach. First you create all the vertices by sending messages to them. Each message contains one neighbor of that newly created vertex. Once the vertices are created and assigned to the worker nodes in the cluster, each vertex, in parallel, begins iterating over the received messages and creating edges connecting it to its neighbors. Listing 5-21 is a modification of the previous example that does just that.

Listing 5-21. GenerateTwitterParallel.java

```java
import org.apache.giraph.GiraphRunner;
import org.apache.giraph.graph.BasicComputation;
import org.apache.giraph.graph.Vertex;
import org.apache.giraph.edge.EdgeFactory;
import org.apache.hadoop.io.DoubleWritable;
import org.apache.hadoop.util.ToolRunner;
import org.apache.hadoop.io.Text;
import java.io.IOException;

public class GenerateTwitterParallel extends
  BasicComputation<Text, DoubleWritable, DoubleWritable, IntWritable> {

  private static final String[] twitterMembers = { "",
          "John", "Peter", "Mark", "Anne", "Natalie", "Jack", "Julia" };
  private static final byte[][] twitterFollowership = {{0},
          {2},   {},      {1, 4}, {2, 7}, {1, 2, 4}, {3, 4}, {3, 5}};

  @Override
  public void compute(Vertex<Text, DoubleWritable, DoubleWritable> vertex,
                    Iterable<IntWritable> messages) throws IOException {

    if (getSuperstep() == 0) {
      for (int i = 1; i < twitterFollowership.length; i++) {
        Text destVertexID = new Text(twitterMembers[i]);
        for (byte neighbour : twitterFollowership[i]) {
          sendMessage(destVertexID, new IntWritable(neighbour));
        }
      }

      removeVertexRequest(new Text("seed"));
    } else {
      for (IntWritable m : messages) {
        Text neighbour = new Text(twitterMembers[m.get()]);
```

```
        vertex.addEdge(EdgeFactory.create(neighbour,
                                    new DoubleWritable(0)));
    }

    vertex.voteToHalt();
  }
 }
}
```

Of course, for a tiny graph like this one, the difference between sequential and parallel versions of the same code is negligible. But it does highlight an important design point that every Giraph application needs to consider: parallel graph processing is always limited by the amount of sequential code that cannot be parallelized. This is also known as *Amdahl's law of parallelism*.

The latest example makes it possible for edges in the graph to be created in parallel as opposed to being created in a single loop running at superstep 0. This is a good thing, but there's still one more optimization possible.

Consider the fact that for each edge, you have to create a dedicated message. Given that all you need to communicate in that message is an integer ID, the overhead can be pretty high. It would be much more efficient if you could somehow aggregate or combine neighbor IDs into a single message. One way of accomplishing this is to change the data type of the messages being communicated into a byte array. Another way is to use Giraph's combiners.

You implement a combiner for messages sent between vertices whenever there's a natural way to aggregate multiple messages into a single one. Giraph provides a few useful combiners to get you started. SimpleSumMessageCombiner, for example, sums individual messages into one; and MinimumIntMessageCombiner finds a minimum value in messages containing individual integers and creates a single message containing that value. The next example, however, creates a custom combiner that combines all of your messages in a bitmap array. Each integer value sent in a message is represented as a bit set in a bitmap. For example, if a message contains the integer 3, the third bit from the right is set to 1. Listing 5-22 shows the code for BitArrayCombiner.

Listing 5-22. BitArrayCombiner.java

```
import org.apache.giraph.combiner.MessageCombiner;
import org.apache.hadoop.io.IntWritable;

public class BitArrayCombiner
  extends MessageCombiner<IntWritable, IntWritable> {

  // the guts of each combiner is a method that defines how to
  // combine a new message into an overall message accumulator
  @Override
  public void combine(IntWritable vertexIndex,
                      IntWritable originalMessage,
                      IntWritable messageToCombine) {

    originalMessage.set((1<<messageToCombine.get()) |
                      originalMessage.get());
  }
```

```
// this a method that defines how to create an initial message accumulator
// in our case this simply creates a bit-map array with all bits set to 0 (no messages
   seen)
@Override
public IntWritable createInitialMessage() {
  return new IntWritable(0);
}
}
```

In general, combiners tend to be dead simple: all you have to do is to define the function that returns an initial message. This message is used as an accumulator for the messages emitted as part of compute()execution. The job of accumulating the data is performed by the code defined in the combine() method. As you can see, this method treats originalMessage as a mutable accumulator; in this case it sets a bit corresponding to the neighbor integer ID in an overall integer bit array.

Combiners can be useful, but they are not transparent. First, you have to tell Giraph that it needs to use a combiner. This is typically done by passing a class name to the –c command-line option, just as you pass an input format. Second, you must modify the code from the previous example so it can unpack the bit array that is now the payload of each message. This issue, plus the fact that you can only work with fewer than 32 vertices (because integers in Java are 32-bit) makes BitArrayCombiner a bit impractical and only good as an illustration of how easy it is to write a combiner.

Speaking of limitations, if you tried running the previous example, you probably noticed a subtle bug in its implementation: it's impossible to generate the vertices with zero neighbors. The usual way to reduce the likelihood of this type of bug is to develop code hand-in-hand with unit tests verifying its correctness. You have been developing code in this chapter without any kind of testing or verification, and it is time to change that.

Unit-Testing Your Giraph Application

Developing complex applications against any framework typically requires a suite of unit tests guaranteeing against accidental changes in semantics during code changes and refactoring. Fortunately, the event-driven nature of Giraph applications makes it extremely easy to test the semantics of various methods (such as compute()) in isolation. You can also create a miniature execution environment for full-fledged application testing.

As mentioned earlier, the previous example had a subtle issue. With that in mind, let's create a unit test verifying that the resulting graph topology is indeed what you expect it to be; see Listing 5-23.

Listing 5-23. TestGiraphApp.java

```
import org.junit.Test;
import org.junit.Assert;
import org.junit.Ignore;
import org.apache.giraph.conf.GiraphConfiguration;
import org.apache.giraph.io.formats.TextDoubleDoubleAdjacencyListVertexInputFormat;
import org.apache.giraph.io.formats.AdjacencyListTextVertexOutputFormat;
import org.apache.giraph.utils.InternalVertexRunner;

public class TestGiraphApp  {
  final static String[] graphSeed = new String[] { "seed\t0" };
  final static int EXPECTED_VERTICES = 7; // The number of vertices we expect to
                                          // be created in the resulting graph
```

```
@Test
public void testNumberOfVertices() throws Exception {
  GiraphConfiguration conf = new GiraphConfiguration(); // Giraph configuration
      // object: it will configure our test Giraph run just like a combination of
      // static configuration files and command line options we used in all of our
      // previous examples
  conf.setComputationClass(GenerateTwitterParallel.class);
  conf.setVertexInputFormatClass(
    TextDoubleDoubleAdjacencyListVertexInputFormat.class);
  conf.setVertexOutputFormatClass(
    AdjacencyListTextVertexOutputFormat.class);

  // The following is all that is required for our test to simulate
  // a full Giraph application run: the conf is how we configure the
  // simulation and graphSeed is an array of strings simulating input
  // graph data. The result it returns back allows us to iterate over
  // the simulated output in order to check assumptions about what
  // we expect after the GenerateTwitterParallel code is done executing.
  Iterable<String> results =
      InternalVertexRunner.run(conf, graphSeed);

  // Iterating over the simulated output is how you check the
  // resulting graph structure. In our case we're simply counting
  // the number of lines in the simulated output.
  int totalVertices = 0;
  for (String s: results ) {
    totalVertices++;
  }
  // We expect a certain number of lines (vertex descriptions)
  // to be presented in the simulated output.
  Assert.assertEquals(EXPECTED_VERTICES, totalVertices);
  }
}
```

Even though this implementation is only a few lines long, a number of remarkable things happen behind the scenes courtesy of the InternalVertexRunner implementation. The run() method used for this example requires that you supply Giraph's configuration and a String array simulating the content of the input. What you get from calling run() is a String array simulating the output that would be written to a filesystem during a normal run of the application. Although the input and output files are simulated, everything else is not. The conf object is constructed exactly as it would be during the actual run of the application by combining default settings with settings defined in giraph-site.xml and those specified as part of the giraph command line. If you don't want to be constrained by the syntax of the command line, you can write a main() method to explicitly set configuration values.

Running the unit test is a simple matter of either hooking it up to your build system (Ant, Maven, or Gradle) or passing it to the JUnit executor. If you are using Maven to build the examples, all you need to do is save the test case under src/test/java/TestGiraphApp.java in your project file tree and add JUnit dependency to the pom.xml file as shown in Listing 5-24. After that, executing mvn test will run the unit tests for you (and so will the mvn package command).

Listing 5-24. Maven Project Object Model Definition: pom.xml After Adding JUnit Dependency

```xml
<?xml version="1.0" encoding="UTF-8"?>
<project>
  <modelVersion>4.0.0</modelVersion>

  <groupId>giraph</groupId>
  <artifactId>book-examples</artifactId>
  <version>1.0.0</version>

  <dependencies>
    <dependency>
      <groupId>org.apache.giraph</groupId>
      <artifactId>giraph-core</artifactId>
      <version>1.1.0</version>
    </dependency>

    <dependency>
      <groupId>org.apache.hadoop</groupId>
      <artifactId>hadoop-core</artifactId>
      <version>1.2.1</version>
    </dependency>

    <dependency> <!- This is the only section we have added to the original pom file ->
      <groupId>junit</groupId>
      <artifactId>junit</artifactId>
      <version>4.12</version>
    </dependency>
  </dependencies>

  <build>
  </build>
</project>
```

Either way, the assertion fails because the resulting graph is one vertex short of what you expect. Making the unit tests pass is left as an exercise for you; the next section covers the somewhat orthogonal subject of extensions to the BSP compute model that Giraph provides.

Beyond a Single Vertex View: Enabling Global Computations

So far, the Giraph applications you've seen have assumed no shared state between different vertices. Sending messages works well for communicating data between two given vertices (or even between a given vertex and a set of message recipients), but it doesn't really work if the entire computation needs to keep a running tally that can be updated and accessed by all compute() methods working on all the vertices in a graph.

For example, imagine a situation where you need to run a parallel search on a graph. Let's say your graph represents a map, with vertices being cities, edges being roads, and edge labels being distances between cities. Ideally you could find an absolute shortest path between the two vertices; but more often than not, a close approximation of a shortest path will do. The question then becomes, how do you keep track of the current shortest path distance, and how do you abort the computation when the path gets to be small enough (but perhaps not as small as the absolute shortest path)?

Giraph offers an efficient way to solve this: named aggregators. The next two sections talk about aggregators and how they work with master compute.

Using Aggregators

As mentioned earlier, aggregators really shine when you need to track graph statistics that don't naturally belong to any of the vertices (or even a subset of vertices). One such statistic is the total number of connections (edges) in the graph. Listing 5-25 offers a very simple way to count the number of edges in the graph by using an aggregator. That aggregator happens to be a summing aggregator, but you don't know that unless you look at Listing 5-26 and see that the aggregator implementation is registered under the name TotalNumberOfEdgesMC.ID.

Listing 5-25. TotlaNumberOfEdges.java

```
import org.apache.giraph.graph.BasicComputation;
import org.apache.giraph.graph.Vertex;
import org.apache.hadoop.io.Text;
import org.apache.hadoop.io.DoubleWritable;
import org.apache.hadoop.io.NullWritable;
import org.apache.hadoop.io.LongWritable;

public class TotalNumberOfEdges extends
  BasicComputation<Text, DoubleWritable, DoubleWritable, NullWritable> {

  @Override
  public void compute(Vertex<Text, DoubleWritable, DoubleWritable> vertex,
                      Iterable<NullWritable> messages) {

    aggregate(TotalNumberOfEdgesMC.ID, new LongWritable(vertex.getNumEdges()));

    vertex.voteToHalt();
  }
}
```

The idea here is simple: every compute() method pushes the number of outgoing edges associated with a given vertex into an aggregator registered under the name TotalNumberOfEdgesMC.ID. As you can see, the compute() methods don't care what happens to the values sent to the aggregator, nor do they care what kind of aggregator they are sending values to. Setting up an aggregator is, by itself, an example of a global action that must be done once before the first superstep, so you cannot do that in a vertex-centric compute() method. You have to use a bit of code that runs once before each superstep, and that is what the master compute implementation shown in Listing 5-26 allows you to do.

Listing 5-26. TotalNumberOfEdgesMC.java

```
import org.apache.giraph.aggregators.LongSumAggregator;
import org.apache.giraph.master.DefaultMasterCompute;

public class TotalNumberOfEdgesMC extends DefaultMasterCompute { // master compute
    // implementations are usually extending the DefaultMasterCompute abstract class
    // just the same way that our basic BSP compute implementations extended
    // BasicComputation. Unlike BasicComputation you have to provide implementations
    // for two methods: compute() and initialize()

  public static final String ID = "TotalNumberOfEdgesAggregator"; // all registered
    // aggregators are referenced by unique (within a given application) but arbitrary
    // strings
```

```
@Override
public void compute() {
    // this is a global master compute method that will be executed once
    // before every superstep. In our case we're simply outputting the
    // running tally of total number of edges in a graph
    System.out.println("Total number of edges at superstep " +
                       getSuperstep() + " is " +
                       getAggregatedValue(ID));
}


@Override
    // this method gets called during overall initialization of the Giraph
    // BSP machinery. It is an ideal place to register all required
    // aggregators.
public void initialize() throws InstantiationException,
                                IllegalAccessException {
    registerAggregator(ID, LongSumAggregator.class); // this is how we associate
        // an aggregator ID with a particular implementation of an aggregator:
        // in our case we are using a built-in LongSumAggregator that simply
        // sums all of the values sent to it
}
}
```

Master compute allows you to execute global actions using the entire graph topology and all the aggregator values available to you. Because the master compute initialize() method runs once very early during the initialization of Giraph, this is an ideal place to register all the aggregators you use in a given application. Once the aggregators are registered, vertex-centric compute() methods can send values to them, and the master compute() method can inspect the state of the aggregators before each superstep. In this case, you are using the compute() method of the master compute implementation to print out the current value of the summing aggregator. In real-world applications, you can use it to do any kind of global computations, based on the results of which you can even terminate your application early (provided that a globally computed heuristic of your choice has been satisfied).

Executing the example in Hadoop local mode provides the expected output, but for cluster execution you are better off recording the state of the aggregators to external storage at given intervals during execution of your application. Think of it like having the graph's final state recorded at the end of the run. Of course, you can replace the System.out.println() call in master compute, but there's a better way. Giraph provides the notion of an *aggregator writer*. An aggregator writer class is expected to implement an AggregatorWriter interface and provide an implementation for the writeAggregator() method that, at the end of each superstep, is passed a map of all the registered aggregator names, their values, and the numeric ID of a superstep (the end of a computation is signaled by passing the ID LAST_SUPERSTEP). As usual, you're free to implement your own aggregator writer or use the default TextAggregatorWriter. For now, let's stick with the default.

Putting it all together, Listing 5-27 shows how you can run your edge-counting Giraph application from the command line.

Listing 5-27. Running the edge-counting application

```
$ rm -rf aggregatorValues*
$ giraph target/*.jar TotalNumberOfEdges -mc TotalNumberOfEdgesMC  \
    -aw org.apache.giraph.aggregators.TextAggregatorWriter         \
    -ca giraph.textAggregatorWriter.frequency=1                    \
    -vip src/main/resources/2                                      \
    -vif \
```

```
org.apache.giraph.io.formats.TextDoubleDoubleAdjacencyListVertexInputFormat \
   -w 1 -ca giraph.SplitMasterWorker=false,giraph.logLevel=error
Total number of edges at superstep 0 is 0
Total number of edges at superstep 1 is 12

$ cat aggregatorValues*
superstep=0      TotalNumberOfEdgesAggregator=0
superstep=-1     TotalNumberOfEdgesAggregator=12
```

There are a couple of interesting points to note about the giraph command line presented in Listing 5-27. First, because you are using master compute, you tell Giraph which class contains that implementation as a third argument (-mc TotalNumberOfEdgesMC). Second, you tell Giraph to use an aggregator writer that outputs the values of the aggregators into the default Hadoop filesystem (-aw org.apache.giraph. aggregators.TextAggregatorWriter). Finally, you specify that you want the aggregator values to be recorded at every superstep (-ca giraph.textAggregatorWriter.frequency=1). Note that you specify the frequency by providing a property that the TextAggregatorWriter implementation recognizes. You can use the same technique for configuration properties in your own implementations.

The last command in Listing 5-27 demonstrates that TextAggregatorWriter did, indeed, record the state of your aggregator at each superstep. Those values are the same as the ones produced by your System.out.println statements.

The next section provides an additional level of detail about how the aggregator and master compute machinery works. But first, let's take a moment to look at the implementation of LongSumAggregator, shown in Listing 5-28. Even though this aggregator is available to you out of the box, it is useful to know how simple it is to create specialized aggregators of your own.

Listing 5-28. Implementation of LongSumAggregator

```
import org.apache.hadoop.io.LongWritable;

// Most custom aggregators are extending the BasicAggregator class and
// focus on implementing two key methods createInitialValue() and
// aggregate(). The type variable used for BasicAggregator defines the
// type of the values that the aggregator will be receiving from compute()
// methods and also providing to its consumers.
public class LongSumAggregator extends BasicAggregator<LongWritable> {
  @Override
  public LongWritable createInitialValue() {
    // this method is expected to create an initial state for the aggregator,
    // which in our case happens to be the value of 0
    return new LongWritable(0);
  }

  @Override
  public void aggregate(LongWritable value) {
    // this method is expected to fold incoming values into the aggregator
    // created by a previous method
    getAggregatedValue().set(getAggregatedValue().get() + value.get());
  }
}
```

The implementation here is as simple as it gets, but it is not a toy example. This code was lifted verbatim from the Giraph code base. Hopefully this provides enough inspiration for you to create your own aggregators for cases when Giraph doesn't come with the ones you need.

Aggregators and Master Compute

As you have seen, aggregators are referenced via simple flat namespace of String names. Each name is associated with a class implementing the Aggregator interface. Once this association is established, the compute() method of every vertex can provide a value to be aggregated via the call to the aggregate(name, value) method or query the current state of the aggregator by calling getAggregatedValue(name) and expecting the value of the same type as the one supplied in corresponding calls to the aggregate() method. The aggregator implementation defines how multiple values supplied by calls to aggregate() coming from different vertices are aggregated into a single value of the same type. This is similar to the pattern you saw with combiners.

For example, an aggregator may sum numeric values or find a min/max value. Giraph provides about a dozen aggregators, and users can add to the collection by implementing the Aggregator interface directly or, better yet, subclassing a BasicAggregator abstract class. The latter provides a reasonable starting point for the most practical aggregators, whereas the former leaves the implementation unconstrained. There can be as many aggregators as you require: they are all independent in terms of data type and the aggregation performed on multiple values coming from different vertices.

Finally, aggregators can be regular or persistent. The difference is that the value of a regular aggregator is reset to the initial value in each superstep, whereas the value of a persistent aggregator exists throughout the application run. You can use the same aggregator implementation as either regular or persistent, depending on whether you register it by calling registerAggregator() or registerPersistentAggregator().

The functionality of aggregators serving as rendezvous points between vertices is useful, but their real power for affecting graph computation comes from combining them with master compute. The idea behind master compute is a slight extension of a straight BSP model. All you are doing is introducing a hook to run a special compute() method of a master compute object at the beginning of every superstep (before invoking compute() for individual vertices). This gives you a centralized location to affect every aspect of the rest of the graph computation. Master compute provides an ideal opportunity to register and initialize aggregators and also to inspect their state between consecutive supersteps. Aggregators, in turn, provide a means of communication between the workers and the master compute methods. The values of all the aggregators on all the worker nodes are always consistent in a given superstep, because their values are broadcast at the beginning of each superstep. On the flip side, at the end of each superstep, the values are gathered and made available to the next invocation of master compute. This means aggregator values used by workers are consistent with aggregator values from the master from the same superstep, and aggregators used by the master are consistent with aggregator values from the workers from the previous superstep.

A Real-World Example: Shortest Path Finder

So far, you have been focusing on artificially small examples that highlight certain aspects of Giraph's APIs. You may think that a real-world Giraph application is much more complex and involved. The truth, however, is that more often than not, even applications used in very large-scale graph-processing jobs tend to be pretty compact.

In general, Giraph applications follow a Unix philosophy of small tools that do one job and do it well. In Unix, it is common to use pipelines with the next tool operating on the output of the previous one, and so it is with Giraph. Giraph pipelines are strung together with the next Giraph application operating on the Graph representation serialized by the previous one.

Out of the box, Giraph comes with a few well-known, useful algorithms. You can get a list of built-in algorithms by giving Giraph the command-line option -la.

The example shown in Listing 5-29 is a complete implementation of one of the quintessential algorithms in graph theory: it finds the shortest paths in a graph (also known as *Dijkstra's algorithm*). Here's what it does: given a graph representation and a vertex, it finds the shortest path between that vertex and every other one in the graph. For example, if you think of your graph as a map with vertices representing cities, edges representing roads, and integer edge labels representing driving distances, then the results are the shortest driving distances between a chosen city and the rest of the cities on the map.

The idea behind this algorithm is fairly straightforward. You start by assigning a driving distance of 0 to a chosen city (after all, you don't have to drive to get there) and a distance of infinity to all the others (the worst-case scenario is that there is no path between all the other cities and your starting city). You then proceed, in parallel, to calculate for every vertex a sum of vertex's value (the current shortest distance to it) plus the distance to all the vertex's neighbors. You can think of it as each vertex broadcasting its best-known shortest past to all of its neighbors and the neighbors having a chance to update their values based on that information. If such an update offers a shorter distance, the neighbor would be foolish not to consider it the new shortest path. The algorithm converges when no more updates happen on the graph.

Listing 5-29 is the unabridged source lifted directly from the Giraph code base. It is a good example of how even the smallest, simplest graph-processing algorithms can be indispensable in practice. It also gives you an opportunity to review quite a few API calls that you learned about in this chapter.

This code operates on a graph with every vertex having a LongWritable ID and a DoubleWritable distance to the source vertex as its vertex data. All the edges have FloatWritable labels associated with them, representing the distance between two connected vertices. A source vertex is given to the implementation via the SimpleShortestPathsVertex.sourceId command-line option; it is expected to be a LongWritable ID. You measure the distance from this source vertex to all the other vertices in a graph. Initially you assign a distance metric of Double.MAX_VALUE to all the vertices except the starting one. Each vertex expects to receive messages from the neighbors it is connected to, announcing the distance from the source vertex to each neighbor plus the distance between a neighbor vertex and a vertex that is receiving. You then pick the smallest value (because you are interested in the shortest paths), and if it happens to be smaller than the current distance, you assume that a shorter path through one of the neighbors was uncovered; you then send messages to the neighbors announcing that fact. Finally, an unconditional call to voteToHalt() guarantees that the computation continues as long as there are unprocessed messages. This makes sense, because when there are no messages left, it means every vertex has the shortest distance and all that is left is to record the state of the graph.

Listing 5-29. SimpleShortestPathsComputation.java

```java
import org.apache.giraph.graph.BasicComputation;
import org.apache.giraph.conf.LongConfOption;
import org.apache.giraph.edge.Edge;
import org.apache.giraph.graph.Vertex;
import org.apache.hadoop.io.DoubleWritable;
import org.apache.hadoop.io.FloatWritable;
import org.apache.hadoop.io.LongWritable;
import org.apache.log4j.Logger;
import java.io.IOException;

public class SimpleShortestPathsComputation extends
  BasicComputation<LongWritable, DoubleWritable,
                    FloatWritable, DoubleWritable> {

  // This is how we will be passing an ID of a starting
  // vertex: via command line argument given to the giraph
  // execution utility.
```

```
public static final LongConfOption SOURCE_ID =
  new LongConfOption("SimpleShortestPathsVertex.sourceId", 1,
                     "The shortest paths id");

private static final Logger LOG =
  Logger.getLogger(SimpleShortestPathsComputation.class);

private boolean isSource(Vertex<LongWritable, ?, ?> vertex) {
  return vertex.getId().get() == SOURCE_ID.get(getConf());
}

@Override
public void compute(
  Vertex<LongWritable, DoubleWritable, FloatWritable> vertex,
      Iterable<DoubleWritable> messages) throws IOException {
  if (getSuperstep() == 0) {
    vertex.setValue(new DoubleWritable(Double.MAX_VALUE));
  }
  double minDist = isSource(vertex) ? 0d : Double.MAX_VALUE;
  for (DoubleWritable message : messages) {
    minDist = Math.min(minDist, message.get());
  }
  if (LOG.isDebugEnabled()) { // a good way to handle debug output
    LOG.debug("Vertex " + vertex.getId() + " got minDist = " +
            minDist +    " vertex value = " + vertex.getValue());
  }
  // the following is the guts of the shortest path algorithm
  if (minDist < vertex.getValue().get()) {
    vertex.setValue(new DoubleWritable(minDist));
    for (Edge<LongWritable, FloatWritable> edge : vertex.getEdges()) {
      double distance = minDist + edge.getValue().get();
      if (LOG.isDebugEnabled()) {
        LOG.debug("Vertex " + vertex.getId() + " sent to " +
                  edge.getTargetVertexId() + " = " + distance);
      }
      sendMessage(edge.getTargetVertexId(),
                  new DoubleWritable(distance));
    }
  }
  vertex.voteToHalt();
  }
}
```

Summary

This chapter covered these topics:

- Building your first "Hello World" Giraph application and writing a unit test for it

- Discovering various ways of specifying graph-definition data

- Exploring the details of the Giraph graph computation model

- Manipulating the topology of the graph structure

- Looking into advanced uses of messaging

- Taking advantage of Giraph extensions to the simple BSP compute model such as aggregators and master compute

The Giraph framework offers a set of powerful APIs catering to variety of graph-processing algorithms. One thing that all Giraph implementations have in common is that they must provide an implementation of the compute() method that is called for every vertex at least once. Each implementation of compute() has full access to a variety of utility methods:

- Every Giraph application uses voteToHalt() when it needs to signal that a vertex is done computing. When all vertices call voteToHalt() and there are no more pending messages, the entire computation stops.

- You can send messages from any vertex to any other vertex (regardless of whether they are connected) by using methods like sendMessage() and sendMessageToAllEdges(). If the destination vertex doesn't exist, it is created.

- On the I/O side of things, you have access to a wide array of built-in input and output formats for parsing various graph representations and an option to roll your own if needed. You saw IntIntNullTextInputFormat and TextDoubleDoubleAdjacencyListVertexInputFormat in this chapter, but many more come with Giraph and allow you to slurp data not just from files but also from HBase, Hive, and other data stores.

- Another commonly used set of APIs provides ways to query graph topology (getEdgeValue(), getVertexId(), and so on) and also to modify the graph during the application run (addVertexRequest(), addEdge(), and so on).

- A subset of APIs (supporting aggregators and master compute) goes beyond the basic BSP graph computation model.

- If the cost of sending a message ever becomes a concern, Giraph offers efficient tools to deal with it by applying combiners to multiple messages and effectively compressing the traffic.

This chapter laid the foundation for the rest of the book in terms of practical usage of Giraph APIs, but you still haven't seen the implementation details of the Giraph framework and how it uses lower-level mechanisms such as Hadoop MapReduce and Apache ZooKeeper. This is the focus of the next chapter.

CHAPTER 6

■ ■ ■

Giraph Architecture

The previous chapter introduced a gamut of Giraph APIs and showed how you can use them for various graph-processing applications. This chapter pops the hood on Giraph implementation and shows you what transpires behind the scenes when a graph-processing application is executed and APIs are called.

For the rest of this chapter, you use the example Giraph applications introduced in Chapter 5. Assume that the size of the graph on which the application operates is measured in billions of lines of input. With this assumption in place, the chapter traces every step of the application run on a Hadoop cluster. You start by learning about the functional roles of the different services that form a running Giraph application, how these services are assigned to compute nodes in the cluster, and how they coordinate their activity to perform an overall graph-processing task. You also look into steps that are needed to partition the input and load the initial graph topology in such a way that each vertex is assigned to a worker that performs compute operations on it. You then dive deep into the implementation details of each service and close the chapter by examining failure scenarios that a Giraph implementation needs to cope with.

Genesis of Giraph

So far, you have only used Giraph on tiny data sets that fit comfortably into the disk and memory space of a single host. Although Giraph is certainly applicable for use with small and medium-size graphs, its true power comes from its scalability. To give you a taste of the level of scalability required from a Giraph implementation, consider its use at Facebook: analyzing the social graph of the social network's members. In 2013, the total number of vertices was estimated at around 1.1 billion, and the number of edges was well into the trillions. Even if storing each chunk of data associated with either a vertex or an edge takes only hundreds of bytes, you are approaching petabyte scale for the graph representation alone. Storing datasets of this size on a single node is impossible; and analyzing them using the CPU resources of a single node would be prohibitively time consuming. The only way to scale for both storage and compute requirements is to exploit clusters of commodity hardware.

It should come as little surprise, then, that a running Giraph application is a collection of distributed, networked services running in parallel on different nodes in the cluster. Each of these services provides a particular function for the rest of the Giraph application, has its own lifecycle, and interacts with the rest of the services running on different nodes via remote API calls. Because the majority of these interactions are network-based, all the complications of network programming (collectively known as *fallacies of distributed computing*) need to be dealt with. For example, the Giraph implementation spends a lot of time tuning the lifecycles of its internal services to enable the most efficient recovery in case of an individual service failure.

FALLACIES OF DISTRIBUTED COMPUTING

Back in 1994, a few smart engineers working at Sun Microsystems started to realize the power of distributed computing that has been unlocked by advances in local network design. Around the same time, one of those engineers, John Gage, coined the phrase "the network is the computer" to capture the company's new direction. That was the good news. The bad news was compiled by another Sun engineer (Peter Deutsch) and delivered to future generations of engineers as a list of false assumptions to watch out for when designing this next-generation computer architecture. The list is known as the fallacies of distributed computing, and it is as relevant today as it was 20 years ago:

- The network is reliable.
- Latency is zero.
- Bandwidth is infinite.
- The network is secure.
- Topology doesn't change.
- There is one administrator.
- Transport cost is zero.
- The network is homogeneous.

Keep this list in mind when reading this chapter—it will help you appreciate quite a few of the architectural decisions made by the Giraph implementation.

Giraph Building Blocks and Concepts

Regardless of whether you are running the example application on a single host (as you did in Chapter 5) or on a 1,000-node cluster, everything Giraph does boils down to making sure a piece of Java code that acts as a particular type of a network service can be executed on a given number of hosts. Before moving on, let's quickly define the three types of network services that serve as fundamental building blocks for Giraph architecture: masters, workers, and coordinators.

Every Giraph computation is a result of the coordinated execution of these services, operating on subsets of vertices that are split into partitions and processed in parallel. Partitioning is a key optimization, allowing parallel execution of graph-processing tasks; it is *dynamic*, meaning the assignment of partitions to workers can change based on the characteristics of the running application. At every superstep, however, the overall functional structure of a Giraph application looks like Figure 6-1. All the services represented in this figure are internal to the Giraph implementation and are not visible to the end user, but it is still important to know their architecture—if for no other reason than to be able to debug failures in Giraph application runs.

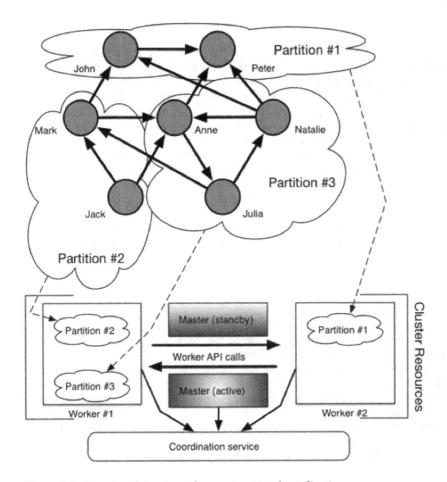

Figure 6-1. *Functional structure of a running Giraph application*

Masters

There are always one active master service and a few standby masters bidding to become active if the current master fails. The standby masters are dormant and don't play an active role in the lifecycle of a running Giraph application. Once a master becomes active, its job is to coordinate computation. This task consists of doing the following:

- Transitioning the workers from one superstep to the next in a coordinated manner

- Before each superstep, assigning partitions to active workers

- Running the master compute code

- Monitoring the health and statistics of all the workers

Workers

Workers represent the majority of Giraph services. Their primary function is to manage the state of graph partitions assigned to them. Each worker exposes a set of network APIs to let remote workers manipulate the data in partitions assigned to it. Again, note that these network APIs are internal to the Giraph implementation and are not expected to be available to end users. Workers transition from superstep to superstep as directed by the active master. During each superstep, workers iterate over the graph partitions they own and execute the compute() method for all the vertices belonging to these partitions. Workers are also responsible for checkpointing their state from time to time as a means of recovery from a worker failure.

Note that the assignment of graph partitions to workers is not permanent and is subject to master-driven rebalancing before every superstep. A class implementing the MasterGraphPartitioner interface provides the implementation of the partitioning logic. An instance of that class belonging to an active master manages the current mapping of partitions to workers and provides the generateChangedPartitionOwners() method for reevaluating the mapping based on the various statistics coming from the workers. The master updates the mapping via the coordination service. Workers, on the other hand, look up which partitions no longer belong to them and re-create the state of those partitions on a target worker by issuing network API calls.

Coordinators

Nodes running the coordination service provide the nervous tissue for the rest of the Giraph services. They don't participate in performing any graph-processing work; instead, they provide distributed configuration, synchronization, and naming registry services for the rest of Giraph.

There are several coordination service implementations you can use, but the default choice in the Hadoop ecosystem has always been Apache ZooKeeper. The ZooKeeper service is so ubiquitous that it is hard to imagine a production Hadoop cluster deployment running without it. It has become synonymous with the coordination of any application in the Hadoop ecosystem.

A collection of nodes running a coordination service is collectively known as a ZooKeeper *ensemble*. All nodes in the ensemble are considered peers, and each node has a replica of a ZooKeeper state as a means of achieving high availability and fault tolerance. As long as the majority of the nodes are up, the service will function correctly.

Giraph provides two options for managing coordination services. The default option is to spin up ZooKeeper ensemble nodes on demand as part of a Giraph application run. That way, coordination services are no different from masters or workers: they are fully managed by Giraph, and they vanish once the application exits. The second option is to use a stand-alone ZooKeeper ensemble that is centrally maintained as part of the cluster. In that case, Giraph applications act as pure clients, and there is no need to have the coordination service running on nodes used by the Giraph application.

The second approach is recommended for production use of Giraph. It allows for easier postmortem and real-time debugging (because the coordination service stays alive after the Giraph application exits). Also, because it uses a cluster-wide, centralized, highly tuned, well-maintained ZooKeeper ensemble, your application's performance may be better.

Bootstrapping Giraph Services

Consider any of the examples from Chapter 5, but this time assume that you have to deal with a much bigger input dataset. The size of the dataset makes it prohibitively expensive to execute the application on a single host and requires you to use a cluster. The good news is that you don't have to change the application to make it run on a cluster of compute resources: the Giraph architecture is flexible enough to scale elastically. All you have to do to begin using a cluster is to tell Giraph how many compute resources it can use for which purpose.

Setting the configuration property `giraph.maxWorkers` (or passing the same value via the command-line option -w) tells Giraph how many compute resources should be used to execute worker services. Another configuration property, `giraph.zkServerCount` (which has a default value of 1), tells Giraph how many compute resources need to be allocated for running master and coordination services. Finally, the total number of cluster resources used by an application is determined based on the configuration property `giraph.SplitMasterWorker`. If that property is set to false, all services can run on all cluster resources, and the total number of resources used is equal to the value of `giraph.maxWorkers`. On the other hand, if it is set to true (its default value), the total number of utilized resources is the sum of `giraph.maxWorkers` and `giraph.zkServerCount`.

For the rest of this chapter, assume a cluster of at least ten nodes so you can execute the example application with `giraph.maxWorkers` set to 8 and `giraph.zkServerCount` set to 2. Also assume the default value (true) for `giraph.SplitMasterWorker` so the total number of utilized cluster resources is exactly ten (eight plus two). Those resources will be used to run exactly eight worker services, two master services, and two coordination services (with the master and coordination services co-located together).

With these settings in mind, let's look into the underlying mechanism that pushes the required bits of Java code for each service to the desired cluster resources and makes them behave like a set of eight worker services, two master services, and two coordination services. See Figure 6-2.

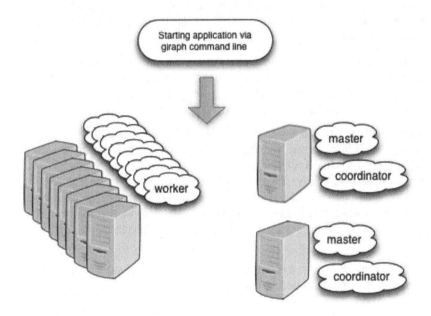

Figure 6-2. *Running an example on a cluster of 10 hosts*

Although it is nice to know that given the right combination of command-line options, Giraph can start using the power of a Hadoop cluster, the question you answer in this chapter is how that happens. One way to distribute services on a cluster of compute resources requires building a custom Giraph-specific cluster-management layer, purposefully designed with a graph-processing application in mind. There is nothing wrong with doing that; Apache Hama (a graph-processing framework very similar to Giraph) made exactly that type of design decision. It allows for a greater degree of control, but the price is maintaining low-level cluster-management code that has nothing to do with Graph processing.

Unlike Hama, Giraph decided to use existing cluster-management solutions to do all the heavy lifting. Apache Hadoop is the most popular solution and also the one that comes by default with the Giraph distribution. It is not the only option, though. Even within the Hadoop project, Giraph has two alternatives: a more classic MapReduce framework (also available in prior versions of Hadoop) and a brand-new collection of resource-management APIs known as YARN. The YARN back end is still experimental, and this chapter assumes that the example application is executed using the Hadoop MapReduce framework.

WHEN MAPREDUCE IS NOT REALLY MAPREDUCE

The MapReduce framework for distributed data analysis was first described by two Google engineers (Jeffrey Dean and Sanjay Ghemawat) in the seminal 2004 paper "MapReduce: Simplified Data Processing on Large Clusters." The core idea behind the framework was simple enough, but it required a radical shift in application design. The paper proposed that every data-processing application should be built around two things: a piece of code called a *mapper* and another piece of code called a *reducer*. Both pieces of code are transparently instantiated on a large cluster of compute resources, thus allowing the application to process as much data as there are compute resources. In other words, thousands of mapper copies run in parallel, all supplying data to hundreds of reducers, also running in parallel. The programmer is free from low-level cluster-management plumbing and can focus on implementing the mappers and reducers.

Apache Hadoop was the first highly popular, open source implementation of a MapReduce framework. Its availability led to explosive growth of data-processing applications that can be neatly decomposed into mappers and reducers. Graph processing is not one of those applications; it is typically much more iterative in nature, and it requires a different kind of data partitioning.

Thus, when running with a MapReduce back end, Giraph doesn't have a chance to exploit the spirit of the framework. All it needs from Hadoop is a way to push a bit of its own code to a given number of compute resources. The trick it plays is that it constructs a fake MapReduce application with a fixed number of mappers and no reducers. This is known as a *map-only application*. Because it is disconnected from all the Hadoop data pipelines from Hadoop's standpoint, it "runs forever" (as opposed to a normal MapReduce application, which runs as long as unprocessed data remains). Because of that, the Giraph implementation has to deal with ingesting the initial data and limiting the application's runtime.

Even though a running Giraph application looks to a Hadoop cluster like a MapReduce application, it has very little to do with the behavior of a typical MapReduce job.

How does Giraph use MapReduce to spin up all the required internal services? It involves encapsulating Giraph's business logic in the GraphTaskManager class implementation and expecting it to be instantiated on every node available to the Giraph application. The MapReduce Hadoop framework expects GraphTaskManager to provide a way to execute a mapper. Giraph doesn't need to worry about the details of how an instance of GraphTaskManager is created; rather, it focuses on the job it needs to perform in the overall mesh of Giraph services. When Giraph uses MapReduce as a back end, the job of instantiating a GraphTaskManager object belongs to a mapper. This is the only thing this unusual mapper ever does.

At the highest level, the entire process of executing a Giraph application on a cluster of machines consists of the following steps (as illustrated in Figure 6-3):

1. An underlying cluster-management framework (most likely Hadoop's MapReduce or YARN, although Giraph can be plugged in to other ones as well) instantiates a `GraphTaskManager` class implementation on a number of nodes. The exact number is determined by Giraph's configuration.

2. The same framework then calls the `GraphTaskManager.setup()` method, which takes care of the initialization and also determines which internal Giraph service (master, worker, or coordination service) this node is supposed to provide to the rest of the Giraph application.

3. The `GraphTaskManager.execute()` method gets control. The Giraph application run commences, with each node assuming the role of one of the three internal services: master, worker or coordinator.

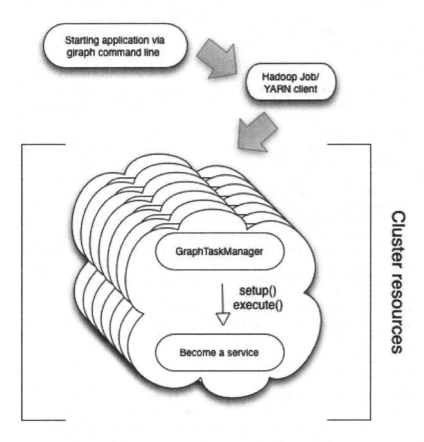

Figure 6-3. *Bootstrapping a Giraph application on a cluster of nodes*

Despite its simplicity, this architecture means Giraph doesn't have to be concerned with the details of finding appropriate cluster resources, scheduling containers to run there, restarting containers on different hosts if they fail, and otherwise allocating low-level resources. This is a very elastic approach that lets the

Giraph application have as many workers as needed; the cluster-management framework is responsible for maintaining the worker pool. This is all good news, except for one detail: containers can be instantiated on any host in a cluster, and how can they communicate with each other if they don't know the host names (or IP addresses) of other services?

The answer to this question lies in Giraph's use of a coordination service that every container can communicate with. Different Giraph services use the coordination service as a bulletin board or directory where all containers can post information about themselves (such as their network coordinates, operational status, and so on) and inspect information posted by others. Thus, instead of every container keeping track of the network address and state of every other container, the only information needed to bootstrap the Giraph framework is the location of a coordination service.

An interesting consequence of a coordination service playing the role of central information radiator is that it bears the brunt of communicating with every container. To avoid overwhelming it, the Giraph implementation tries to minimize the amount of traffic between each container and the coordination service. The approach is to manage a small amount of metadata via the coordination service and opt for direct communication between containers whenever possible.

As long as the Giraph services are running as they should, a cluster-management framework such as Hadoop stays out of the way. If services fail (due to software bugs or hardware failure), it is up to the cluster-management framework to spin up new instances. The coordination service plays a vital role in allowing the Giraph implementation to detect when services fail and when they are being brought back online (perhaps on a different host). That way, an active master always has a way to rebalance the current graph computation between the active workers.

Of course, detecting service failure and bringing back new instances solves only half the problem. When a service goes down, it also brings down its in-memory state. It would be unfortunate if Giraph had to restart an entire computation simply because one service out of thousands went down. An elegant solution to that problem is to use periodic checkpoints. At a given interval (requested via the `giraph.checkpointFrequency` configuration property), Giraph records the state of all the services in a permanent location in HDFS (under the home directory of a current user). Suppose `giraph.checkpointFrequency` is set to 5. This means after every five supersteps, the state of all the services is checkpointed. Should any service fail, the computation must redo at most five recent supersteps, because it is restarted from a checkpointed state. The exact value the checkpoint frequency is set to is highly application specific. On one hand, more frequent checkpointing increases the overall runtime of the application; on the other, it speeds up recovery if a worker fails. Experiment with different values, and pick the one that works for you. Just keep in mind that the default value for this property is 0: by default, checkpointing is disabled.

Now that you understand the mechanics of how Giraph uses cluster compute resources, let's proceed with a detailed overview of what each Giraph service does and how they interact.

Anatomy of Giraph Services

Once fully bootstrapped and running, every Giraph application consists of a network of services that, collectively, accomplish graph processing by communicating with each other via network API calls. The two services implemented by Giraph (masters and workers) share a common design based on the `CentralizedService` interface and the `BspService` abstract class implementing it. The coordination service is implemented separately by a stand-alone Apache ZooKeeper project and is not discussed in this book.

WHAT IS APACHE ZOOKEEPER?

Apache ZooKeeper is an effort to develop and maintain an open source server that enables highly reliable distributed coordination. It was developed in response to the proliferation of highly distributed applications, all trying to implement similar functionality in an ad hoc fashion. You can find information about the design and implementation of ZooKeeper at `http://zookeeper.apache.org` or in a book written by two of its authors, Flavio Junqueira and Benjamin Reed: *ZooKeeper: Distributed Process Coordination* (O'Reilly, 2013).

All the functionality offered by masters is implemented in a `BspServiceMaster` class. On the worker side, `BspServiceWorker` is the corresponding implementation. Both classes extend the common functionality of `BspService` by subclassing it and implementing more specific interfaces, as shown in the class hierarchy in Figure 6-4.

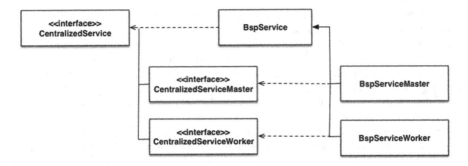

Figure 6-4. *Giraph services class hierarchy*

Both services act as network servers and clients at the same time by using the classes implementing the following interfaces:

- For masters: `MasterClient` and `MasterServer`.
- For workers: `WorkerClient` and `WorkerServer`.

Currently, the only implementation available for all four of these interfaces is based on the Netty framework. Netty is a new I/O (NIO) client/server framework that enables quick and easy development of network applications such as protocol servers and clients. Don't worry if you are not familiar with Netty; as long as you understand that it implements NIO, you should have no trouble understanding how Giraph uses it. Giraph abstracts that away by wrapping the code that provides general-purpose, Netty-enabled client/server interactions into two stand-alone classes: `NettyClient` and `NettyServer`. That part has little to do with the Giraph graph-processing model. The part that does is then wrapped in the four classes shown in Figure 6-5.

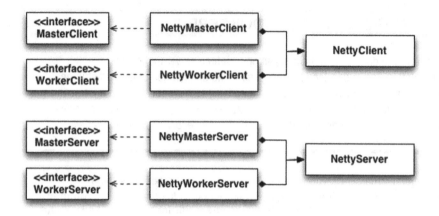

Figure 6-5. *Netty-enabled client/server architecture*

As you can see, each of the classes implements the appropriate interface by delegating all non-Giraph-specific client/server interactions to a NettyClient or NettyServer object that it owns.

Determining which node should provide which Giraph service is handled by GraphTaskManager.setup(). That determination is permanent during the Giraph's application run and is driven by the configuration parameters given to the Giraph framework. For example, if an external ZooKeeper ensemble has been specified via the giraph.zkList configuration option, then there is no need for Giraph to assign this service to the nodes it manages. The implementation has the flexibility to assign any combination of services to a given node, with one exception: a coordination service cannot be the only service assigned to a node. In other words, nodes running coordination services always have an additional master or all three services running there as well. The only two impossible combinations of node service assignments are these:

- A node exclusively dedicated to running a coordination service

- A node running a coordination service and a worker service

Now that you have enough background about the generic aspects of how Giraph implements its master, worker, and coordination services, it is time to dive deeper into their detailed architecture.

Master Services

The master service always runs as an extra Java thread implemented by the MasterThread class. An instance of that class owns a reference to the BspServiceMaster object and implements its functionality by orchestrating calls to the APIs exposed by BspServiceMaster. The master algorithm is pretty simple, but it needs to be designed in a way that is resilient if the active master fails and a different master (typically running on a different host) assumes the responsibilities of an active master. Figure 6-6 shows the high-level steps that every master service goes through.

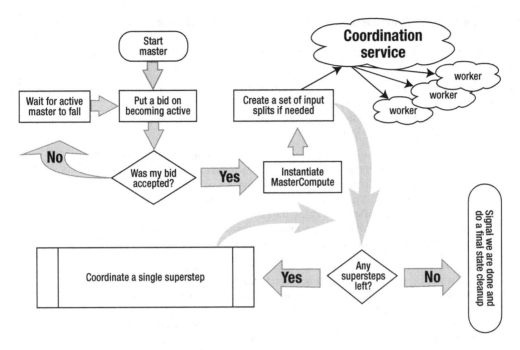

Figure 6-6. *A single master service lifecycle*

The first thing a master does is place a bid on becoming an active master via a `BspServiceMaster.`
`becomeMaster()` call. Note that although all masters issue this call, all but one will block. The one that
succeeds will determine the active master; the others will stay blocked until the active master fails. At that
point, one of the blocked calls will return and thus determine the next active master. This synchronization
barrier is implemented by using the functionality of a coordination service, as you see later.

On becoming an active master, the master thread instantiates a `MasterCompute` object, provided it
was requested as part of the Giraph application configuration. This, in turn, has a chance to register the
application aggregators on the active master.

Before entering a superstep loop, an active master tries to perform one more important function:
creating a set of input splits for the graph data (both vertex and edge data) to be loaded by the appropriate
input formats. Because Giraph applications typically operate on a set of huge files stored in a distributed
filesystem, effectively spreading the work of loading that data among all the available workers is a key
scalability requirement. This functionality is implemented in the `BspServiceMaster.createInputSplits()`
method, and it works as follows. First it consults a coordination service to see if the mapping of which worker
is supposed to load which file has been established. If it hasn't, the map is generated and externalized into
the coordination service. This accomplishes two things at once. First, because every worker is watching the
coordination service for updates, once the mapping is made available, the workers can load their portion
of the input data (the details of this step are laid out in the section "Coordination Services" below). Second,
persisting the mapping to the coordination service as a transaction makes it possible for the standby master
to effectively take over in case the active master dies.

Once the input splits' mapping has been made available to all the workers, the active master transitions into the superstep loop that runs for as long as supersteps remain. The bulk of the functionality driving the superstep loop is implemented in the BspServiceMaster method coordinateSuperstep() and consists of the following steps:

1.   Determine the set of healthy workers by observing the self-reporting available in the coordination service. The master observes the healthy workers for the duration of the superstep by listening to the coordination service events and takes corrective actions if they fail.

2.   Assign graph partitions to the healthy workers. This is a dynamic assignment for a given superstep and may change in the next superstep based on worker load and other considerations. If any partition is assigned to a worker that doesn't have the graph data for that partition in its local memory, the worker that owns the partition is in charge of sending graph data by issuing API calls re-creating the partition on a target worker. The partition data may also come from the checkpoint file if this superstep was restarted from a checkpoint.

3.   Finalize the state of the aggregators from the previous superstep, and send them to the new worker owners. This is the last step in the two-phase distributed aggregator management. At this point the master has done the final aggregation and is ready to send the values back for further aggregation during the current superstep. See the section "Worker Services" below for the worker side of the two-phase distributed aggregator implementation.

4.   If the current superstep is an input superstep (a fake superstep that is needed to make workers load the initial graph data from input splits), coordinate the loading of vertex and edge data. This is accomplished by partitioning the data files into input splits and writing the mapping of input splits to workers to the coordination service. The workers watching updates in the coordination service will notice that the mapping belonging to them is available and will interpret that as a signal to start loading the data.

5.   Wait for all the workers to finish processing the current superstep. If any of the workers fail while processing the current superstep, attempt to restart from a checkpointed state. If that is successful, the next iteration of the superstep loop will assume that the previous superstep was the one restored from a checkpointed state.

6.   If the checkpoint frequency is met, wait for all the workers to checkpoint their state, and then finalize the checkpointing process globally (including checkpointing the aggregators).

7.   At this point, assume that all the workers have successfully completed the current superstep and it is safe to collect the values of the aggregators assigned to them.

8.   Once all the aggregator values are collected and the global state of the aggregators (as stored in the active master's memory) is brought up to date, call a master compute's compute() method.

9.   Determine whether it is time to stop the entire computation. To do so, first check whether an external flag was set in a coordination service, signaling to halt the computation for an arbitrary reason. If it wasn't, check whether all the vertices voted to halt and there are no more messages pending to be delivered to the vertices. Finally, check whether more supersteps have happened than a

maximum number of supersteps allowed for this application (you can set a value for this using the `giraph.maxNumberOfSupersteps` property). If any of these conditions are true, record the fact that the computation has been halted in the global statistics for this job, and bail out of the superstep loop.

10.   Publish the aggregated application state into the coordination service.

11.   Depending on the value of `giraph.textAggregatorWriter.frequency`, check whether it is time to output the state of all the registered aggregators. If it is, call the supplied `AggregatorWriter` implementation.

Once the superstep loop is over, the only thing left for the master to do is to call the cleanup routine `BspServiceMaster.cleanup()`. All the master processes should signal that they are done by posting a special kind of note to a coordination service. When the number of notes equals the number of partitions for workers and masters, the master cleans up the global state associated with a job. This is how a global barrier is implemented with the help of a coordination service.

Getting back to the example application, there are always exactly two master services running: one active and one standby. Should the active master fail, the standby resumes coordination activities. As long as the master is active, its superstep loop goes through the steps outlined in Figure 6-7.

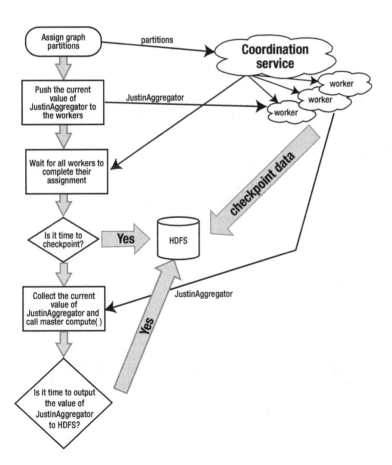

Figure 6-7. *Example application's active master superstep loop*

Worker Services

The key function of a worker service in a Giraph application is to manage the state of a few graph partitions assigned to it by the master. Just as with the master, the lifecycle of a worker is managed by the GraphTaskManager.execute() method orchestrating calls to the API methods exposed by the object of the BspServiceWorker class, which provides the bulk of the implementation.

Each worker service operates under the same resiliency constraints that masters operate under. Worker failure is a norm in a Giraph application, and the worker algorithm should be designed with that possibility in mind.

The algorithm consists of two phases: the setup phase and a superstep loop phase. The setup phase is mostly concerned with loading initial graph data into the memory of all the workers. The key steps of how it happens are outlined in Figure 6-8.

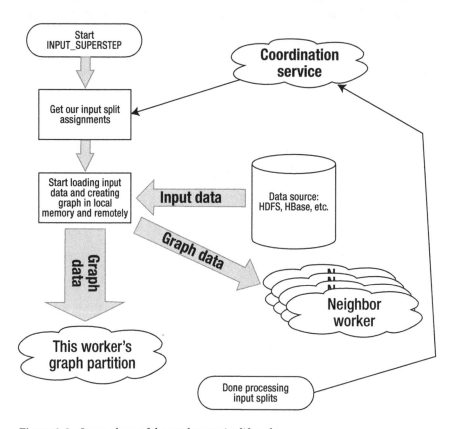

Figure 6-8. *Setup phase of the worker service lifecycle*

Not surprisingly, BspServiveWorker.setup() provides the implementation of a setup phase, which, before it loads the data, checks whether setup() was called as part of restarting the failed superstep. If that's the case, it loads the state from the checkpoint and bails, because the reset of the setup() machinery only applies when the entire Giraph application is initialized for the first time:

1. Start a fake superstep that does no computation but instead waits for the master to calculate input splits and then loads the slice of input data that is assigned to this worker. Internally, this superstep is called `INPUT_SUPERSTEP` and has a numeric ID of -1. An important note about loading input splits is that in general, you shouldn't expect the input data from a split assigned to the worker to generate vertices belonging to one of the graph partitions assigned to the same worker. Thus, a worker loading vertex and edge data is virtually guaranteed to generate network traffic, sending messages to other workers and requesting them to create vertices in their partitions.

2. Keep loading data and creating new vertices or edges as long as unprocessed input splits remain. The two classes `VertexInputSplitsCallable` and `EdgeInputSplitsCallable` provide an implementation of this functionality.

3. When there are no splits left to process, signal that this worker is done by posting a message to the coordination service, and wait for the other workers to finish. This is necessary because it allows a worker that is finished to still receive network API calls and update its partitions with the vertices loaded by workers still processing their input splits. The fact that all workers are finished is recognized by the master and reflected in the coordination service.

4. When the master has signaled that all workers are finished loading their data, do a few housekeeping tasks: create the remaining partitions owned by this worker (the ones for which no input data has arrived), and finalize mutations deriving from requests to add edges to the vertices belonging to partitions managed by this worker.

5. Wait for the master to signal that the fake `INPUT_SUPERSTEP` has completed and all the vertices are ready to move into the superstep loop.

Once all workers are ready to transition to the superstep loop, they do so at once and enter the main computational loop of the Giraph application. This worker service superstep loop cycles through the following steps as long as supersteps remain:

1. Prepare for the current superstep by calling `BspServiceWorker.startSuperstep()`. This combines all API requests received during the previous superstep and mutates the graph partitions accordingly (adding, removing, and updating vertices, and so on). It also waits until master is finished rebalancing the partitions between workers and returns this worker's partition assignment.

2. Based on the current partition assignment, the worker may have to push some of its partitions to the workers they were assigned to. It does so by calling `BspServiceWorker.exchangeVertexPartitions()`. The method makes sure not to return until all of the worker's dependencies are finished sending their data to it. As usual, the worker signals completion of pushing its own data by sending a message to the coordination service; it relies on the master to aggregate those messages and signal when the entire worker collective is done exchanging graph partitions.

3. Check whether this superstep is supposed to be loaded from a checkpointed state or whether, instead, it is time to save the current state into a checkpoint (as signaled by the checkpointFrequencyMet() method). The implementation of both actions is provided by two complementary methods defined in BspServiceWorker: loadCheckpoint() and storeCheckpoint().

4. Call the BspServiceWorker.prepareSuperstep() method, which is currently mostly concerned with synchronizing the state of the aggregators owned by this worker and the rest of the workers. In this step, the worker fetches the state of the aggregators it manages from the master and blocks until they are pushed to all the other workers.

5. Call compute() for all the vertices belonging to partitions managed by this worker. Iterate over all the vertices, but only call compute() for those that either are not halted or have messages available. This is done in a multithreaded fashion, with threads using the ComputeCallable class as an implementation for the task that needs to be added to the task queue and later executed.

6. Call the BspServiceWorker.finishSuperstep()method, which does all the necessary work to complete the current superstep and then blocks, waiting for the master to signal that all the workers are ready for the next superstep to begin. As part of this step, all the aggregators that ended up having a value assigned to them as part of this worker are sent to their rightful owners (the owners are picked dynamically by the master). Then, wait to receive values of the aggregators that are managed by this worker from all the other workers. Once that is done, aggregate them one final time and push them back to the master. The last bit of housekeeping, performed by the finishSuperstep() method, is to reflect the statistics from the current superstep in a coordination service. This is needed to keep track of what's happening on what worker for observability and load-balancing reasons.

7. Because the previous step makes the global Giraph statistics available to each worker, the superstep loop continues to the next superstep, provided there are active vertices left in the graph.

Once again, if you turn to the example application, each worker (once it enters the superstep loop phase) goes through the set of transitions outlined in Figure 6-9.

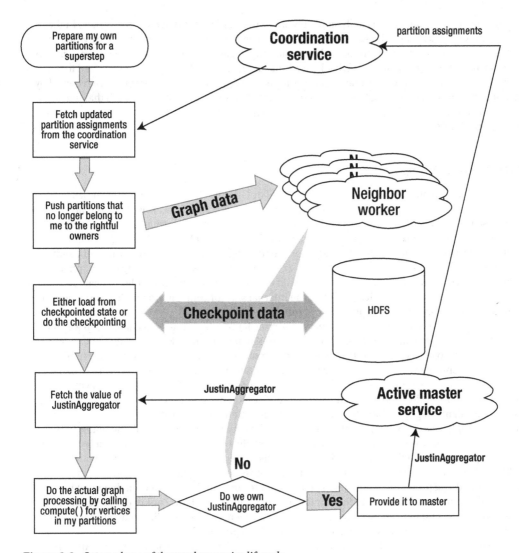

Figure 6-9. *Setup phase of the worker service lifecycle*

The worker superstep loop is the central process driving the execution of every Giraph application. Most of the steps are pretty easy to follow, and the only part worth reiterating is the implementation of sharded aggregators. Steps 4 and 6 are the key to understanding how sharded aggregators work.

First, during each superstep, values provided to the aggregators are partially aggregated by the workers locally. No network API calls are involved. Once the superstep is done, all of these partially aggregated values coming from different workers need to be aggregated once more, and the resulting value must be made available to MasterCompute.compute(). One option is to let the master manage all the aggregators because their value needs to end up there anyway. This simple approach unfortunately doesn't scale if an application uses a lot of aggregators; it turns the master into a bottleneck from both the computation and network communication standpoints.

The approach that Giraph takes is to dynamically shard the aggregators between the workers in the cluster. Here's how it works: when the superstep ends, each aggregator is assigned to one of the workers, and that worker becomes responsible for receiving partially aggregated values from its neighbors. It is also expected to push the final aggregated value back to the master. That way, the master doesn't need to perform any aggregation, and it always receives one final value for each aggregator. Once the values are received by the master and it is finished executing `MasterCompute.compute()`, the only worker to which that master sends each value is the owner of that aggregator. As a final step, the owners are expected to distribute these values to their neighbors.

Coordination Services

As noted earlier, a coordination service is not implemented by Giraph. The implementation comes from a stand-alone Apache ZooKeeper project. Still, it is important to understand the APIs it offers and how masters and workers use them. Most of the data that Giraph puts into a ZooKeeper is permanent, so knowing the details of how Giraph externalizes its state can be an invaluable tool for debugging failed applications. And if you want to make it even easier to observe and diagnose running Giraph applications, run them with an external ZooKeeper ensemble (by setting the `giraph.zkList` property) instead of relying on Giraph to spin up coordination services on demand. By running with an external ZooKeeper ensemble, you can inspect the internal state of the coordination service (and thus the internal state of your Giraph application) whenever you like, as opposed to only while the application is running.

One of the key properties of ZooKeeper is that the service it provides is highly available. The way it is implemented requires that all nodes that are part of the ZooKeeper ensemble know the network addresses (hostnames and ports) of every other node that is part of the same ensemble. That way, they can make the best effort to synchronize the common state between as many nodes as possible. ZooKeeper clients (remember, Giraph is only a ZooKeeper client) talk to only one ZooKeeper node at a time. But even the client has to know the network addresses of as many nodes in the ZooKeeper ensemble. After all, if the node the client is currently talking to goes down, the client has to know where to reconnect.

As long as Giraph uses an external ZooKeeper service, the network addresses of all the nodes in the ensemble are passed to the worker and master services as a static, comma-separated list specified via the `giraph.zkList` property. If, on the other hand, Giraph needs to spin up a coordination service on demand, it can no longer know in advance what nodes in the cluster those services will be instantiated at. It is a curious chicken-and-egg problem: Giraph needs ZooKeeper to store all the information that must be shared between services running on different nodes, but the network addresses of the ZooKeeper nodes are exactly the kind of information that need to be stored in ZooKeeper. It seems as though Giraph needs ZooKeeper to be able to bootstrap ZooKeeper. How do you avoid this infinite regression?

Giraph uses another storage substrate that all Giraph services know how to access: HDFS. Although most of the time HDFS is an extremely poor choice for keeping small files that need to be frequently updated, it is OK to use it on a case-by-case basis. That is exactly how Giraph uses it. It records the network addresses of the ZooKeeper ensemble nodes as names of HDFS files. The exact location where it stores these files is set by the `giraph.zkManagerDirectory` property; the default value is `_bsp/_defaultZkManagerDir` (note that because the default value has a relative, not absolute, path name, it is rooted under the home directory of the user running the job).

Now that you know how bootstrapping the ZooKeeper ensemble works and how master and worker services are enabled to connect to it as clients, let's see what ZooKeeper client APIs they use. From a client perspective, ZooKeeper offers a filesystem-like hierarchical API with a namespace structured as a tree of nodes (each node in a tree, including inner nodes, is called a *znode*), with every node answering to a small set of API calls. Each znode in the hierarchy can hold some data, be a parent node to other nodes, or, unlike in any of the filesystems, do both. ZooKeeper is designed for coordination, which typically requires small chunks of data stored in znodes (the amount of data is capped at 1MB). All APIs are atomic, which means clients receive either all the data or none of it. When a client creates a znode, it may elect to declare the znode to be ephemeral and thus guarantee that it is visible only as long as the client that created it is connected to ZooKeeper. If the node is not ephemeral, it is considered to be persistent, and an explicit delete call would be required to remove it from the tree. Both types of znodes can have an additional property of being sequential (thus ZooKeeper znodes fall into one of four categories: ephemeral, ephemeral sequential, non-ephemeral, or non-ephemeral sequential). A sequential znode's name is derived from the initial name supplied as part of the create API call and the sequence number ZooKeeper assigns it. So, if multiple clients try to create a sequential znode with the same path name, each of them will end up creating a unique znode with ordering guaranteed by the ZooKeeper service. Finally, clients can register watches that are triggered when other clients perform certain types of operations on the tree of znodes. Events such as creation, deletion, and modification can be tracked that way.

INSPECTING THE ZOOKEEPER ZNODES TREE

By default, Giraph doesn't provide any tools to inspect the tree of znodes it maintains in the ZooKeeper service. It is highly recommended that you download and install Apache ZooKeeper separately and use the command-line utility zkCli.sh or zkCli.cmd to browse the file tree. The only command-line argument you need to supply is -server. This is the comma-separated list of ZooKeeper ensemble nodes that the client connects to. Once the client is started, it behaves similarly to an ftp client. Type help to get a quick overview of supported commands.

Giraph keeps all of its coordination metadata and internal state data rooted under the _hadoopBsp znode. By default, it is created directly at the root of ZooKeeper's hierarchy, but you can also instruct Giraph to root it at a subtree by setting the giraph.zkBaseZNode property. For example, setting giraph.zkBaseZNode to a value of /giraph/examples results in a running Giraph application creating the set of znodes shown in Listing 6-1.

Listing 6-1. Inspecting the Znode Tree of a Running Giraph Application Using the zkCli.sh Command-Line Utility

```
$ zkCli.sh -server zknode.cluster
[zk: localhost:2181(CONNECTED) 0]  ls -R /giraph/examples

/giraph/examples/_hadoopBsp/job_local1896501187_0001
/giraph/examples/_hadoopBsp/job_local1896501187_0001/_applicationAttemptsDir
/giraph/examples/_hadoopBsp/job_local1896501187_0001/_applicationAttemptsDir/0
....
/giraph/examples/_hadoopBsp/job_local1896501187_0001/_masterElectionDir
/giraph/examples/_hadoopBsp/job_local1896501187_0001/_masterElectionDir/
localhost_00000000000
/giraph/examples/_hadoopBsp/job_local1896501187_0001/_masterJobState
/giraph/examples/_hadoopBsp/job_local1896501187_0001/_vertexInputSplitDir
/giraph/examples/_hadoopBsp/job_local1896501187_0001/_vertexInputSplitDir/0
....
/giraph/examples/_hadoopBsp/job_local1896501187_0001/_vertexInputSplitDoneDir
/giraph/examples/_hadoopBsp/job_local1896501187_0001/_vertexInputSplitDoneDir/localhost_0
/giraph/examples/_hadoopBsp/job_local1896501187_0001/_vertexInputSplitsAllDone
/giraph/examples/_hadoopBsp/job_local1896501187_0001/_vertexInputSplitsAllReady
```

As you can see, Giraph exposes a treasure trove of information in the ZooKeeper state it maintains. An obvious implication of this is help with debugging; a less obvious one is integration with external applications that may need to track the progress of running Giraph applications. Keep in mind that the location and content of Giraph znodes are considered internal APIs of Giraph and may change without notice between releases. Figure 6-10 gives you a complete view of a hierarchy of znodes used by various Giraph services for coordination and metadata management.

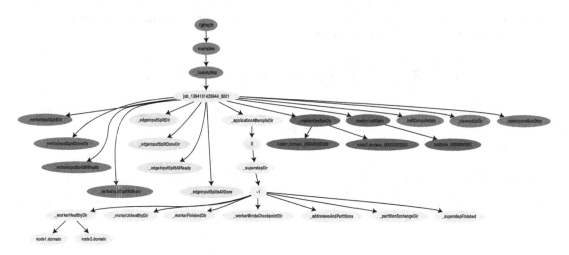

Figure 6-10. *Tree of znodes used by various Giraph services for coordination*

All the znodes from the example are rooted under /giraph/examples/_hadoopBsp/ job_1394131425944_0001. This common prefix was determined by the giraph.zkBaseZNode setting and the ID of a graph-processing job currently run by Giraph. Under that common root are roughly five different classes of znodes, color-coded according to the function they serve. Magenta znodes (names starting with _vertexInputSplit) are used to coordinate workers processing splits. Yellow znodes (names starting with _edgeInputSplit) are used for the same purpose when edge-split processing is requested. For both of these, the postfix of the znode name determines what it is used for:

- _vertexInputSplitDir and _edgeInputSplitDir contain paths to the input splits written by the master.

- _vertexInputSplitDoneDir and _edgeInputSplitDoneDir contain paths to the vertex input splits that have been fully processed.

- _vertexInputSplitsAllReady and _edgeInputSplitsAllReady contain paths to the vertex input splits ready to be processed by workers.

- _vertexInputSplitsAllDone and _edgeInputSplitsAllDone contain paths to the finished vertex input splits, to notify the workers that they should proceed.

Using these four znodes, you can create a fault-tolerant work-sharing scheme with an active master coordinating the processing and assignment of edge and/or vertex input splits as part of creating the initial graph topology. Green znodes are used for electing the active master (_masterElectionDir) and also for keeping the global, master-specific job state in a safe location (_masterJobState) in case an active master dies. Violet znodes are used to signal from the client side to stop the current graph-processing job and also to track worker cleanup activities:

- _haltComputation: A client setting this global flag can signal Giraph to halt the computation regardless of whether it should be halted naturally. This is the quickest way to abort the job without attempting to save anything.

- _checkpointAndStop: A client setting this global flag can signal Giraph to do the checkpointing of its current state and then halt the computation. This is a graceful way to stop the job.

- _cleanedUpDir: This znode denotes which workers have been cleaned up.

Finally, blue znodes are the heart of graph processing. Each graph-processing application goes through a number of attempts to run a given graph-processing job the same way a Hadoop MapReduce execution goes through a certain number of attempts to run a mapping phase before declaring a failure. All the attempts are tracked under the znode path _applicationAttemptsDir/<attempt ID>. Within each attempt's znode, _superstepDir tracks supersteps by ID (starting from -1). During each superstep, the master coordinates the progression of work among workers and rebalances the graph partitions. The following znodes are used to do that:

- _workerHealthyDir and _workerUnhealthyDir are used to partition workers (nodes) in the cluster into a set that can be assigned work and a set that needs to be skipped for work assignment. Each worker can be assigned to one or the other.

- _workerFinishedDir and _workerWroteCheckpointDir contain znodes of workers that are either done with the current superstep or done checkpointing the state of the current superstep, if checkpointing is enabled.

- _addressesAndPartitions is used for tracking master and worker addresses and partition assignments.

- _partitionExchangeDir helps coordinate the partition exchange between workers.

- _superstepFinished flags this superstep as finished.

Most of the time, the Giraph implementation doesn't interact with ZooKeeper client APIs explicitly, and instead wraps all API calls into methods provided by the ZooKeeperExt class. This class not only acts as a façade for ZooKeeper client APIs but also provides extended functionality such recursive path creation and non-atomic operations.

The majority of znodes created by Giraph fall under the category of regular (non-ephemeral, non-sequential) znodes; working with them is very similar to working with files in a regular filesystem. There are, however, a few exceptions when Giraph needs to use the APIs provided by the ephemeral and sequential flavors of znodes:

- Depending on whether they consider themselves healthy, workers create ephemeral, non-sequential znodes (named according to the hostname the worker is running on) under either the _workerHealthyDir or _workerUnhealthyDir znode. Giraph offers a hook for defining the criteria of what makes a node unhealthy however default implementation always returns health status for any node as healthy. Because these znodes are ephemeral, ZooKeeper guarantees that when a worker dies, a znode associated with it will disappear as well, and an active master will be notified because it maintains a watch over the two locations.

- Workers processing splits assigned to them by the master put the claim on the work that needs to be done by creating an ephemeral, non-sequential znode with a name corresponding to the input split assigned to them. When a worker fails, its work reservation znode disappears, allowing other workers to claim its previously read splits.

- Different master services place a bid to become an active master by creating an ephemeral, sequential znode under _masterElectionDir. Once the bids are placed, ZooKeeper imposes total ordering on the file names. The master whose bid has the smallest ID in that order proceeds under the assumption that it can now act as an active master. The rest of the masters block, watching over the children of the _masterElectionDir znode. Whenever an active master dies, its ephemeral znode is automatically removed by ZooKeeper. This, in turn, wakes up all the candidate masters. The candidate master with the smallest sequential ID associated with its bid assumes the duties of an active master.

- An active master service maintains periodic snapshots of the global state of the entire Giraph job by writing it out to a persistent (non-ephemeral), but at the same sequential, znode named _masterJobState/jobState_XXX. This guarantees that the master never overwrites the previous state, but keeps creating additional snapshots with increasing IDs. This snapshot of the internal state is used during recovery from a master failure.

This concludes the review of the design and implementation of each of the three services: masters, workers, and coordinators. So far, the chapter has focused on the normal execution path and has not spent much time covering fault-tolerance considerations. This is the subject for the rest of the chapter.

Fault Tolerance

As mentioned earlier, much of the Giraph architecture is dictated by the need to effectively address the issues captured by the fallacies of distributed computing. The biggest one often is the fact that at Giraph's typical operating scale, nothing can be assumed to be reliable.

HOW OFTEN DOES IT BREAK?

Everybody knows Google is an authority on managing the world's information. What's less well known is that based on its track record of building and operating large datacenters, Google has become one of the leading authorities on datacenter buildout and maintenance. The company is extremely careful about the design of everything that goes into its datacenters, but even Google, according to Google Fellow Jeff Dean (in his "Software Engineering Advice from Building Large-Scale Distributed Systems" talk), sees a lot of things go wrong.

In each cluster's first year, it's typical for 1,000 individual machine failures to occur; along with thousands of hard drive failures. One power distribution unit will fail, bringing down 500 to 1,000 machines for about 6 hours; 20 racks will fail, each time causing 40 to 80 machines to vanish from the network; 5 racks will "go wonky," with half their network packets missing in action; and the cluster will have to be rewired once, affecting 5% of the machines at any given moment over a 2-day span, Dean says. And there's about a 50% chance that the cluster will overheat, taking down most of the servers in less than 5 minutes and requiring one or two days to recover.

So far in this chapter, you have seen many techniques employed by Giraph to withstand various failure scenarios. This section looks at those scenarios and reviews the steps Giraph must go through to recover from disk failures, node failures, and network failures.

Disk Failure

As mentioned, Giraph is not concerned with managing storage. It does, however, interact with various storage layers to accomplishing the following goals:

- Reading the initial graph data and possibly outputting the final graph data at the end of the application run. Most of the time, the data comes from HDFS; but with an increased number of Giraph input/output formats available for other storage frameworks, the HDFS may be out of the picture.

- Storing checkpoints of each vertex state during the computation. Unlike with input/output formats, which provide unlimited flexibility for connecting Giraph to various data sources, when it comes to checkpoints, flexibility is very limited. Currently, the Giraph implementation can only use Hadoop-compatible filesystems, but most of the time stock HDFS is used.

For both use cases, what happens when a disk fails depends a great deal on what storage substrate is being used. In the case of a correctly deployed and configured HDFS, Giraph will not notice a failure rate of a few disks per day. HDFS is specifically designed to cope with those types of failures behind the scenes and mask them from the client. This is accomplished by making copies of all the data on at least three nodes (one on the same server rack and one on a different server rack). As far as metadata (a mapping between file names and blocks) is concerned, it is also typically stored in a few alternative locations. This makes Giraph checkpoints and the input/output formats operating on storage frameworks riding on top of HDFS (HBase, Hive, and so on) extremely reliable and oblivious to the disk failure rate of a typical datacenter. On a correctly configured HDFS deployment, the failure of a Giraph application due to a disk failure requires a very improbable set of events (all three disks hosting the same block failing at the same time).

Of course, however improbable it is, it can still happen. In addition, there are input/output formats for non-HDFS backed storage frameworks. These frameworks may not be as tolerant to individual disk failures as HDFS is. Either way, if a disk fails and a storage framework cannot silently recover, the error is propagated to the level of Giraph. At that point, if the error was encountered while an input/output format was reading or writing the data, the application will fail with an error message.

Node Failure

Because nodes in a cluster typically host a collection of different services, a useful way to analyze what happens when a node fails is to look at what happens when a particular service fails. That way, a failure of a single node can be viewed as a complex event consisting of a simultaneous failure of all the services hosted by it.

Giraph Master Failure

A master service can be in one of two states: it can be blocked on a bid to become an active master, or it can function as one. The master service that is blocked on a bid doesn't have any internal state. Thus, when it fails, the only event of any significance is that the znode advertising its bid for becoming an active master disappears from the coordination service. There is no internal state to recover (because a blocked master doesn't have any state), and additional master services can be spun up (and wait to become active) very easily.

A failure of an active master is more complex. If an active master fails, its ephemeral, sequential znode disappears as well, and that event wakes up all the master services blocked on a bid to become active. One of those master services will discover that its znode is now the first child of _masterElectionDir and will proceed to become an active master (the rest remain blocked on a bid). Because an active master keeps most of its state in a coordination service, a newly appointed active master doesn't need to recover the state of the failed master. Even the aggregator values don't need to be recovered, because they are fetched from the workers they are assigned to at the end of every superstep.

All in all, regardless of where an active master fails, a new active master can start from the beginning of coordinating the current superstep, fetch the state from the other services, and retrace all the steps just before the previous master failed.

Giraph Worker Failure

The purpose of the worker service is to manage the graph partitions and the aggregators assigned to it. This means, unlike a master service, a failed worker brings down quite a bit of unique state associated with it. The state is unique to each worker and is not redundant, and it would be undesirable for Giraph to have to restart an entire computation because of one failed worker. Fortunately, as mentioned earlier in this chapter, you can protect a long-running computation against this risk by using checkpoints.

Before you go any further with checkpoints, recall that a worker service goes through two phases: a split-loading phase and a superstep loop. During the split-loading phase (designated as a fake superstep with ID -1), each worker is tasked with loading initial graph data, turning it into vertices and/or edges, and sending those objects to an owning worker via network API calls. Each worker tries to load as many input splits as possible. The master monitors overall progress and waits for all the input splits to be loaded and (possibly) checkpointed on the receiving worker. If any worker fails during the split-loading phase, the master aborts the entire application run. This approach means there's no need for a special recovery mechanism during the split-loading phase.

After the split-loading phase, each worker transitions into the superstep loop. If checkpoints are enabled, you are guaranteed to have the initial graph data safely stored in HDFS. From that point on, whenever a healthy worker fails, its ephemeral znode disappear from _workerHealthyDir and an active master is notified by the coordination service. On receiving this notification, the active master declares the current superstep as failed, immediately restarts its own superstep loop from the last checkpointed one, and instructs all the healthy workers to load the state corresponding to that last checkpointed superstep. Once that is done, the computation continues, and the current superstep is the last checkpointed superstep plus one. Of course, if checkpoints are not enabled, the loss of even a single worker will mean an entire application run will terminate immediately and be marked as a failure.

ZooKeeper Service Failure

ZooKeeper is architected to be resilient to the failures of the individual nodes participating in its ensemble by reliably replicating its state to the majority of nodes. It also guarantees a reliable process for nodes to rejoin the ensemble after they go down or are cut off from the rest of the service. As long as the majority of the nodes in the ensemble are available, the service will be available and consistent.

From a client perspective, a single ZooKeeper node failure triggers a connection loss event. But because clients are supplied with a list of all the nodes in the ensemble, they can try to reconnect to the nodes that are still functioning.

If Giraph uses an external ZooKeeper service, the number of nodes in the ensemble is predetermined and can't be changed. If Giraph is managing its own coordination service nodes, you can control the size of the ensemble by setting the giraph.zkServerCount configuration property. The default value is 1, which doesn't allow for any kind of recovery. If you have to run with a Giraph-managed coordination service, be sure to set it to at least 3. In general, it is wise to set it to an odd number, because an even number doesn't add much to the reliability of the overall service compared to the lower odd number. For example, if you set it to 4, ZooKeeper will tolerate at most one node failure (because if more than one node fails, the remaining one or two nodes won't constitute a majority). This is no different from setting it to 3, which can also tolerate at most one node failure.

Network Failure

A network failure occurs when one node can no longer communicate with a subset of nodes in a cluster. A particularly nasty type of a network failure, known as *network partitioning*, happens when two subsets of cluster nodes can no longer communicate with each other but retain perfect communication abilities between the nodes in each subset.

Network failures affect Giraph in two major ways. First, network errors may prevent communication between the services (masters, workers, and coordinators) that form a running Giraph application. As you've seen in this chapter, the entire graph computation is done by master and worker services communicating via a set of network-enabled API calls. Masters and workers communicate with each other by using the Netty-based implementation of NettyClient and NettyServer. In general, the NettyClient implementation tries to be as asynchronous as possible. The remote API calls made in a superstep are non-blocking and go into a queue of outstanding requests for asynchronous processing. The superstep is considered finished when the queue of outstanding requests is drained. This is accomplished by relying on the synchronization method waitAllRequests(), which blocks until all the outstanding API calls are finished. It doesn't just passively block, though; it keeps polling for the status of outstanding requests with a frequency determined by the giraph.waitingRequestMsecs property (default value 15 seconds). Each API request sits in a queue of outstanding requests for not more than the time determined by the giraph.maxRequestMilliseconds property (default value 10 minutes). Any requests that take longer than that are considered to have failed and are resent. The size of the queue of outstanding requests is determined by the giraph.maxNumberOfOpenRequests property (default value 10,000).

Because communication with the coordination service is not Netty based but relies on native ZooKeeper Java APIs, it works slightly differently and relies on a different set of timeouts and polling intervals. As you've seen, an object of class ZooKeeperExt manages a connection to a coordination service for both masters and workers. As part of its initialization, it expects to receive values for the following network-related ZooKeeper client settings:

- How long a connection can be in a failed state before the entire session with a ZooKeeper ensemble is declared to be timed out (set by the configuration property giraph.zkSessionMsecTimeout with a default value of 1 minute).

- How many times to retry connecting to a ZooKeeper ensemble before giving up (set by the giraph.zkOpsMaxAttempts property with a default value of 3).

- How much time to wait before retrying due to a connection loss. This timeout interval is controlled by two properties. The giraph.zkOpsRetryWaitMsecs property (default value 5 seconds) on the client side determines the timeout for the masters' and workers' connections to the ZooKeeper ensemble. The giraph.zkServerlistPollMsecs property controls the same timeout for some internal operations and is also used for spinning up coordination service nodes on demand (instead of using a permanent ZooKeeper ensemble). Its default value is 3 seconds.

Finally, unbeknownst to Giraph, network-related failures may affect the internal state of the services it interacts with. For example, if you are using an external ZooKeeper ensemble and a network partition splits that ensemble in two, different masters and workers may end up interacting with different subsets of a single ZooKeeper ensemble. It is up to the ZooKeeper service to make sure it does whatever is necessary to present a consistent view of the world to all Giraph masters and workers. The last thing you want is an active master that has a different view of the workers than was reported by the workers themselves. This

161

could happen in a scenario where a few workers report back their state to part of a ZooKeeper ensemble that is currently partitioned away from the subset of the ensemble that the master is connected to. The reason it doesn't happen has to do with how ZooKeeper combats network partitioning; describing the ZooKeeper implementation is outside the scope of this book.

Remember that whenever Giraph services are talking to external distributed services, they rely on those services to either report an error or maintain a fully consistent model of their world. In general, Giraph is not architected for interacting with eventually consistent services.

Summary

This chapter covered these topics:

- High-level architectural overview of the Giraph framework

- A detailed walkthrough of internal services that together form the Giraph implementation

- How Giraph uses Apache Hadoop for all low-level cluster-management tasks while remaining flexible enough to allow alternative cluster-management frameworks to be plugged in by putting Giraph business logic into a stand-alone GraphTaskManager class

- How Giraph uses the Apache ZooKeeper implementation for all its coordination tasks, and the overall structure of the znode tree that various Giraph services use to communicate with each other

- Review of failure scenarios and scalability challenges and how Giraph deals with them

The key takeaway from this chapter is that the Giraph architecture is fundamentally built around three types of internal network services (masters, workers, and coordinators) running in a distributed fashion and collectively orchestrating the work that needs to be done for each superstep:

- Workers are defined by two phases: a setup phase and a superstep loop phase, each of which deals with working on a particular partition of either input data or graph data assigned to every worker by the active master.

- Masters play a central role in coordinating worker partition assignments (for both input data and graph data) and also coordinating worker transitions from phase to phase and from superstep to superstep.

- Coordinators form a ZooKeeper ensemble and let masters and workers coordinate with each other while also tracking the overall composition of roles in the Giraph service assignment.

Throughout this chapter, you have looked at various "knobs" specific to each of the services and how different setting for these knobs affects the overall execution of a Giraph application (such as setting the number of masters via giraph.zkServerCount or specifying a checkpoint interval by setting a non-zero value for giraph.checkpointFrequency). Overall, the Giraph implementation is extremely tunable; to understand the ways you can change its behavior, you have to dig into Giraph's source code (start with GiraphConstants) or consult the manual at http://giraph.apache.org/options.html.

Although this chapter is heavy on technical details, it provides an important overview of the internals of the Giraph implementation—something that comes in handy any time you need to debug or tune a Giraph application. One thing you haven't spent much time on is how Giraph interacts with external storage frameworks by using its flexible input/output format extensions. This is the focus of the next chapter.

CHAPTER 7

■ ■ ■

Graph IO Formats

This chapter covers

- Graph representations

- Reading input graphs in different formats

- Saving the result of your analysis

So far, you have familiarized yourself with the concept of graphs and their constituents—vertices and edges—through a variety of use cases. In Chapter 5, you learned how Giraph uses the corresponding Vertex and Edge objects to represent a graph and to programmatically manipulate Vertex and Edge objects and compute useful graph metrics. But before Giraph creates Vertex and Edge objects for you to use, a graph is stored on a storage system in different formats: for instance, plain text files or a binary format. Similarly, you may require the result of a Giraph computation to be stored in different output formats.

In this chapter, you learn about the tools Giraph provides for reading input graphs and writing output, called *input formats* and *output formats*, respectively. Giraph uses input formats to convert an input graph into Vertex and Edge objects that you can then manipulate programmatically; it uses output formats to save the result of your analysis. When you launched your first Giraph job in Chapter 5, you were already using specific implementations of the input and output formats. Here, you get to know in detail how input and output formats work.

More important, you learn how to write your own input and output format implementations. Because graphs may be stored in different ways, it is difficult to provide an input format suitable for every possible storage system or graph format. Giraph exports a programming interface that you can implement to build support customized to your specific scenario. This chapter first discusses the different ways graphs can be represented and then teaches you how to implement input and output format interfaces to read graphs and output results in various formats. At the end of this chapter, you will have completed your knowledge of the basic features of the Giraph API and will be ready to move to more advanced features: reading graphs, computing, and outputting the result of your computation.

Graph Representations

Before diving into the details of the Giraph API, you first need to understand the ways you can represent graphs to store them in a storage medium such as a disk. Until now, this book has discussed graphs abstractly and with the help of two-dimensional figures. Let's borrow the example of the Twitter social graph from Chapter 3, shown here in Figure 7-1, and enrich it with extra information to make it a bit more interesting. This example graph is used throughout the chapter.

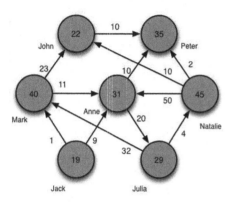

Figure 7-1. *An example Twitter graph. Vertices represent users, and an edge from vertex A to vertex B denotes that user A follows user B. The vertex value represents the age of a user, and the edge value represents the number of mentions from user A to user B*

In the example graph, every vertex corresponds to a Twitter user and is labeled with a name: the vertex ID. An edge from user A to a user B signifies, in this case, that A follows B on Twitter. Although vertices and edges are the basic constituents of a graph, in most cases you want to associate interesting information, or *metadata*, with vertices and edges. In the case of a social network, a vertex may be associated with user profile information. Here, the vertices have an associated user age. As you may have already observed, this corresponds to the Vertex value you learned about in Chapter 5. Every edge also has an associated number, which denotes how many times a user mentions another user; this corresponds to an Edge value.

It is easy to conceptualize a graph for this image; but to store this graph on a disk, you must break it into pieces of information in such a way that you can later reconstruct the original graph. The typical way to store a graph is by storing information about the vertices—a *vertex-based representation*—or by storing information about the relations among the vertices—an *edge-based representation*.

In a vertex-based representation, a graph is defined by a collection of per-vertex information. The *adjacency list* is one of the most common vertex-based graph representations. In its simplest form, an adjacency list is a collection of lists, where each list corresponds to a vertex and contains the vertex's neighbors. For example, Figure 7-2 shows the adjacency list that represents the Twitter graph from Figure 7-1. Each list corresponds to user, and the vertices in the list correspond to the Twitter accounts the user follows.

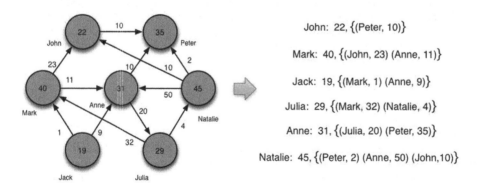

Figure 7-2. *A vertex-based representation of the Twitter graph*

Notice that the adjacency list provides exactly the same information as Figure 7-1, but in a different form. With this form, you break down the graph into per-vertex information that you can easily store.

Often, a graph is described not by its individual vertices, as in the previous example, but by the relations that exist between vertices. In this edge-based representation, a graph is defined by the set of edges, possibly accompanied by edge metadata. Figure 7-3 shows how you can represent the same Twitter graph by breaking it into per-edge information. Notice that an edge-based representation does not contain any information about the vertices themselves. If you try to reconstruct the Twitter graph from the edge information, in the end you will be missing the vertex information. In general, an edge-based representation is more suitable when you want to describe the structure of the graph through the relations among vertices.

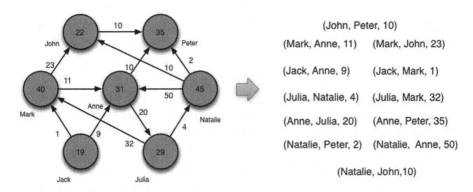

Figure 7-3. *An edge-based representation of the Twitter graph*

You are likely to come across various representations in your encounters with graph data. In some cases, you choose the representation to use to store the graph data, and in other cases, you have to access graph data that's organized for you. For instance, a vertex-based representation may be preferred over an edge-based one because it takes up less space on disk. Other times, you want to store vertex-specific information, such as user profile information in a social network, in which case a vertex-based format fits more naturally than an edge-based format.

Different representations must be handled in different ways when it comes to creating Vertex and Edge objects from them. For example, in a vertex-based description, you can reconstruct a Vertex object just by looking at an individual piece of information such as a table row in a database describing a user in social network. In contrast, in an edge-based representation, you need to look for and put together various pieces of information to reconstruct a Vertex object. This seems easy in pictures, but it becomes challenging when you have to deal with terabytes of distributed graph data—imagine trying to do this manually. The following sections examine how Giraph makes it easier to handle graph representations without worrying too much about such details.

Input Formats

Regardless of the representation, a graph may be stored in different storage systems, and possibly distributed in different formats. Giraph provides tools called *input formats* that simplify the task of reading an input graph. An implementation of an input format is essentially a way to specify how to read data from a storage system and then how to convert the data to the familiar Vertex and Edge objects. Giraph exports a simple API that you can implement to support new types of storage systems and formats. Giraph, in turn, is responsible for doing the hard work for you.

The Giraph API for vertex-based graph representations is VertexInputFormat, and the API for edge-based graph representations is EdgeInputFormat. These are the most basic abstract classes you extend to implement new input formats. Giraph also provides more specialized APIs that support common storage systems and formats; for instance, one common format is adjacency lists stored as text files on the Hadoop Distributed File System (HDFS) or on HBase tables.

In general, you can either extend the basic API to write your own input format implementation that is customized to your application scenario, or you can extend an existing specialized API. In many cases, you do not have to implement an input format from scratch; it is very likely that one of the implementations provided by Giraph already serves your needs.

■ **Tip** You're encouraged to look into the available input and output format implementations before creating your own. The Giraph code base is constantly being enriched, and you may find what you need there and save some time.

In other cases, you may have to extend one of the specialized text-based input formats that already implements most of the functionality for you. For instance, the TextVertexInputFormat handles details like locating files on HDFS and reading them line by line. Implementing this specialized API only requires you to specify how to parse a text line into a Vertex object.

Figure 7-4 shows a small sample of the vertex-based input format classes in the Giraph code base. It also shows how you can build functionality starting from the basic API. White boxes represent abstract classes that you can extend, and shaded boxes represent concrete implementations that already exist in the Giraph source code.

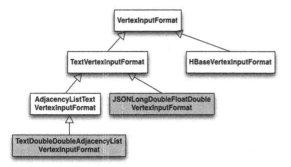

Figure 7-4. *Input format class hierarchy. Starting from the basic VertexInputFormat, every child class adds functionality and adds a specific type of format*

For instance, TextVertexInputFormat is an extension of VertexInputFormat specifically for input graphs stored as text files on HDFS. The AdjacencyListTextVertexInputFormat abstract class, in turn, can handle text formats where the graph is in the form of an adjacency list. You can find all these under the org.apache.giraph.io.formats package in the Giraph code.

In the rest of this section, you learn how to create input formats from scratch by implementing the basic APIs. This will give you the necessary knowledge to create your own input format when necessary. This chapter uses text-based formats as examples; in Chapter 9, you learn how to use the same API to implement support for more advanced systems and formats, such as HBase and Hive.

Vertex-Based Input Formats

You start by seeing how to implement input formats for vertex-based graph representations. VertexInputFormat is the most basic API you can implement. Let's look at how to extend this API step by step to implement an input format that can be used to read graphs in the form of an adjacency list stored as plain text files on HDFS, one of the most common input formats. Listing 7-1 shows what the adjacency list representing the Twitter graph would look like in a text file.

Listing 7-1. Example Text File Describing the Twitter Graph

```
John 22 Peter 10
Mark 40 John 23 Anne 11
Jack 19 Mark 1 Anne 9
Julia 29 Mark 32 Natalie 4
Anne 31 Julia 20 Peter 35
Natalie 45 Peter 2 Anne 50 John 10
```

In this text input, every line contains a user, the user's age, the people they follow, and how many times they have mentioned those followers. This implementation is called SimpleTextVertexInputFormat.

Giraph uses VertexInputFormat to split the input data into parts and then process each part to generate Vertex objects. Listing 7-2 shows the VertexInputFormat API.

Listing 7-2. VertexInputFormat Abstract Class

```
public abstract class VertexInputFormat<
    I extends WritableComparable,
    V extends Writable,
    E extends Writable> extends GiraphInputFormat<I, V, E> {

  public abstract void checkInputSpecs(Configuration conf);

  public abstract List<InputSplit> getSplits(JobContext context,
      int minSplitCountHint) throws IOException, InterruptedException;

  public abstract VertexReader<I, V, E> createVertexReader(
      InputSplit split,
      TaskAttemptContext context) throws IOException;
}
```

Notice first that VertexInputFormat uses Java generics to declare as parameters type I of the vertex ID, type V of the vertex value, and type E of the edge value. Any implementation of this API, unless abstract, must specify these three types. Recall from Chapter 5 that Vertex objects and Computation implementations also require the same parameters. Before loading the graph, Giraph uses the information about the types to ensure that the types of the vertices your VertexInputFormat implementation creates match the types of the Computation.

In the example in Listing 7-3, the vertex IDs are names, so you use type Text to represent them. Vertex and edge values are the age and number of mentions, respectively, so you use IntWritable to represent them. Next, let's get into the details of the API method-by-method through our example.

Listing 7-3. An Example Text-Based VertexInputFormat Implementation

```
public class SimpleTextVertexInputFormat
    extends VertexInputFormat<Text, IntWritable, IntWritable> {

  public void checkInputSpecs(Configuration conf) { ... }      #1

  @Override
  public List<InputSplit> getSplits(JobContext context,
                                    int minSplitCountHint)
    throws IOException, InterruptedException {

    List<FileStatus> files = getFileList(context);          #2
    List<InputSplit> splits = new ArrayList<InputSplit>();
    for (FileStatus file : files) {                         #3
      Path path = file.getPath();                           #3
      long length = file.getLen();                          #3
      splits.add(new FileSplit(path, 0, length,             #3
          new JobConf(context.getConfiguration())));        #3
    }
    return splits;                                          #4
  }

  public VertexReader<I, V, E> createVertexReader(
      InputSplit split, TaskAttemptContext context) throws IOException {
    return new SimpleTextVertexReader();                    #5
  }
}
```

#1 Check the validity of the input.
#2 Get the list of files in the input path.
#3 For every input file, create a FileSplit object and add it to a list.
#4 Return the list of FileSplit objects created.
#5 Return an implementation of the VertexReader class.

The first method you have to implement is checkInputSpecs(). Giraph calls this method before it starts using the input format. Typically, the role of this method is to check whether the job configuration contains all the necessary information for the job to read the input graph. For instance, in this method you may check whether an input directory has been set in the configuration or whether the input directory exists.

The next method you must implement is getSplits(). It takes as input a JobContext object, which contains configuration information about the job such as the input directory, and returns a list of InputSplit objects. An InputSplit is a logical representation of a part of the input. For instance, a FileSplit implements the InputSplit interface to represents a file on HDFS; the InputSplit contains information about the file's path, size, and location. In this example, the getSplits() method lists all the files in the input directory passed as a parameter in the job configuration and creates a FileSplit for every file.

TYPES INHERITED FROM THE HADOOP API

Just as in Chapter 5, notice that certain types, such as InputSplit, JobContext, and TaskAttemptContext, are inherited from the Hadoop API. But in most cases you don't need to know the Hadoop details.

The most important method of the class is createVertexReader(), which returns an object extending the VertexReader abstract class. VertexReader is responsible for performing the actual processing of the data described by InputSplit. For every InputSplit returned by the previous method, Giraph creates a VertexReader object that processes the corresponding data and creates the Vertex objects that eventually form your graph. Figure 7-5 illustrates this entire process.

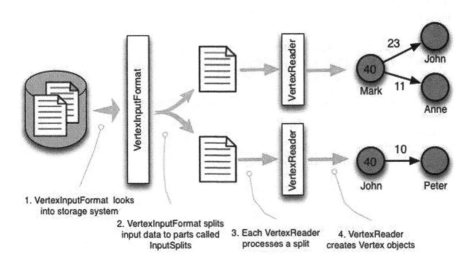

1. VertexInputFormat looks into storage system

2. VertexInputFormat splits input data to parts called InputSplits

3. Each VertexReader processes a split

4. VertexReader creates Vertex objects

Figure 7-5. *Putting all the pieces together.* VertexInputFormat *splits the input into parts.* VertexReader *processes a part and creates* Vertex *objects*

Now let's look at the details of VertexReader and how to implement one. Listing 7-4 shows the methods defined in the VertexReader abstract class.

Listing 7-4. The VertexReader Abstract Class

```
public abstract class VertexReader<I extends WritableComparable,
    V extends Writable, E extends Writable> extends
    DefaultImmutableClassesGiraphConfigurable<I, V, E>
    implements WorkerAggregatorUsage {

  public abstract void initialize(InputSplit inputSplit,
                                  TaskAttemptContext context)
    throws IOException, InterruptedException;

  public abstract boolean nextVertex() throws IOException,
      InterruptedException;

  public abstract Vertex<I, V, E> getCurrentVertex()
    throws IOException, InterruptedException;

  public abstract void close() throws IOException;
}
```

In Listing 7-5, you implement SimpleTextVertexReader, which is responsible for reading the lines in a text file one by one and creating a Vertex object for each line. This example shows how to implement each of the VertexReader methods.

Listing 7-5. An Example Text-Based VertexReader

```
public class SimpleTextVertexReader
 extends VertexReader<Text, IntWritable, IntWritable> {

  private RecordReader<LongWritable, Text> lineRecordReader;
  private TaskAttemptContext context;

  @Override
  public void initialize(InputSplit inputSplit, TaskAttemptContext context)
     throws IOException, InterruptedException {
     this.context = context;
     lineRecordReader = new LineRecordReader();                        #1
     lineRecordReader.initialize(inputSplit, context);                 #1
  }

  @Override
  public final boolean nextVertex()
       throws IOException, InterruptedException {
     return lineRecordReader.nextKeyValue();                           #2
  }

  @Override
  public final Vertex<Text, IntWritable, IntWritable> getCurrentVertex()
     throws IOException, InterruptedException {

     Text line = lineRecordReader.getCurrentValue();                   #3
     Vertex<Text, IntWritable, IntWritable> vertex =
         getConf().createVertex();                                     #4
     String[] words = line.toString().split(' ');                      #5
     Text id = new Text(words[0]);                                     #6
     IntWritable age = new IntWritable(Integer.parseInt(words[1]));    #7

     List<Edge<TextWritable, IntWritable>> edges = new ArrrayList();
     for (int n = 2; n < tokens.length; n=n+2) {
       Text destId = new Text(tokens[n]);                              #8
       IntWritable numMentions =                                       #8
         new IntWritable(Integer.parseInt(tokens[n+1]);                #8
       edges.add(new DefaultEdge(dstId, numMentions));                 #9
     }
     vertex.initialize(id, age, edges);                                #10
     return vertex;
  }

  @Override
  public void close() throws IOException {
    lineRecordReader.close();                                         #11
  }
}
```

#1 Create a line record reader to read text files.
#2 Check whether there are more lines to read.
#3 Get the current line.
#4 Create an empty vertex object.
#5 Split the line into tokens.
#6 The first token is the vertex ID.
#7 The second token is the vertex value.
#8 Parse the destination IDs and number of mentions.
#9 Create edge objects.
#10 Initialize the vertex object.
#11 Close the input file.

The first method you implement is initialize(), which Giraph calls immediately after creating but before it starts using VertexReader. Giraph passes as input to this method the InputSplit that the vertex reader will process and a TaskAttemptContext object that contains configuration information such as the HDFS input directory. In general, you use this method to set up VertexReader before for execution. In the example, you use this to initialize the local variables and create a LineRecordReader. LineRecordReader is a helper class that you borrow from the Hadoop API and use to read a text file on HDFS line by line. To keep the example simple, it omits the LineRecordReader details; you can always use it as is by passing it the InputSplit and the context information.

The next two methods, nextVertex() and getCurrentVertex(), resemble an iterator interface. The nextVertex() method must return true if there are more vertices to read and advance the iterator to the next vertex. The getCurrentVertex() method returns the current vertex the iterator points to. Giraph calls these two methods repeatedly as a pair until there is no more data in the input to read.

Now let's see how you implement these methods. In the example, nextVertex() only has to check whether there are more lines in the file and, if so, read and buffer the next line. This is the operation LineRecordReader performs for you automatically, so you do not need to worry about the details.

The getCurrentVertex() method, in turn, must create a Vertex object from a text line. First it gets the current text line read. Then it creates an empty Vertex object that you need to fill with information: the ID, the value of the class, and its edges. Recall that the input graph is in the form of an adjacency list, as follows:

```
John 22 Peter 10
Mark 40 John 23 Anne 11
Jack 19 Mark 1 Anne 9
Julia 29 Mark 32 Natalie 4
...
```

After you split the text line into words, the first word in the line describes the vertex ID of a user. The next word describe the vertex value—that is, the age of the user—which you parse into an IntWritable object. The next words describe the IDs this user follows on Twitter and how many times the user mentions them. You use these to create the vertex edges: you parse every such pair in the line and create Edge objects. After you have parsed the entire line, you use the initialize() method of the Vertex class to fill the empty Vertex object with the necessary information.

Finally, after reading an InputSplit, Giraph calls the close() method of VertexReader. In the example, you use this to close the file opened upon initialization. Again, LineRecordReader takes care of this for you.

This concludes the operation of VertexInputFormat. Using this simple yet very useful example, you can now begin writing your own vertex-based input formats.

Edge-Based Input Formats

As described in the previous section, edge-based graph representations are common as well. Recall that an edge-based representation looks like Figure 7-6 for the example Twitter graph.

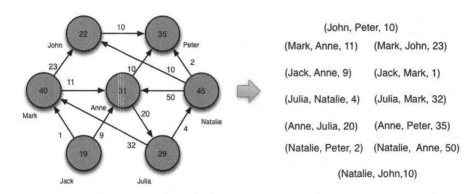

Figure 7-6. Edge-based representation of the Twitter graph. Edge values represent number of mentions

Listing 7-6 shows how you store such a representation in a text file. Each line in the file represents an edge in the graph. Recall that an edge from A to B denotes that user A follows B and is accompanied by the number of times A mentions B. Each line in the file includes this information as well, next to each edge.

Listing 7-6. An Example Edge-Based Input File for the Twitter Graph

```
John Peter 10
Mark John 23
Mark Anne 11
Jack Mark 1
Jack Anne 9
Julia Mark 32
Julia Natalie 4
Anne Julia 20
Anne Peter 10
Natalie Peter 2
Natalie Anne 50
Natalie John 10
```

Giraph provides an easy way to read such representations, called EdgeInputFormat. Similar to VertexInputFormat, Giraph uses EdgeInputFormat to split the input data into parts and then processes the data in each split. The basic difference is that EdgeInputFormat does not create Vertex objects, but Edge objects. At this point you may wonder how Vertex objects are created; after all, when you implement a Computation, Giraph passes a Vertex object for you to manipulate. The answer is that Giraph handles the creation of the Vertex objects for you. This is one of the conveniences that EdgeInputFormat provides.

Let's get into the details of the API, illustrated in Listing 7-7. Notice the similarities to VertexInputFormat. The first two methods serve exactly the same purpose: checking the validity of the configuration and returning a list of InputSplit objects describing the input parts. One difference to note is that EdgeInputFormat does not require you to specify the vertex value V, but only the vertex ID type I and the edge value type E. Keep in mind that the vertex ID type and edge value type still have to match those of your Computation.

Listing 7-7. EdgeInputFormat Abstract Class

```
public abstract class EdgeInputFormat<I extends WritableComparable,
    E extends Writable> extends GiraphInputFormat<I, Writable, E> {

  public abstract void checkInputSpecs(Configuration conf);

  public abstract List<InputSplit> getSplits(JobContext context,
      int minSplitCountHint) throws IOException, InterruptedException;

  public abstract EdgeReader<I, E> createEdgeReader(
      InputSplit split,
      TaskAttemptContext context) throws IOException;
}
```

The most important difference in the API is the createEdgeReader() method, which returns an object that extends the EdgeReader abstract class, shown in Listing 7-8. EdgeReader is responsible for reading the data described by InputSplit and creating the Edge objects.

Listing 7-8. EdgeReader Abstract Class

```
public abstract class EdgeReader<I extends WritableComparable,
    E extends Writable> {

  public abstract void initialize(InputSplit inputSplit,
                                  TaskAttemptContext context)
    throws IOException, InterruptedException;

  public abstract boolean nextEdge()
    throws IOException, InterruptedException;

  public abstract I getCurrentSourceId()
   throws IOException, InterruptedException;

  public abstract Edge<I, E> getCurrentEdge()
    throws IOException, InterruptedException;

  public abstract void close() throws IOException;
}
```

Let's look at the methods one by one and how to implement them to read the example file in Listing 7-9. This implementation is called SimpleTextEdgeReader.

Listing 7-9. Example EdgeReader for Text-Based Formats

```
public class SimpleTextEdgeReader extends EdgeReader<Text, IntWritable> {
  private RecordReader<LongWritable, Text> lineRecordReader;

  @Override
  public void initialize(InputSplit inputSplit,                    #1
      TaskAttemptContext context)                                 #1
        throws IOException, InterruptedException {                #1
    lineRecordReader = new LineRecordReader();                    #1
```

```
      lineRecordReader.initialize(inputSplit, context);              #1
    }

    @Override
    public final boolean nextEdge()                                  #2
      throws IOException, InterruptedException {                     #2
        return lineRecordReader.nextKeyValue();                      #2
    }

    @Override
    public final Text getCurrentSourceId() throws IOException,
          InterruptedException {
      Text line = getRecordReader().getCurrentValue();               #3
      String[] words = line.toString().split(' ');                   #3
      return new Text(words[0]);                                     #3
    }

    @Override
    public final Edge<Text, IntWritable> getCurrentEdge()
      throws IOException, InterruptedException {
      Text line = getRecordReader().getCurrentValue();
      String[] words = line.toString().split(' ');
      Text targetVertexId = new Text(words[1]);                      #4
      IntWritable edgeValue = Integer.parseInt(words[2]);            #5
      return new DefaultEdge(targetVertexId, edgeValue);             #6
    }

    @Override
    public void close() throws IOException {                         #7
      lineRecordReader.close();                                      #7
    }
  }
}
```

#1 This method is called at the very beginning to set up EdgeReader.
#2 Check whether there are text lines.
#3 Get the current text line, and parse the source vertex ID.
#4 Parse the target vertex.
#5 Parse the edge value.
#6 Create an Edge object with the parsed values.
#7 This method is called at the end to close the opened files.

First, notice that you have to implement initialize() and close() methods, as with VertexReader. These are called right before Giraph starts using the reader and at the end of execution. Here you use them to open and close the file described by InputSplit the same way you did previously. Their implementation in this case is exactly the same, so we won't elaborate. Instead, let's get into the most interesting part of EdgeReader.

EdgeReader implements an iterator-like interface similar to VertexReader. The nextEdge() method returns true if there are more edges to read in the input file and advances the iterator to the next edge. Although the details aren't shown, if a line is available, LineRecordReader reads and buffers it. After Giraph calls this method, it can use the following two methods. First it calls the getCurrentSourceId() method. This method must return the source ID of the edge the iterator points to currently. Recall that Giraph creates Vertex objects automatically for you; this is how Giraph knows which Vertex to attach the edge to. In this implementation, all you have to do is read the first word in a line and return it as a Text object. Finally, Giraph calls the getCurrentEdge() method to create an Edge object and add it to the vertex. An edge is defined by a target vertex ID and an edge value. To obtain these, the implementation parses the second and third words of each line and then creates and returns an Edge object.

Note the creation of the Edge objects. In Chapter 5, you were only concerned with accessing Edge objects through a Vertex; you never had to create one. In reality, Edge is an interface; under the hood, Giraph uses specific types that implement the Edge interface. DefaultEdge is one of these Edge implementations provided by Giraph. Most of the time, the default type will serve your needs, but you are free to write your own Edge implementations.

USING THE BUILT-IN EDGE IMPLEMENTATIONS

Giraph provides different implementations of the Edge interface that serve different purposes. You can find these in the org.apache.giraph.edge package. For instance, the EdgeNoValue edge type is suitable when edges have no associated value, because this implementation handles this scenario in a more memory-efficient way than the default implementation. As a convenience, Giraph provides the EdgeFactory utility class with methods to simplify the creation of Edge objects.

Combining Input Formats

You have learned that graphs can be stored in either vertex-based or edge-based representations. But in several scenarios, a graph may be stored in multiple representations at the same time. One common case is to store a graph by separating the graph structure: that is, by separating the connecting edges from the per-vertex information. There are various practical reasons for doing so. For instance, different analytical applications may require different data; you can separate the data for performance reasons so that no application needs to read all the data every time. In the Twitter graph scenario, you may want to keep the who-follows-whom relations stored in text files on HDFS in an edge-based format, and the user profile information for each user in a vertex-based format in a key-value store. Figure 7-7 shows an example of this separation.

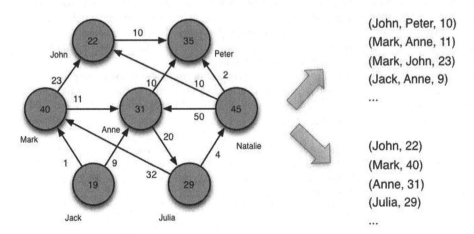

Figure 7-7. *Graph information split into two separate representations*

This separation may be common practice, but it is also common for an application to combine data sources to analyze a graph in a meaningful way. An application must be able to combine the two sources on the right in Figure 7-7 to reconstruct the original graph.

You may wonder whether you need a vertex-based or an edge-based input format to handle this scenario. But you will quickly realize that no one type of input format alone is sufficient for this. For instance, in the previous subsection, the `EdgeInputFormat` does not provide a means to specify a vertex value, so you cannot use it to read profile information. Similarly, you cannot use a `VertexInputFormat` because information about a single vertex is scattered across different places in the input. As a workaround, you could preprocess the data to merge the two data sources into one input data set that is in a vertex-based format before reading. But this approach can be cumbersome and inefficient, because you have to read the same data twice every time you want to do your analysis.

Fortunately, Giraph provides a more convenient and efficient way to handle this scenario by allowing you to combine two different graph representations. Specifically, you can use an `EdgeInputFormat` that is responsible for creating the edges of a vertex by reading the edge representation, and at the same time a `VertexInputFormat` that is responsible for setting the value of a vertex by reading the profile information.

Recall that in the example in Chapter 5, when you launched the job you specified a vertex input format using the `-vif` command-line parameter to define the input format class name and a vertex input path using the `-vip` parameter. Similarly, you can use an edge input format by replacing this with the `-eif` parameter and specifying the edge input format class name and the `-eip` parameter to indicate the edge input path. Combining two types of input formats is as simple as including all four parameters in your command line. Giraph takes care of the rest.

Input Filters

Filtering a graph is a common operation in graph analysis. *Filtering* means removing some of the vertices or some of the edges from the original graph and performing your analysis on the remaining graph. There are various reasons to do this. Certain graph-mining algorithms give good approximations even if executed on a random sample of the graph. This is often preferred as a way to speed up the computation.

In other cases, filtering is an implicit requirement of an application. A common scenario is when the weight of the edges in the graph signifies the strength of the connection between vertices, and you want to perform the analysis only on users with strong connections. Consider the Twitter scenario, and assume you want to recommend new people for Mark to follow by looking at who the users he follows—Anne and John—follow. For instance, John follows Peter, so Peter may be a good recommendation for Mark too. But to make the recommendation more relevant, perhaps you want to include in this analysis only neighbors with whom Mark has a strong connection. In this case, you need to filter out edges with low weights. For instance, you could filter out neighbors that a user mentions fewer than 20 times, excluding Anne in this case, indicating that Julia may not be a relevant recommendation for Mark.

You could even perform this kind of filtering based on per-vertex information. For instance, you may want to perform this analysis on Twitter users in a specific age range. This way, you ensure that you do not recommend friends to a youngster based on who older people follow. To address this, you could filter from the graph all vertices that have an age above an age threshold. These are only a few examples where filtering might be necessary.

Giraph provides an elegant way for you to integrate this filtering process into your analysis when necessary. It allows you to specify *vertex filters* and *edge filters* to be used along with a `VertexInputFormat` and an `EdgeInputFormat`. Vertex and edge filters provide a way for you to specify whether a vertex or an edge should be added in the final graph during the loading of the input data. You can specify filters by implementing the `VertexInputFilter` and `EdgeInputFilter` interfaces, shown in Listing 7-10 and Listing 7-11, respectively.

Listing 7-10. VertexInputFilter

```
public interface VertexInputFilter<I extends WritableComparable,
    V extends Writable, E extends Writable> {

  boolean dropVertex(Vertex<I, V, E> vertex);                    #1
}
```

#1 Decide whether to drop a vertex.

Listing 7-11. EdgeInputFilter

```
public interface EdgeInputFilter<I extends WritableComparable,
  E extends Writable> {

boolean dropEdge(I sourceId, Edge<I, E> edge);
}
```

In the case of a vertex-based input format, during the loading of the graph, Giraph uses a VertexInputFilter object to decide whether a vertex should be filtered. The VertexInputFilter interface has a single method named dropVertex() with a boolean return value. Recall from the previous sections that Giraph uses VertexReader to create Vertex objects. For every Vertex object created by VertexReader, Giraph calls dropVertex() and passes it the Vertex object as an input parameter to decide whether to keep the corresponding vertex. If the method returns true, then Giraph drops the vertex.

Similarly, for edge-based input formats, Giraph uses an EdgeInputFilter to decide whether an edge that is read should be included in the graph. For every Edge object created, Giraph calls the dropEdge() method and passes it the ID of the edge source vertex and the edge object itself. If the method returns true, Giraph drops the edge. Listing 7-12 shows an example EdgeInputFilter implementation that decides whether to include an edge based on the edge weight. In the example of the Twitter graph, you could use such an edge filter to filter out weak connections identified by a low number of mentions.

Listing 7-12. An Example Edge Input Filter

```
public class WeightInputFilter extends EdgeInputFilter<Text,IntWritable> {

  public static int THRESHOLD = 10;

  boolean dropEdge(Text sourceId, Edge<Text,IntWritable> edge) {
    if (edge.getValue().get()<THRESHOLD) {
      return true;
    }
    return false;
  }
}
```

As you may have expected, the vertex ID and edge value types must match those of the input format implementation you use.

Once you have implemented your filter, you can specify that you want to use it through the command line. You can do this by adding the giraph.vertexInputFilterClass custom argument in the execution command and setting your filter implementation class name as its value.

Note that you could implement the filtering logic in your application if you wished, using the mutation API you learned about in Chapter 5. This way, you would dedicate the first superstep of the computation to filtering. However, filters offer a couple of benefits. First, they allow you to decouple the filtering logic from your main application logic, leading to clean, reusable code. Second, using input filters, Giraph has an opportunity to do the filtering at the early stage of reading the graph, leading to more efficient graph loading. It is faster and requires less memory.

Alternatively, you could preprocess your graph with separate scripts. However, this can be a tedious task, because you have to write code to read data from different storage systems and formats. This functionality is already provided by Giraph and its input formats. It can also be inefficient, because you have to read the graph twice every time you want to analyze it: once for preprocessing and once for the actual analysis. Input filters simplify these tasks.

Output Formats

You have learned how to read an input graph and how to perform useful analysis on it. What you are missing to complete the picture is a way to output the result of your analysis. After all, the result is no good if you cannot somehow save it to use later. Output formats are the tools that Giraph provides for you to achieve this. At the end of a computation, the Vertex and Edge objects in the graph contain all the useful information you want to store. For instance, in the shortest-paths application you wrote in Chapter 5, at the end of the computation the vertices contained the value of the shortest distance. Giraph uses an output format to save the contents of Vertex and Edge objects to a storage system.

As with input formats, there are both vertex-based and edge-based output formats. As you may have guessed, a vertex-based output format is used to save information about individual vertices, and an edged-based output format is used to save information stored in Edge objects. In the examples in Chapter 5, you primarily used vertex-based output formats. The type of output format you need will depend on your particular scenario. For instance, when you compute a metric per vertex, such as a distance or a rank, you naturally need a vertex-based output format. But suppose you read an input graph in an edge-based input format and you simply want to transform it. (You saw an example of such a transformation in Chapter 4, where you converted an undirected graph to a directed one.) In this case, you may want to save the transformed graph in an edge-based format as well.

Similar to when reading an input graph, you may wish to store the result information to different storage systems and in different formats. To accommodate this, Giraph exports two basic abstract classes VertexOutputFormat and EdgeOutputFormat that you can extend. For instance, the GraphvizOutputFormat that you used in Chapter 5 is an implementation of VertexOutputFormat provided for you. In this section, you learn how to write your own output formats. This chapter focuses on text-based formats, but in Chapter 9 you discover how to output data to any storage system and in any format.

Vertex-Based Output Formats

First you learn how to implement a vertex-based output format. Let's assume that you run the shortest-distances application, and you want to save the distance computed for every vertex in text files and store them to HDFS—the typical scenario. To do this, you extend the VertexOutputFormat abstract class as shown in Listing 7-13.

Listing 7-13. VertexOutputFormat Abstract Class

```java
public abstract class VertexOutputFormat<
    I extends WritableComparable, V extends Writable,
    E extends Writable> {

  public abstract void checkOutputSpecs(JobContext context)
    throws IOException, InterruptedException;

  public abstract VertexWriter<I, V, E> createVertexWriter(
    TaskAttemptContext context) throws IOException, InterruptedException;

  public abstract OutputCommitter getOutputCommitter(
    TaskAttemptContext context) throws IOException, InterruptedException;

}
```

The first thing to notice in the VertexOutputFormat is that the vertex ID, vertex value, and edge value types are parameters of the class. This means your implementation, unless abstract, must specify these types, and they must match the corresponding types in your Computation.

Next, Listing 7-14 describes the VertexOutputFormat class methods and implements them. This implementation is called SimpleTextVertexOutputFormat.

Listing 7-14. An Example Text-Based VertexOutputFormat

```java
public class SimpleTextVertexOutputFormat
    extends VertexOutputFormat<Text, IntWritable, IntWritable> {

  @Override
  public void checkOutputSpecs(JobContext context)
    throws IOException, InterruptedException {
    textOutputFormat.checkOutputSpecs(context);
  }

  @Override
  public VertexWriter<Text, IntWritable, E> createVertexWriter(
    TaskAttemptContext context)
      throws IOException, InterruptedException {

    return new SimpleTextVertexWriter();                          #1
  }

  @Override
  public OutputCommitter getOutputCommitter(TaskAttemptContext context)
    throws IOException, InterruptedException {
    return textOutputFormat.getOutputCommitter(context);
  }
}
```

#1 This returns an object extending the VertexWriter class.

Giraph calls checkOutputSpecs() at the beginning of a job and passes it a JobContext object that contains configuration information. Its role is to perform checks similar to the one you learned for input formats. For instance, you can check whether the output directory has been set in the job configuration and, if so, whether it already exists and contains data that cannot be overwritten. In such a case, this method may throw an exception.

Like VertexReader objects, Giraph uses VertexWriter objects to do most of the work of writing Vertex information in objects to output. Specifically, after your computation finishes, Giraph creates a VertexWriter object on every worker machine in your cluster and uses these objects to write information about every vertex to the output. It does so by calling the createVertexWriter() method of the VertexOutputFormat class, which returns an object extending the VertexWriter abstract class. Every such object is responsible for processing a set of vertices in the graph. Figure 7-8 shows this process.

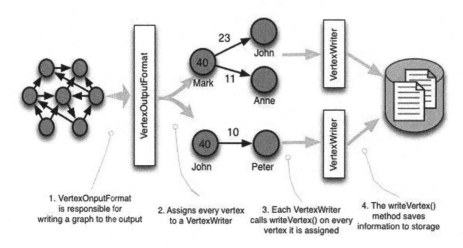

Figure 7-8. Saving a graph using a VertexOutputFormat and VertexWriter objects

Notice that Giraph follows a process that is the inverse of the one it performed when it was reading the graph. Giraph started by reconstructing a graph from text files, and now it must save the graph back to the same type of format. But let's look what happens in a VertexWriter in more detail. Listing 7-15 shows the VertexWriter abstract class methods you implement.

Listing 7-15. VertexWriter Abstract Class

```
public abstract class VertexWriter<I extends WritableComparable,
    V extends Writable, E extends Writable> {

  public abstract void initialize(TaskAttemptContext context)
    throws IOException, InterruptedException;

  public abstract void writeVertex(Vertex<I, V, E> vertex)
      throws IOException, InterruptedException;

  public abstract void close(TaskAttemptContext context)
    throws IOException, InterruptedException;

}
```

The example output format returns a VertexWriter called SimpleTextVertexWriter, which is designed to write vertex information to text files. Listing 7-16 shows the implementation of SimpleTextVertexWriter and its methods.

Listing 7-16. An Example Text-Based VertexWriter

```
public class SimpleTextVertexWriter
      extends VertexWriter<Text, IntWritable, IntWritable> {

  private RecordWriter<Text, Text> lineRecordWriter;

  @Override
  public void initialize(TaskAttemptContext context)
    throws IOException, InterruptedException {
    lineRecordWriter = createLineRecordWriter(...);          #1
  }

  @Override
  public final void writeVertex(Vertex vertex) throws
      IOException, InterruptedException {
    String line = vertex.getId()+" "+vertex.getValue();      #2
    lineRecordWriter.write(new Text(line), null);            #3
  }

  @Override
  public void close(TaskAttemptContext context)
    throws IOException, InterruptedException {
    lineRecordWriter.close(context);                         #4
  }
}
```

#1 Create a line record writer object.
#2 Make a string containing the vertex information to write.
#3 Use a record writer to write the string.
#4 Close the record writer.

Before it begins using the VertexWriter, Giraph calls initialize(), passing it a TaskAttemptContext as input. The initialize() method typically sets up the output files using configuration information from the TaskAttemptContext object. This involves, for instance, creating the output text file and obtaining a handle on the file. To write output to HDFS, you use the LineRecordWriter class, which takes care of these tasks for you. After the initialization of a VertexWriter, Giraph calls the writeVertex() method for every vertex in the graph. In this method, you must specify what information to output and how to write it. In this case, you write the vertex value to the output text files. The implementation first makes a string that contains the vertex ID and the value of the vertex; it then uses the LineRecordWriter to append the string to the output files.

USING THE LINERECORDWRITER

Notice that you use the write() method of LineRecordWriter to append text to the file. The write() methods accepts two input parameters: a key and a value. The LineRecordWriter then writes both the key and value parameters to the output file. Here you only need to write a single string, so you only need to pass it as the key parameter and leave the second parameter as null. The LineRecordWriter knows to ignore the null parameter.

After writing every vertex, Giraph calls the close() method, which in this case closes the files you opened upon initialization. Again, the LineRecordWriter takes care of this task for you.

One new feature here is OutputCommitter. An *output committer* is a component borrowed from Hadoop; its primary job is to set up the output directory of a job. This involves setting up a directory where temporary output files can be written before the job finishes and moving the files from the temporary directory to a final one upon successful completion of the job. We do not go into the details of the OutputCommitter methods here. For HDFS output, you can simply reuse the standard OutputCommitter provided by the Hadoop API, which handles all the tasks involved in setting up the output directories for you. Chapter 9 revisits OutputCommitter, when it discusses storage systems.

Edge-Based Output Formats

Edge-based output formats are very similar to their vertex-based counterparts. You can immediately see the similarities in Listing 7-17, which shows the EdgeOutputFormat abstract class.

Listing 7-17. EdgeOutputFormat Abstract Class

```
public abstract class EdgeOutputFormat<
    I extends WritableComparable, V extends Writable,
    E extends Writable> {

  public abstract void checkOutputSpecs(JobContext context)
    throws IOException, InterruptedException;

  public abstract OutputCommitter getOutputCommitter(
    TaskAttemptContext context) throws IOException, InterruptedException;

  public abstract EdgeWriter<I, V, E> createEdgeWriter(
    TaskAttemptContext context) throws IOException, InterruptedException;
}
```

The basic difference is that now you have to implement an EdgeWriter instead of a VertexWriter. Listing 7-18 shows the EdgeWriter abstract class. The most important method in this class is writeEdge(); Giraph calls this method for every edge in the graph.

Listing 7-18. EdgeWriter Abstract Class

```
public abstract class EdgeWriter<
    I extends WritableComparable, V extends Writable,
    E extends Writable>

  public abstract void initialize(TaskAttemptContext context)
    throws IOException, InterruptedException;

  public abstract void writeEdge(I sourceId, V sourceValue, Edge<I,E> edge)
      throws IOException, InterruptedException;

  public abstract void close(TaskAttemptContext context)
    throws IOException, InterruptedException;
}
```

Let's look at an example of how to implement an EdgeWriter that saves the edges along with their values into text files stored on HDFS. The implementation is called SimpleTextEdgeWriter and is shown in Listing 7-19.

Listing 7-19. An Example Text-Based EdgeWriter

```
public class SimlpeTextEdgeWriter<
      extends EdgeWriter<Text, NullWritable, IntWritable> {

  private RecordWriter<Text, Text> lineRecordWriter;

  @Override
  public void initialize(TaskAttemptContext context)              #1
    throws IOException, InterruptedException {                    #1
    lineRecordWriter = createLineRecordWriter(...);              #1
  }

  @Override
  public final void writeEdge(Text sourceId,
      IntWritable sourceValue, Edge<Text, IntWritable> edge)
      throws IOException, InterruptedException {

    String line = sourceId+" "+                                   #2
              edge.getTargetVerexId()+" "+                        #2
              edge.getValue();                                    #2
    lineRecordWriter.write(new Text(line), null);                 #3
  }

  @Override
  public void close(TaskAttemptContext context)
    throws IOException, InterruptedException {
    lineRecordWriter.close(context);                              #4
  }
}
```

#1 Initialize the EdgeWriter by creating a line record writer.
#2 Construct the string to write to the output.
#3 Use the line record writer to write the string.
#4 Close the output files at the end.

By now, most of the methods you need to implement for the `EdgeWriter` should not surprise you. You use `initialize()` to set up the `EdgeWriter` by creating the familiar `LineRecordWriter`. Similar to `VertexWriter`, Giraph calls `writeEdge()` for every edge in the graph. Giraph passes to this method the ID of the source vertex of the edge and its vertex value as well as the edge object itself. In this implementation, you construct a string that contains the source and destination vertex IDs and the edge value. You then use the `LineRecordWriter` to write the string to the output file.

Output formats provide you with all the tools you need to output information stored on your graph: the vertex and edge values. But recall that a Giraph program allows you to keep information in different objects, as well. These are the most useful aggregators.

Aggregator Writers

Aggregators are used to hold information, such as counters and aggregate statistics, about the entire graph. For example, in Chapter 5, you saw an example of how to use an aggregator to keep track of the Twitter user with the most followers. Often, at the end of a program, aggregators hold useful information—in this case, the most popular Twitter user—that you want to output to a file in the same way you output information held in the graph's vertices and edges.

To achieve this, Giraph provides a tool called an *aggregator writer*. Giraph uses aggregator writers to save the values of aggregators to a storage medium and in a format of your choosing: for instance, text files on HDFS. In Chapter 5, you used an aggregator writer implementation provided in the Giraph code base that saves aggregator values in a file. Here, you learn how to implement your own aggregator writer.

To specify how to save the values of aggregators, you must implement the `AggregatorWriter` interface shown in Listing 7-20. Giraph uses an `AggregatorWriter` at the end of every superstep to save the values stored in aggregators.

Listing 7-20. The `AggregatorWriter` Interface

```
public interface AggregatorWriter
    extends ImmutableClassesGiraphConfigurable {

  void initialize(Context context, long applicationAttempt)
     throws IOException;

  void writeAggregator(Iterable<Entry<String, Writable>> aggregatorMap,
     long superstep) throws IOException;

  void close() throws IOException;
}
```

Let's look at an example of how to implement these methods to customize the saving of aggregator information. Listing 7-21 shows a simple aggregator writer called `SimpleAggregatorWriter` that exports aggregator values to a text file on HDFS.

Listing 7-21. Example Aggregator Writer

```
public class SimpleAggregatorWriter implements AggregatorWriter {
private FSDataOutputStream output;

  @Override
  public void initialize(Context context, long applicationAttempt)
    throws IOException {
    Path p = new Path("aggregatedValues_" + applicationAttempt);   #1
    FileSystem fs = FileSystem.get(context.getConfiguration());
    output = fs.create(p, true);                                   #2
  }

  @Override
  public void writeAggregator(
    Iterable<Entry<String, Writable>> aggregatorMap,
    long superstep) throws IOException {
    for (Entry<String, Writable> entry : aggregatorMap) {          #3
      String text = entry.getKey()+" "+entry.getValue();
      output.writeChars(text);                                     #4
    }
    output.flush();
  }

  @Override
  public void close() throws IOException {
    output.close();                                                #5
  }
}
```

#1 Name the output file based on the application attempt.
#2 Create a file on HDFS.
#3 Iterate over all aggregators.
#4 Append the aggregator name and value to the file.
#5 Close the output file.

The first method you implement is initialize(). Giraph calls this method right before it starts using the aggregator and typically uses it to set up the aggregator writer. The first argument it passes is a Context object that contains configuration information about the job. The second argument indicates the number of the application attempt. An application attempt occurs whenever Giraph elects a new master worker; in most cases this happens only once, but sometimes, such as after a master worker failure, a new master must be elected. The example uses this method to create the HDFS file where you output the aggregator values. Notice how you name the output file based on the application attempt so that you can distinguish between attempts.

Next, you implement the writeAggregator() method. Giraph calls this method at the end of every superstep and passes two input arguments. The first is a Map that contains a key-value pair for each aggregator registered, where the key is a String representing the name of an aggregator and the value is the aggregator's associated value. The second parameter is the number of the superstep that just finished. Note that Giraph uses the special superstep number -1 to denote that this is the last superstep. You can use this information to decide whether you want to perform some action at every superstep or only at the end. This example iterates over all aggregators; for every aggregator, you write to the output file a line containing the name and the value of the aggregator.

In this method, you can implement arbitrary logic. For instance, you may want to select only specific aggregators to export to the file, or you may decide not to export an aggregator value unless it is the last superstep. You may also want to output different aggregators to different file names or even output the values to a different storage system or format. You have the ability to accommodate your particular application scenario.

Finally, you implement the `close()` method. Giraph calls this method at the end of a successful execution. Here, you use this method to close the file that you opened upon initialization.

Summary

Reading an input graph is the first step you need to perform to use Giraph, and saving the output of your analysis is equally necessary. These sound like simple tasks, but they can be challenging once you start thinking about the various types of formats and storage systems where your graph may be stored. Giraph provides the necessary tools to simplify this process.

In this chapter, you learned the following:

- Understanding how your input graph is represented, the details of the storage system where your graph is stored, and its format is the first step in managing your input and output data.

- You can use `VertexInputFormat` and `EdgeInputFormat` implementations to read input data in vertex-based and edge-based representations, respectively. Their role is to convert raw input data into `Vertex` and `Edge` objects for you to use.

- `VertexOutputFormat` and `EdgeOutputFormat` implementations, in turn, allow you to save output information in a vertex- or edge-based format.

- Giraph provides a library of input and output formats that support common cases and may cover your needs. Be sure to check the existing code base, because it may save you time.

- As you start using Giraph in a more advanced way and integrating it into your particular data-analytics architecture, you may need more than the basics. Extending the input and output format APIs allows you to add more functionality—for instance, supporting new storage systems.

- You may frequently need to combine different graph representations. The ability to combine a `VertexInputFormat` with an `EdgeInputFormat` becomes very handy in these cases.

In the last few chapters, you have learned how to use the basic Giraph APIs. Although this gives you the ability to perform sophisticated analysis on graphs, the Giraph API is even more powerful. The following chapters explore advanced features provided by Giraph; you learn how to add functionality to your applications, make them more efficient, and write cleaner and more reusable code.

CHAPTER 8

■ ■ ■

Beyond the Basic API

This chapter covers

- Graph mutations
- The Aggregator API
- Vertex coordination
- Writing modular applications

In the previous chapters, you learned how to use the basic parts of the Giraph programming API to implement various graph algorithms. While the basic API already gives you enough flexibility to build a wealth of useful graph mining algorithms, this chapter covers features of the Giraph API beyond the basic ones that enable you to write more sophisticated applications. The Giraph API is rich with features that allow you to add more functionality to your applications, write algorithms more efficiently, and even make your life easier from a programming perspective.

While we typically think of the input graph as an immutable data structure, this chapter discusses scenarios where you want to alter the input graph by adding or removing vertices and edges. Giraph provides an API that allows you to perform such mutations to the input graph. Examples are used to describe the different ways that you can use this API.

Next, you revisit aggregators, a tool that allows you to compute global statistics across the entire graph. In previous chapters, you saw examples of typical aggregators and learned how to use them inside of your applications. Here you look at the Java API that Giraph provides for writing custom aggregators; it is explained by implementing the aggregator.

Further, one of the most important features in distributed algorithms is coordination. So far you learned how to think in a vertex-centric way, writing programs where vertices communicate with each other through messages in a distributed fashion. Giraph took care of the coordination of the execution of the algorithm. Here, you learn how to intervene in this coordination process when you wish to further customize the execution of your algorithms.

This chapter also covers the Giraph API features that make it easier for you to write sophisticated applications from a programming perspective. You learn how to break down a potentially complex algorithm into modules that perform distinct logical operations, resulting in code that is cleaner and easier to understand. Inversely, you also learn how to combine these modules into different applications, improving code reuse and saving you programming effort.

Graph Mutations

In this subsection, you explore the ability to mutate the graph structure during the execution of an algorithm. In general, mutating a graph means adding or removing vertices and edges. While for the most part, you have thought about analyzing an input graph, here you see that mutating a graph is often a part of this analysis.

There are different reasons graph mutation might be necessary. In some scenarios, you may simply want to transform an input graph; for instance, as a preprocessing step. You already saw an example of graph transformation in Chapters 3 and 5, where you learned how to convert a graph from directed to undirected. In such cases, graph mutation is a natural requirement.

In other scenarios, the output of your analysis may be a graph that is completely different from your input graph. A common case is when you want to divide an input graph into communities or clusters and subsequently analyze the connections among the communities themselves. Community detection is a common application in social network analysis. In an online social network, you may want to find communities of users with similar profiles and output how the communities themselves connect with each other. In this case, the output of the community detection algorithm is a graph representing the communities, a graph structure totally different from your input graph. This implies that during the execution of your algorithm, somehow you must be able to create a graph; that is, create new vertices and connect them with new edges.

Graph generators make for another application scenario that requires graph mutation. Graph generators are useful tools that allow users to construct synthetic graphs that conform to some model, such as the popular "small-world" model. In fact, the Giraph code already contains implementations of a couple of graph generators. Graph mutations are natural in this case; these tools may start from an empty graph or an initial seed graph and gradually add vertices and edges until they build the final graph.

Before going into the API that allows you to change the graph, it is important to understand what it means to mutate the input graph. At this point, you have to remember that Giraph loads the input graph in-memory and maintains a copy in an internal representation. It is this copy of the graph that Giraph manipulates and can mutate if desired. The original copy of the graph remains intact. For instance, if you had stored your graph in an HDFS directory on your Hadoop cluster, and then executed an algorithm that uses the mutation API, the graph data on HDFS would be exactly the same after the execution.

Of course, it is possible for a user to mutate this internal copy of the graph and output a new copy to some external storage system. In the case of converting the graph from directed to undirected, you would be creating a new copy of the graph in an undirected form. This would be the output of the Giraph job.

■ **Note**   It is always possible for a user to modify the original graph from within a vertex computation function. Imagine that the graph is stored in a table store, like HBase. Nothing prevents a user from writing a vertex computation that connects to HBase and modifies the graph. However, such an operation would be outside the scope of Giraph. Giraph makes no guarantees about the consistency of these changes. For instance, in the event one of the Giraph workers fails, Giraph guarantees that it will produce a correct result, but does not guarantee the correctness of such external changes. This is both a capability and risk that comes with the flexibility of the Giraph programming API. If you choose to take advantage of this flexibility, you should be aware of all the risks and take extra care to design your program such that it does not cause any problems, especially if used in a production environment.

Now that we have discussed the usefulness of graph mutations, let's explore the actual API that gives you this ability. While Chapter 5 went thought the mutation API in brief, here it is explained in detail through a variety of examples.

The Mutation API

Giraph provides three ways to alter the graph structure during execution: (i) direct mutations on a vertex, (ii) mutation requests, and (iii) mutation through messages. Each of these may allow different types of graph mutations and have different semantics with respect to when the mutations are realized. Next, you review the APIs for each of these types of mutation and illustrate their use.

Direct Mutations

To illustrate the use of the direct mutation API, let's use an example that you have already seen in Chapter 5. In this example, you want to convert an input graph from a directed to an undirected one. Figure 8-1 shows this transformation.

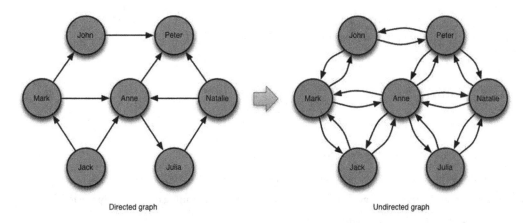

Directed graph   Undirected graph

Figure 8-1. *Transforming a directed graph into an undirected one*

In Figure 8-1, you see that such a transformation requires you to change the graph structure. In particular, you need to add some extra edges between existing vertices.

The Vertex interface contains a number of methods that allow the vertex computation function to directly change the structure of the vertex—that is, change its edges.

- They can only change the vertex from which they are called.

- The effect of such a call is immediate.

Let's look at the Vertex methods that allow the mutation of its structure:

- addEdge(Edge<I, E> edge): Adds an edge to the vertex. The ID and value of the edge are specified through the passed Edge object.

- removeEdges(I targetVertexId): Removes from the vertex all the edges with the specified target ID.

- setEdges(Iterable<Edge<I, E>> edges): Sets the outgoing edges of the vertex by iterating over the passed Iterable and adding the edges, one by one.

In Listing 8-1, let's revisit the example code from Chapter 5; the way it uses the mutation API is explained in detail next.

Listing 8-1. Converting a Graph from Directed to Undirected

```
public class ConvertToUndirected extends
  BasicComputation<Text, DoubleWritable, NulleWritable, Text> {

  static final NullWritable DEFAULT_EDGE_VALUE = NullWritable.get();

  @Override
  public void compute(Vertex<Text, DoubleWritable, DoubleWritable> vertex,
                      Iterable<Text> messages) {

    if (getSuperstep() == 0) {                                          #1
      sendMessageToAllEdges(vertex, vertex.getId());                    #1
    } else {
      for (Text m : messages) {                                        #2
        if (vertex.getEdgeValue(m)==null) {                            #3
          vertex.addEdge(EdgeFactory.create(m, DEFAULT_EDGE_VALUE));   #4
        }
      }
    }
    vertex.voteToHalt();
  }
}
```

#1 In the first superstep, every vertex sends its own ID to all its neighbors.

#2 In the second superstep, a vertex receives messages from every vertex that has an edge to it.

#3 For every such message the vertex gets, it checks whether the corresponding edge already exists.

#4 If the edge does not exist, the vertex adds it.

Note that the addEdge method takes an implementation of the Edge interface as input. While you can implement your own Edge types, Giraph already provides default implementations that should suffice in most cases. In particular, you can use the create() method of the EdgeFactory utility class to create Edge objects.

Mutation Requests

Next, you are going to look at an example showing why direct mutations might not always be suitable for modifying the graph structure and learn how mutation requests provide more flexibility. Let's consider that you want to analyze a Twitter-like social network and explore, not the individual user connections as you have done so far, but how users from different countries are connected. Starting from your original social graph, you want to create a graph where every vertex corresponds to a country, and edges imply that there are users among these countries that are connected. This allows you to create a high-level view of the Twitter graph and observe the social influence at the country level as shown in Figure 8-2.

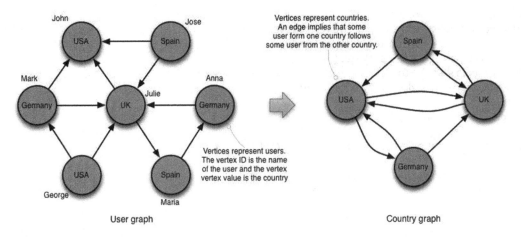

Figure 8-2. *Analyzing how users from different countries are connected*

It is obvious that while you are analyzing the original graph, you are also creating a new one. In fact, the new graph is the actual output of the algorithm. However, since the vertices of this new graph do not exist in advance, you cannot really use the direct mutations as before. Let's see how mutation requests solve this problem.

Mutation requests are methods of the Computation interface. Just like direct mutations, you can call these methods from within your vertex computation function. However, unlike direct mutations, a mutation request has the following properties:

- It can modify any part of the graph, regardless of the vertex that makes the request. For instance, vertex A can request the addition of an edge between vertices B and C.

- The effect of a mutation request is visible only after the current superstep finishes and right before the next one begins.

■ **Caution**   The second property might seem a bit unnatural at first, but it is really meant to simplify your programs. This property allows you to separate the programming logic that handles the creation or removal of vertices and edges from the rest of the programming logic. A vertex request has no effect on the remaining computation of the vertex that does the request, even if it modifies the same vertex. It has no effect on the computation of other vertices that may occur after the request is made, but within the same superstep.

Next, let's see this API in action.

The Listing 8-2 shows the implementation of this algorithm. Although the logic may not be readily apparent, all the parts are explained in a moment.

Listing 8-2. Finding Country Connections in the Twitter Graph

```
public class CreateCountryGraph extends
  BasicComputation<Text, Text, NullWritable, Text> {

  static final NullWritable DEFAULT_EDGE_VALUE = NullWritable.get();
```

```
@Override
public void compute(Vertex<Text, Text, NullWritable> vertex,
                    Iterable<Text> messages) {
  Text myCountry = vertex.getValue();
  if (getSuperstep()==0) {                                          #1
    sendMessageToAllEdges(vertex, myCountry);                       #1
  } else {
    for (Text m : messages) {                                       #2
addEdgeRequest(m,                                                   #3
  EdgeFactory.create(myCountry, DEFAULT_EDGE_VALUE));               #3
    }
    removeVertexRequest(vertex.getId());                            #4
    vertex.voteToHalt();
  }
 }
}
```

#1 To every user I am following, I send the name of my own country.

#2 At superstep 1, every message contains the country of a user that follows me.

#3 For every message, create a new edge from the other country to my country.

#4 Each user vertex removes itself from the graph, so what we are left only with country vertices.

In this scenario, the ID of the vertices is of type Text, representing the name of the user, and the value of a vertex is also of type Text, representing the country of the user. The general way the algorithm works should be familiar. Somehow, you need to know about what pairs of countries hide inside the followership graph. The natural way to do this is to have every vertex tell its neighbors about their own country. The first superstep of the algorithm implements exactly this logic. Every vertex sends to the vertices it follows its own vertex value—that is, its corresponding country (#1).

In the second superstep, each vertex now knows what countries its followers are from. In other words, it knows what an edge in the new country graph should be. At this point, it can simply request the creation of such an edge from the follower's country to its own country, using the addEdgeRequest() method (#3). The addEdgeRequest() method makes a request to add an edge at the vertex with the specified source ID. The edge value and destination ID are specified through the passed edge object.

■ **Note**   You do not have to explicitly create the vertices that represent the countries. By requesting the creation of an edge between two countries, Giraph automatically creates the corresponding vertices if they do not already exist.

You are not done yet though. By creating these new edges, there are two types of vertices at the end of the algorithm execution: user vertices that comprise the original graph and country vertices comprising the country graph. If you use an output format, you see that the output contains both of these types of vertices. In this case, though, you are only interested in the country graph, so ideally you would like the final graph to contain only the vertices representing the countries.

The mutation request API provides a solution to this. Apart from modifying the edges of a vertex, the mutation request API also allows you to create or delete vertices. To achieve this, you can make use of the removeVertexRequest() API. At superstep 0, apart from creating new edges and vertices, you can also remove the vertices that you do not need. Each vertex also makes a request to remove itself by calling the removeVertexRequest() method and passing as a parameter its own ID.

ALGORITHM DETAIL

The alert reader must have observed that no vertex halts in superstep 0. Imagine what would happen if vertices with no followers halted at superstep 0. They would never receive a message in superstep 1, and therefore, they would not be activated and run the `compute()` function. By not halting vertices at superstep 0, you allow every vertex to run the `compute()` function at superstep 1, and thus, allow them to remove themselves from the graph through the `removeVertexRequest()` method.

Finally, recall that by calling the `addEdgeRequest()` and `removeVertexReqest()` methods, a vertex only registers these requests with Giraph. The edges are not created immediately and the vertices are not removed immediately. You have to wait until the next superstep before these changes really happen.

Apart from these two methods that you saw in the example, the `Computation` interface provides the following methods as well:

- `addVertexRequest(I id, V value, OutEdges<I, E> edges)`: Makes a request to create a vertex with the specified ID, value, and edges.

- `addVertexRequest(I id, V value)`: Makes a request to create a vertex with the specified ID and value. The created vertex has no edges.

- `removeVertexRequest(I vertexId)`: Makes a request to remove the vertex with the specified ID. Nothing happens if the vertex with the specified ID does not exist.

- `addEdgeRequest(I sourceId, Edge<I, E> edge)`: Makes a request to add an edge at the vertex with the specified source ID. The edge value and destination ID are specified through the passed edge object. Note that this request affects only the vertex with the source ID.

- `removeEdgesRequest(I sourceId, I targetId)`: Makes a request to remove an edge from the vertex with the passed source ID. In particular, it removes the edge that has the passed target ID. Nothing happens if the edge does not exist. This call affects only the source vertex, not the target vertex.

Although this example did not make use of these methods, they have similar semantics. They allow a modification that gets realized, not immediately, but at the beginning of the next superstep.

Mutation Through Messages

Next, let's look at a third alternative way to modify the graph in Giraph. Apart from the explicit graph mutation through the previous APIs, Giraph also allows the creation of a vertex implicitly, by sending a message to a vertex that does not exist. Let's see what this means exactly and how it could be useful.

Let's see how you can use this way of creating vertices through the previous example. In fact, you are going to enrich it with the calculation of more information. Let's assume you now want to know, not only what countries are connected to each other, but also the number of users from one country that follows users from another country. This gives us a more informed view of the country relations; you also get a sense of the strength of the relationship among countries, and which is the most followed country in terms of the total number of followers. Figure 8-3 illustrates the result you wish to get by analyzing the original input graph.

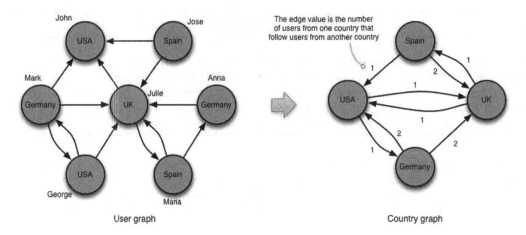

Figure 8-3. *Counting how many users from one country follow users from another country*

Let's now look at the algorithm that computes these more detailed statistics. Unlike the previous example, here we assume that the edges have a value of type IntWritable that is used to hold the number of followers at the end of the execution. Listing 8-3 shows the implementation.

Listing 8-3. Counting the Size of Followership Between Countries

```
public class CreateCountryGraphWithCounts extends
  BasicComputation<Text, Text, IntWritable, Text> {

  @Override
  public void compute(Vertex<Text, Text, IntWritable> vertex,
                      Iterable<Text> messages) {
    Text myCountry = vertex.getValue();
    if (getSuperstep()==0) {                                          #1
      sendMessageToAllEdges(vertex, myCountry);                       #1
    } else if (getSuperstep()==1) {                                   #2
      for (Text m : messages) {                                       #2
        sendMessage(m, myCountry)                                     #3
      }
      removeVertexRequest(vertex.getId());                            #4
      vertex.voteToHalt();
    } else {                                                          #5
      for (Text m : messages) {                                       #6
        IntWritable edgeValue = vertex.getEdgeValue(m);               #7
        if (edgeValue == null) {                                      #7
          vertex.addEdge(EdgeFactory.create(m, new IntWritable(1)));  #7
        } else {                                                      #8
          vertex.setEdgeValue(m, new IntWritable(edgeValue.get()+1)); #8
        }
      }
      vertex.voteToHalt();
    }
  }
}
```

#1 Tell the users that I am following which country I am from.

#2 At superstep 1, every message contains the country of a user that follows me.

#3 For each country following me, send a message to a destination with ID set to that specific country. The message contains this vertex's own country name.

#4 As before, each user vertex removes itself from the graph.

#5 At superstep 2, vertices representing countries are created and receive messages.

#6 A message contains a country name and represents the followership of a single user in that country.

#7 For every such country, check whether this vertex has already an edge to it. If not, add a new edge an initialize its value to 1.

#8 If the country already exists, increment its value by one for every message containing the corresponding country name.

ALGORITHM DETAIL

In the initial user graph, the edge value is not relevant and you simply assume that when the graph is constructed with the use of an InputFormat, it is set to 0.

Recall that the Giraph model allows a vertex to send a message simply by specifying the ID of the destination. In most of the cases, a vertex sends a message to one of its neighbors. It typically choses an ID by looking at its edges, which are constructed during the loading of the input graph. However, the Giraph API is really more flexible and allows you to use any ID in the sendMessage() method. Since you can specify any ID, you may have wondered what happens if you specify an ID that does not really exist in the input graph.

The answer is that Giraph creates the vertex for you. This way, your program can dynamically create a vertex by sending a message to it. In this case, the vertex is initialized with no edges and the vertex value is set to null.

■ **Note** It is your responsibility as a programmer to write code that handles this initialization.

The very first part of this algorithm is the same as before. At superstep 0, every vertex sends its own country to the vertices it follows through a message (#1). Therefore, at superstep 1, every user vertex receives messages containing country names.

Before starting superstep 2, Giraph realizes that there are messages for vertices representing the countries. These vertices do not exist yet. Therefore, Giraph creates them and superstep 2 starts executing as usual. It delivers messages to all vertices and calls the compute() function.

In this case, only the newly created country vertices have messages. Messages contain country names as well. A message that country A receives that contains the country name B means there is a single followership from country A to country B. First, this means that there should be an edge between the vertex representing country A and the vertex representing country B (#7). Second, by counting the messages that contain country B, you know exactly how many users from country A follow users from country B. This count is the value of the edge created.

Now you have been offered three different ways to alter the graph structure during the execution of your algorithm. In general, which one you use depends on your algorithm and how creative you get. Sometimes you find that you can implement an algorithm using either of these ways, while in other scenarios you find that one type of mutation is preferred. For instance, when you only need to add edges to vertices and each vertex is responsible only for adding its own edges; like in the very first example, direct mutations are probably the way to go because they are faster. If, on the other hand, you require the creation of vertices as well, or if a vertex is responsible for adding edges to other vertices, the mutation requests are the solution. In general, you need to think a bit about the design of your algorithm and choose the option that fits best.

Resolving Mutation Conflicts

Even though this was not conveyed in the previous examples, conflicts in mutation requests may occur. For instance, nothing prevents two vertices from making the following conflicting request: one vertex requests the addition of an edge with a certain value, while another vertex may request the creation of the same edge with a different value. What should the value of the edge be in this case? Sometimes even more confusing requests may occur. For instance, a program may issue a request for the addition of an edge to a vertex, while at the same time issue a request for the deletion of the same vertex.

At first glance, these kinds of requests might seem irrational; you may wonder why somebody would ever write a program like this. But you find that when writing distributed programs, such disagreements among the different computations taking place are unavoidable. In fact, they are part of the design of a distributed algorithm as long as there is a way to resolve such conflicts. Think about a simple multithreaded program where two threads try to increment the same, shared counter. The result is undefined if the program is not properly designed, and programming languages typically provide synchronization primitives for you to do design programs correctly. Next, you see how Giraph allows you to resolve this kind of conflict.

First, let's go back to the example of the creation of the Twitter country graph. The code is shown in Listing 8-4. Notice that both the "Mark" vertex and the "Anna" vertex make a request to create an edge for the "Germany" vertex with a target ID equal to "UK".

Every time a set of changes happen on a vertex, like an edge addition or deletion, Giraph uses a VertexChange object to represent such changes and a VertexResolver object that takes a VertexChange as input and defines how to actually handle the changes. Let's start by describing the VertexResolver interface in Listing 8-4.

Listing 8-4. The VertexResolver Interface

```
public interface VertexResolver<I extends WritableComparable,
    V extends Writable, E extends Writable> {

  Vertex<I, V, E> resolve(I vertexId,                    #1
      Vertex<I, V, E> vertex,                            #2
      VertexChanges<I, V, E> vertexChanges,              #3
      boolean hasMessages);                              #4
}
```

#1 The ID of the vertex to resolve.

#2 The original vertex before the changes are applied.

#3 The set of changes to this vertex.

#4 Defines whether the vertex has messages sent to it in the previous superstep.

The VertexResolver interface has a single method. Giraph calls this method for a vertex and passes it the set of changes that have occurred. When you implement this method, you essentially define what the Vertex should look like after the changes. The VertexChange interface shown in Listing 8-5 helps to determine the changes requested for the vertex.

Listing 8-5. The VertexChanges Interface

```
public interface VertexChanges<I extends WritableComparable,
    V extends Writable, E extends Writable> {

  List<Vertex<I, V, E>> getAddedVertexList();    #1
  int getRemovedVertexCount();                    #2
  List<Edge<I, E>> getAddedEdgeList();            #3
  List<I> getRemovedEdgeList();                   #4
}
```

#1 Returns the list of vertex additions requested for this vertex ID.

#2 Returns the number remove requests there were for this vertex ID.

#3 Returns the list of edges that were requested to be added to this vertex.

#4 Returns the list of edges that were requested to be removed from this vertex.

Before implementing your own VertexResolver, let's see what happens if you let Giraph use the default VertexResolver. The default vertex resolver performs the following operations:

1. If there were any edge removal requests, first apply these removals.

2. If there was a request to remove the vertex, then remove it. This is achieved by setting the return Vertex object to null.

3. If there was a request to add the vertex, and it does not exist, then create the vertex.

4. If the vertex has messages sent to it, and it does not exist, then create the vertex.

5. If there was a request to add edges to the vertex, if the vertex does not exist, first create and then add the edges; otherwise, simply add the edges.

The order of this list is important because it defines exactly the way that Giraph resolves any conflicts by default. This means that if, for instance, there is request to remove a vertex and at the same time a request to add it, then because the default resolver checks the vertex creation after it does the deletion, it ends up creating the vertex.

THE DEFAULT VERTEX RESOLVER IMPLEMENTATION

Even though the implementation of the DefaultVertexResolver is not shown here, it is recommended that you look at the source code as an additional example. It gives you more insight into the use of the VertexResolvers.

Now let's assume that you want to change this default behavior. In particular, let's handle the case where there are multiple add requests for the same edge a bit differently. As opposed to just adding the edge to the vertex, you would really like to count the number of requests that exist and set this count as the value of the edge. This essentially gives you an alternative way to count the number of users from one country that follow users from another country. In Listing 8-6, let's look at how you can do this simply by creating a custom VertexResolver.

Listing 8-6. A Custom VertexResolver That Counts the Number of Edge Additions

```java
public class MyVertexResolver implements VertexResolver<Text, Text, IntWritable, Text> {
  @Override
  public Vertex<Text, Text, IntWritable> resolve(
      Text vertexId,
      Vertex< Text, Text, IntWritable> vertex,
      VertexChanges<Text, Text, IntWritable > changes,
      boolean hasMessages) {

    if (changes!=null) {
      if (!changes.getAddedEdgeList().isEmpty()) {          #1
        if (vertex==null) {                                 #2
          vertex = getConf().createVertex();                #2
          vertex.initialize(vertexId,                       #2
            getConf().createVertexValue());
        }
        for (Edge<I, E> edge : changes.getAddedEdgeList()) {
          IntWritable edgeValue =                           #3
            vertex.getEdgeValue(edge.getTargetVertexId()); #3
          if (edgeValue==null) {
            edge.setValue(new IntWritable(1));              #4
            vertex.addEdge(edge);                           #4
          } else {
            vertex.setEdgeValue(edge.getTargetVertexId(),   #5
              new IntWritable(edgeValue.get()+1));          #5
          }
        }
      }
    }
  }
}
```

#1 First, check if there are edge addition requests.

#2 If there are edge additions, and the vertex does not exist already, then create it.

#3 For every edge addition, check if the edge already exists.

#4 If it does not exist, add it and set its value to 1.

#5 If it already exists, just increment its value by 1.

In this example, you assume that you only need to handle edge addition requests. The custom resolver first checks whether there are any such additions (#1). If yes, then you need to check whether the source vertex already exists. Recall that in this case, you create vertices representing countries, which do not already exist. Therefore, if the vertex does not already exist, you need to create it (#2). After this, whenever you see a request for the same edge, you simply increment the value of the edge (#5).

Now that you have implemented your custom resolver, the only thing that is missing is to tell Giraph to use this particular implementation of the vertex resolver in place of the default one. As always, you do this through the familiar command-line custom arguments. You just need to add the following to your command line: -ca giraph.vertexResolverClass=MyVertexResolver.

Overall, mutations are one of the advanced features that can prove very useful once you start thinking about more sophisticated applications. Next, another advanced feature is discussed: how to write your own aggregators.

The Aggregator API

Aggregators are a very useful and easy to use tool that allows you to compute global statistics across the entire graph. You have already seen a handful of examples that use aggregators in Chapters 3 and 5. Giraph already provides a set of common aggregator functions (like sum or max) that you can use in your applications. At the same time, Giraph allows you to implement your own aggregator functions. This section explains the aggregator interface in more detail and shows how to write new ones.

Giraph provides two ways for you to implement custom aggregators. The first way is to implement the Aggregator interface, which gives the most flexibility about how you implement an aggregator. Listing 8-7 shows the interface methods.

Listing 8-7. The Aggregator Interface

```
public interface Aggregator<A extends Writable> {
  void aggregate(A value);              #1
  A createInitialValue();               #2
  A getAggregatedValue();               #3
  void setAggregatedValue(A value);     #4
  void reset();                         #5
}
```

#1 Aggregates the input value to the current value of the aggregator.

#2 Creates the initial value of the aggregator before any aggregation occurs.

#3 Returns the current value of the aggregator.

#4 Sets the current value of the aggregator.

#5 Resets the value of the aggregator.

To understand the implementation of an aggregator, let's first discuss how Giraph uses this interface. The createInitialValue() is called by Giraph before it starts aggregating any values. Giraph calls the aggregate method whenever a vertex wants to add a value to an aggregator. Your implementation of the aggregator is responsible for maintaining the appropriate data structure for this aggregation to happen. For instance, if you are implementing a sum aggregator, your implementation should maintain a partial sum to which you are adding values.

Apart from the Aggregator interface, Giraph also provides an abstract class that implements part of the Aggregator interface and covers the most common aggregator functionality. Most of the time, extending this abstract class covers your needs. Let's look at the details of the BasicAggregator and then an example that extends it to implement a new aggregator.(See Listing 8-8.)

Listing 8-8. The BasicAggregator Abstract Class

```
public abstract class BasicAggregator<A extends Writable> implements
    Aggregator<A> {
  private A value;                          #1

  public BasicAggregator() {                #2
    value = createInitialValue();
  }

  @Override
  public A getAggregatedValue() {           #3
    return value;
  }
```

```java
  @Override
  public void setAggregatedValue(A value) {        #4
    this.value = value;
  }

  @Override
  public void reset() {                            #5
    value = createInitialValue();
  }
}
```

#1 The internal value of the aggregator. It holds the current partial aggregate.

#2 The default constructor sets the internal value to its initial value.

#3 Returns the current value of the aggregator.

#4 Sets the interval value of the aggregator directly.

#5 Resets the value of the aggregator to its initial value.

The only data structure that the BasicAggregator maintains is an internal value of the same type that your aggregate is. This is essentially the current partial value of the aggregator. When you extend the BasicAggregator, your only responsibility is to define what the initial value is through the createInitialValue() method and also how to aggregate a new value to the current internal value of the aggregator through the aggregate() method. Let's look at this with an example. You have already seen the logic of writing a max or a sum aggregator; here you are shown how to write an aggregator that implements a boolean OR function. You may use such an aggregator to detect whether any of the vertices of the graph meets a condition. For instance, imagine a social graph where each vertex is labeled with the age of the user. You also assume that sometimes the age information may be missing, in which case you stop the analysis and print a message. So, you really want know whether any of the vertices—at least one—is missing the age information. A boolean OR aggregator would implement this as "at-least one" logic. In other words, the value of the aggregator should be the boolean value *true* if there is at least one vertex that added the value *true* to the aggregator, and *false* otherwise. Listing 8-9 shows the implementation.

Listing 8-9. A Custom Aggregator That Implements a Boolean OR

```java
public class BooleanOr extends BasicAggregator<BooleanWritable> {

  @Override
  public BooleanWritable createInitialValue() {        #1
    return new BooleanWritable(false);
  }

  @Override
  public void aggregate(BooleanWritable value) {
    boolean currValue = getAggregatedValue().get();    #2
    boolean newValue = currValue || value.get();       #3
    getAggregatedValue().set(newValue);                #4
  }
}
```

#1 Initialize the value of the aggregator to *false*.

#2 Get the current value of the aggregator.

#3 Perform a logical OR operation between the current aggregator value and the aggregated value.

#4 Set the result of the operation as the new value of the aggregator.

THE COMMUTATIVE AND ASSOCIATIVE PROPERTY OF THE OR OPERATION

In Chapter 3 you saw that for Giraph to be able to perform aggregations in an efficient way, the operation must be commutative and associative. It is easy to see that the logical OR operator (||) is indeed both commutative since:

a || b = b || a

and associative since:

(a || b) || c = a || (b || c)

The implementation is quite straightforward. Obviously, if no vertex adds a value to the aggregator, the result should be *false*. Therefore, the createInitialValue() initializes the value of the aggregator to the boolean value *false* using the Writable implementation of the boolean type. Next, through the aggregate method, you essentially perform a logical OR operation between the current aggregator value (#2) and the newly aggregated value (#3). The result replaces the value of the aggregator (#4).

Giraph already provides a rich set of aggregator implementations for the most common operations. As always, the more customized and sophisticated your applications become, the more you want to go beyond the basic functionality and implement your own. This API offers a simple way to implement useful aggregators that Giraph can compute for you in a scalable and efficient manner, without you having to worry what happens under the hood. In the next section, you come across aggregators again, this time looking at how they can help with the coordination of a distributed program.

Centralized Algorithm Coordination

So far you learned that the basic principle of programming in the Giraph API was the "vertex-centric" programming, which required you to think from the perspective of a vertex in the graph. Accordingly, you had to write a compute function that each vertex executes. This leads to algorithms that are naturally distributed, where each vertex does not need to know about the state of the rest of the graph, and only computes for itself.

But in some cases, you find that to implement your algorithm, you need the vertices to coordinate among themselves. This coordination may involve sharing some kind information that is common across the entire graph or by performing some centralized computation that affects the entire algorithm. In fact, you already saw such an example in Chapter 5, where you had to compute the most popular user in Twitter through the use of aggregators. An aggregator offers a form of centralized coordination and computation, since all vertices collectively contribute to the value of a single aggregator, which is then made available back to all vertices. Here, we talk about the MasterCompute, another way to coordinate vertices.

In general, apart from the distributed computations that you execute through the familiar compute() function on a per-vertex basis, you may also need to take some action that depends on or affects the graph and the computation as a whole. As an example, consider that halting your application depends on the value of an aggregator, and not on each vertex individually. Fortunately, Giraph offers the MasterCompute, a mechanism designed exactly for this. You have already seen some examples of how the MasterCompute can be of use, but here it is explained in more detail and you are given more examples of what you can achieve with it.

The MasterCompute provides essentially a centralized location where you can perform actions, such as reading and setting aggregator values, getting statistics about the entire graph, or even stopping the entire job execution. In practice, you can think of the MasterCompute as a piece of application logic that executes only once after the end of each superstep. (See Figure 8-4.)

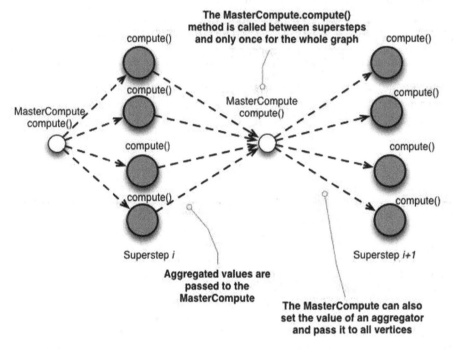

Figure 8-4. *The MasterCompute centralized point of computation. The compute() method executes once before each superstep. Aggregator values are passed from the MasterCompute to the vertices*

Listing 8-10 shows the methods of the MasterCompute abstract class, which are explained shortly, with an example.

Listing 8-10. The MasterCompute Abstract Class

```
public abstract class MasterCompute
    extends DefaultImmutableClassesGiraphConfigurable
    implements MasterAggregatorUsage, Writable {

  public abstract void initialize()        #1
    throws InstantiationException, IllegalAccessException;

  public abstract void compute();          #2
}
```

#1 Called before execution starts.

#2 Called exactly once for the entire graph after each superstep finishes.

The initialize() method of the MasterCompute is called once before the execution of the algorithm starts; it is used typically to set up your MasterCompute implementation. As you will see in a bit, it is very useful to set up the operation of aggregators. The compute() method is the workhorse of the MasterCompute. Giraph calls this method exactly once before the execution of every superstep; that is, before it calls the compute() method on the vertices. More specifically, the first time Giraph calls the compute() method of the MasterCompute is right before the execution of superstep 0. Next, let's look at a few examples of what you can do within the compute function of the MasterCompute.

Halting the Computation

Deciding when to stop the execution is one of the core aspects of a distributed graph algorithm. Think about the PageRank algorithm that you saw in Chapter 4. The PageRank algorithm changes the rank of a web page in each iteration. The decision to stop may depend on the number of iterations that you have performed so far, or on how much the web page ranking has changed since the last iteration. In any case, you need a way to instruct Giraph to stop the computation.

Until now, you have learned that in order to stop a Giraph computation, two conditions must occur: every vertex called the voteToHalt() method and no messages were sent in the current superstep. These conditions depend entirely on your vertex computation. In fact, in Chapter 4 you saw many examples of algorithms and how they implement this halting condition. But you may need to make a decision about halting the computation based on some aggregate information about your graph and your algorithm execution, which does not depend on individual vertices. For example, you may want such a decision to depend on the value of an aggregator. In the case of PageRank, you may want to stop if the aggregate difference of the rankings has not changed much since the last iteration; in other words, PageRank has converged. This type of decision cannot be made by each vertex individually, but it must be done by some centralized entity. As you already have guessed, the MasterCompute is the place to do so.

Giraph gives you an extra way to terminate the computation at a centralized point of control through the MasterCompute API. Recall that the compute() method of the MasterCompute class is called after the end of the superstep and before the beginning of the next. From within the compute() method of the MasterCompute, you can call the haltComputation() method, which terminates the execution. Note that once this method is called, the execution terminates no matter if there are still messages sent by vertices or if not all vertices have called the voteToHalt() method as you knew so far. In other words, a call to haltComputation() from the MasterCompute preempts the usual termination condition.

Let's look at an example of how you could use it. A typical way to decide the termination of an algorithm is based on the number of supersteps executed so far. The PageRank algorithm that you saw in Chapter 4 is such an example. There you put the logic for halting inside the vertex computation itself; here you see how you can do this from inside MasterCompute. One benefit you get from this is that you keep your main computation logic cleaner. Listing 8-11 shows how to do this from inside the MasterCompute implementation.

Listing 8-11. Termination Based on Number of Supersteps

```
public class MyMasterCompute extends DefaultMasterCompute {
  @Override
  public void compute() {
    if (getSuperstep()==10) {                              #1
      haltComputation();                                   #1
    }
  }
}
```

#1 Terminate computation based on the superstep number.

In general, you can implement arbitrary logic inside the `compute()` method of the `MasterCompute`. This logic executes in between supersteps. In this example, you simply check whether you are in superstep 10 and take appropriate action by halting the computation. Otherwise, the `compute()` method does not need to do something else.

Recall that when you call `getSuperstep()` from within `MasterCompute`, it returns the number of the superstep that it is about to execute. In the previous example, the `compute()` method calls `haltComputation()` right after the execution of superstep 9 and before the execution of superstep 10. Therefore, superstep 10 will never execute. This is a small detail that you need to be aware of to avoid programming your termination condition in the wrong way.

Using Aggregators for Coordination

You have already seen how useful aggregators are for computing global statistics. Apart from global statistics, there are other useful things you can use aggregators for, and coordination is one of them. But let's discuss a more concrete example where coordination is necessary. Think about the recommendation algorithms you saw in Chapter 4, used to predict recommendations between users and items. These algorithms iteratively refine a machine learning model that tries to minimize the prediction error. The prediction error is calculated as the aggregate error across all vertices and is computed with the use of an aggregator. These algorithms typically stop execution when the aggregate prediction error has dropped below a specific threshold. One way you can use the aggregators in conjunction with the `MasterCompute` is to coordinate the halting of the execution.

More specifically, you can think of an aggregator as a global variable that all vertices can write and read. You can use aggregators to collect information from the vertices to a centralized location, which is the `MasterCompute`, and from there pass this information back to all their vertices during the execution of the algorithms and have them make decisions according to this information. In this example, you are passing the value of the aggregate error to the vertices, and vertices take actions based on this value. In other words, the aggregators are the way for the master worker to communicate with the vertices. More specifically, the values of the aggregators are broadcast to the workers before vertex `compute()` is called, and collected by the master before master `compute()` is called. Listing 8-12 shows this example.

Listing 8-12. Termination Based on Value of an Aggregator

```
public class MyMasterCompute extends DefaultMasterCompute {
  @Override
  public void compute() {
    if (((DoubleWritable)getAggregatedValue("error")).get()<0.001) { #1
      haltComputation();                                              #2
    }
  }
}
```

#1 Termination is based on the aggregator prediction error falling below a threshold.

#2 When this condition holds, terminate the execution.

In this example, you assume that vertices use the "error" aggregator to sum the total prediction error. After each superstep, the aggregate error is made available to the `MasterCompute`, and you can read it from within the `compute()` method to decide the termination of the execution.

As mentioned, apart from reading the value of an aggregator, inside the `compute()` method of the `MasterCompute` you can also set the value of an aggregator and "broadcast" its value to every vertex in the graph. The value of the aggregator is available to every vertex during the next superstep execution. For instance, imagine that in the previous example, after each superstep, you want to compute the average value of the error across all vertices and then communicate it back to the vertices. The average error then uses each vertex as feedback to improve the prediction.

One way to do this is to first compute the sum of the prediction error across all vertices. You have already seen how to do this using the "error" aggregator. After this, from within the MasterCompute, you can easily compute the average by dividing the aggregator error with the number of vertices in the graph, and then broadcast this to all vertices. Listing 8-13 illustrates this process.

Listing 8-13. Setting the Value of an Aggregator in the MasterCompute

```
public class MyMasterCompute extends DefaultMasterCompute
  @Override
  public void compute() {
    double totalError =
    ((DoubleWritable)getAggregatedValue("error")).get())      #1
    double avgError = totalError/(double)getTotalNumVertices()   #2
    setAggregatedValue("avg.error", new DoubleWritable(avgError));  #3
  }
}
```

#1 Get the aggregator error across all vertices.

#2 Compute the average error by dividing by the number of vertices.

#3 Broadcast the average error to all vertices through a new aggregator.

In this MasterCompute implementation, you first read the value of the "error" aggregator that holds the sum of the prediction errors across all vertices (#1). After this, using the getTotalNumVertices() method, you can easily compute the average error (#2). Finally, using the setAggregatedValue() method of the MasterCompute, you set the value of the "avg.error" aggregator to the average value you just computed (#3). The value of this aggregator is available to all vertices in the next superstep.

In general, the MasterCompute is a very useful tool for coordination among vertices. In this case, it has allowed functionality that was not possible by using only the distributed vertex-centric programming interface. In particular, it allowed you to collect communication among the vertices by aggregating information and making it available to the whole graph. In the following section, you find out about another distinct way to use the MasterCompute—that is, how to compose complex algorithms from simpler blocks.

Writing Modular Applications

So far you have explored those facilities of the Giraph API that allow you to add more functionality in your application or write more efficient algorithms. In this section, you look at a feature of the Giraph API that helps you write better applications from a programming perspective. Specifically, you will look at how Giraph allows you to compose potentially complex algorithms from simpler logical blocks, improving code readability and code reuse. Through examples, you learn how to recognize algorithms that can potentially be decomposed to simpler ones and how to use the Giraph API to simplify their development.

Structuring an Algorithm into Phases

Let's start by discussing the type of algorithms that fall under this category and that could benefit from composability. The programming patterns that make their appearance in many algorithms and are easy to recognize. Note, though, that this is by no means an exhaustive list. These are only meant as common cases for you to understand the composable API and then be more creative with it as new scenarios appear.

One common programing pattern that you may have observed in the Giraph algorithms you have seen so far is that the logic inside your algorithm may depend on the superstep that is executing. This is the basic pattern of graph algorithms in Giraph that you use to break down complex algorithms into simpler ones. Let's revisit one of the algorithms you saw in Chapter 3 that converts a graph from a directed one to an undirected one. In other words, this algorithm ensures that if an edge from vertex A to vertex B exists in the original graph, then the output graph contains an edge from B to A as well. Several algorithms operate on undirected graphs, so this is a common preprocessing step. In Chapter 3, you saw the pseudocode for this algorithm; here you are shown the actual code. (See Listing 8-14.)

Listing 8-14. Algorithm to Convert a Directed Graph to an Undirected

```
public class ConvertGraph
  extends BasicComputation<
    IntWritable, NullWritable, NullWritable, IntWritable> {

  @Override
  public void compute(
    Vertex<IntWritable, NullWritable, NullWritable> vertex,
    Iterable<IntWritable> messages) {
    if (getSuperstep()==0) {
      sendMessageToAllEdges(vertex, vertex.getId())    #1
      vertex.voteToHalt()
    } else {
      for (IntWritable msg : messages) {               #2
        IntWritable id = (IntWritable)msg.get();       #2
        if (vertex.getEdgeValue(id))==null) {          #3
          vertex.addEdge(message)                      #3
        }
      }
      vertex.voteToHalt()
    }
  }
}
```

#1 Send this vertex's ID to all its neighbors.

#2 I got a message with an ID from a vertex that points to me.

#3 If there is no edge to that destination ID, then add it.

In this algorithm, during superstep 0, every vertex sends a message that contains its own vertex ID to all the vertices that it has edges to. This way, during superstep 1, a vertex A receives a message from vertex B if B has an edge to A. If vertex A does not have an edge to B, it can then add it. Figure 8-5 illustrates these two steps.

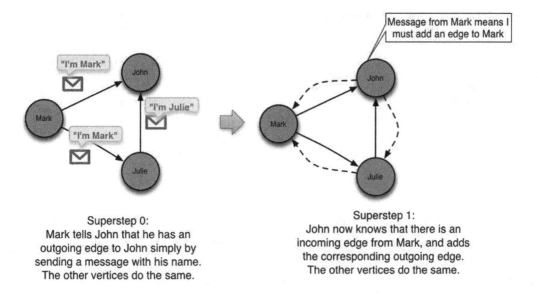

Figure 8-5. *The steps to convert a directed graph to an undirected one*

This is a very representative example of algorithms that benefit from the ability to compose an algorithm from simpler blocks for the following reasons. First, you can immediately observe that it fits the pattern that described earlier; the computation logic depends on the superstep that is executing. In particular, there are two clearly separated phases: one that is executed during superstep 0, and one that is executed during superstep 1.

Second, converting a graph to an undirected one is usually a preprocessing step that is secondary to the main logic of your application. This way, along with your main logic, you would have to include this conversion logic inside your computation function. For instance, assume that you want to run the Connected Components algorithm that you saw in Chapter 4 on an undirected graph. In this case, before you execute the main logic of the algorithm, you have to ensure that the graph is converted to an undirected one by applying the conversion logic.

To illustrate this, let's first look at how you would implement the original connected components algorithm. Chapter 4 described this algorithm with pseudocode; Listing 8-15 shows the actual code.

Listing 8-15. The Connected Components Algorithm

```
public class ConnectedComponents
  extends BasicComputation<
    IntWritable, NullWritable, NullWritable, IntWritable> {

  @Override
  public void compute(
    Vertex<IntWritable, NullWritable, NullWritable> vertex,
    Iterable<IntWritable> messages) {

    if (getSuperstep()==0) {
      sendMessageToAllEdges(vertex, vertex.getId())              #1
      vertex.voteToHalt()
    } else {
      int maxId = vertex.getValue().get()                       #2
```

```
      for (IntWritable msg : messages) {                      #3
        if (msg.get()>maxId) {                                #4
          maxId = msg.get();                                  #4
        }                                                     #4
      }
      if (maxId>vertex.getValue().get()) {                    #5
        vertex.setValue(maxId);                               #5
        sendMessageToAllEdges(vertex, maxId)                  #5
      }
      vertex.voteToHalt();
    }
  }
}
```

#1 In superstep 0, every vertex sends its ID to all its neighbors.

#2 Each vertex maintains the maximum ID seen so far, which represents the ID of the component the vertex belongs too.

#3 In each subsequent superstep, a vertex receives the IDs of all its neighbors.

#4 A vertex finds the maximum ID among its own and the IDs of its neighbors.

#5 If the max ID is greater than the one stored currently in the vertex value, update the value and propagate the new max ID to all the neighbors.

Now that you have the logic for the connected components, let's look at how to modify it to perform the same algorithm, but after converting a directed graph to an undirected one (see Listing 8-16).

Listing 8-16. Applying the Connected Components Algorithm on an Undirected Graph

```
public class ConnectedComponents
  extends BasicComputation<
    IntWritable, NullWritable, NullWritable, IntWritable> {
  @Override
  public void compute(
    Vertex<IntWritable, NullWritable, NullWritable> vertex,
    Iterable<IntWritable> messages) {
    if (getSuperstep()==0) {                                  #1
      sendMessageToAllEdges(vertex, vertex.getId())           #1
      vertex.voteToHalt()                                     #1
    } else if (getSuperstep()==1) {                           #1
      for (IntWritable msg : messages) {                      #1
        IntWritable id = (IntWritable)msg.get();              #1
        if (vertex.getEdgeValue(id))==null) {                 #1
          vertex.addEdge(message)                             #1
        }                                                     #1
      }                                                       #1
      vertex.voteToHalt()                                     #1
    } else if (getSuperstep()==2) {                           #2
      sendMessageToAllEdges(vertex, vertex.getId())           #2
      vertex.voteToHalt()                                     #2
    } else if (getSuperstep()>2) {                            #2
      int maxId = vertex.getValue().get();                    #2
```

```
      for (IntWritable msg : messages) {              #2
        if (msg.get()>maxId) {                        #2
          maxId = msg.get();                          #2
        }                                             #2
      }                                               #2
      if (maxId>vertex.getValue().get()) {            #2
        vertex.setValue(maxId);                       #2
        sendMessageToAllEdges(vertex, maxId)          #2
      }                                               #2
      vertex.voteToHalt();                            #2
    }
  }
}
```

#1 Graph conversion logic is applied in the first two supersteps.

#2 Main algorithm logic is applied in subsequent supersteps.

You can clearly see that this algorithm mixes the two logics: the graph conversion and the connected components algorithm. Again, it uses the superstep number to distinguish between phases. It uses the first two supersteps to convert the graph, and the supersteps afterward to apply the logic of the connected components.

Even though this is a perfectly correct way to structure such an algorithm, you may find that it can lead to unnecessarily complicated code, as it pollutes your computation with code that is not directly related to the core algorithm logic. If your algorithm becomes more complex, you end up placing many different logical blocks inside the same computation. For instance, just like you have a graph preprocessing step here, in some scenarios you may also have a post-processing phase that has to be executed in the supersteps following the main application logic, adding to the complexity of your application. To make matters worse, when you write unit tests for your algorithm, now you have to account for the testing cases that exercise the conversion logic as well, even though they are not directly related to your main logic.

Third, it is obvious that this piece of logic, converting a graph to an undirected one, may repeat itself across different algorithms that must operate on undirected graphs. In fact, you could just copy the logic executed during supersteps 0 and 1 to different algorithms. But this is not a good programming practice for different reasons. For instance, it forces you to modify your main algorithm so that it applies the main logic only for superstep 2 and after. Apart from this, it makes code reuse and code maintenance harder. For instance, if you happen to find a bug in the conversion logic, you have to fix it in all the algorithms that use this logic. Instead, once you start building a library of algorithms, it would be nice if you could just reuse such common blocks of logic and not having to code them every time. Next, you look at how to actually get past all of these problems.

The Composable API

Hopefully, by now you are convinced that writing algorithms can get complicated from a programming perspective. Let's now look at how Giraph can make this task easier for you. At a high level, Giraph gives you the ability to specify different computation classes for different supersteps and decide which computation executes in what superstep in an easy way. This way, you get to separate the logic into different computation classes, but at the same time you can combine them into the execution of a single algorithm. Let's look at an example of how you can do this, followed by an explanation of all the benefits that it gives you.

Specifically, let's use the example of the graph conversion into undirected. As a first step, break down the logic for the graph conversion into two distinct computations. The first one is responsible for sending the ID of the vertex to all of its neighbors. This is the operation that you used to perform in superstep 0 in your original, monolithic application. The second one is responsible for receiving IDs as messages and adding the corresponding edges if they do not exist. This is the operation that you used to perform superstep 1 in the original algorithm. You can see these two computations in Listing 8-17.

Listing 8-17. Computation That Propagates Vertex ID to Neighbors

```
public class PropagateId
  extends BasicComputation<
    IntWritable, NullWritable, NullWritable, IntWritable> {

  @Override
  public void compute(
    Vertex<IntWritable, NullWritable, NullWritable> vertex,
    Iterable<IntWritable> messages) {

    sendMessageToAllEdges(vertex, vertex.getId())    #1
    vertex.voteToHalt()
  }
}
```

#1 Send vertex ID to all neighbors.

In this example, you assume that your input graph has a vertex ID of type integer, and has no associated vertex value and edge value. Therefore, you assume those to be of type NullWritable. In this first computation, each vertex simply sends its ID to all its neighbors using the sendMessageToAllEdges() API and then halts. Let's look at Listing 8-18.

Listing 8-18. Computation That Adds Reverse Edges

```
public class ReverseEdges
  extends BasicComputation<
    IntWritable, NullWritable, NullWritable, IntWritable> {

  @Override
  public void compute(
    Vertex<IntWritable, NullWritable, NullWritable> vertex,
    Iterable<IntWritable> messages) {

    for (IntWritable msg : messages) {
      IntWritable id = (IntWritable)msg.get();
      if (vertex.getEdgeValue(id))==null) {
        vertex.addEdge(message)
      }
    }

    vertex.voteToHalt()
  }
}
```

At this point, you have two different computations that do not seem related whatsoever. That is because so far you have been thinking about algorithms in the context of a single computation class. In a moment, you are going to change this, and discover how you can glue them together in a very easy manner.

Putting together different computation classes consists mainly of deciding which computation class is used at which superstep; in other words, coordinating the use of the computation classes. The word *coordination* must have already hinted to you that you are going to do this through the MasterCompute. Indeed, one of the operations you can do in a MasterCompute implementation is setting the computation to be used at each superstep. You apply the logic of choosing the right function inside the compute() method of the MasterCompute. Recall that Giraph calls the compute method of your MasterCompute implementation right before the execution of a superstep. Inside this method, you have the chance to set the computation used in the next superstep. Listing 8-19 shows an example of how to do this.

Listing 8-19. Setting the Computation Class from the MasterCompute

```
public class MyMasterCompute extends DefaultMasterCompute {

  @Override
  public final void compute() {
    long superstep = getSuperstep();        #1
    if (superstep == 0) {                   #2
      setComputation(PropagateId.class);    #2
    } else {                                #3
      setComputation(ReverseEdges.class);   #3
    }
  }
}
```

#1 Get the number of the superstep to be executed next.

#2 If we are about to execute superstep 0, we set the computation to PropagateId.

#3 If we are about to execute superstep 1, we set the computation to ReverseEdges.

Let's take a step-by-step look at the operations in this method. You first get the number of the superstep that Giraph is about to execute. Based on this, you choose the right computation. In particular, if Giraph is about to execute superstep 0, you want to perform the first phase of the conversion, which is having each vertex send its IDs to its neighbors. This is the logic that the PropagateId computation class implements. You then use the setComputation() API of the MasterCompute to indicate this to Giraph. Notice that you pass as an argument to this method an object of type Class that represents the specific computation class you want to execute. After the first superstep is executed, Giraph calls the compute() method of MasterCompute again. This time, you want to perform the second phase of the conversion, where vertices receive messages from phase one, and add the corresponding edges. This is computation class ReverseEdges. Therefore, if you are about to execute superstep 1, you want to set the computation class to ReverseEdges.class, using again the setComputation() method.

That is it. With the MasterCompute, you have easily coordinated the execution of the algorithm through a simple call to the setComputation() method. As you already have seen in Chapter 5, all you have to do to run this application is to specify the MasterCompute implementation when you run the Giraph job through the –mc command-line option. Giraph takes care of the rest, ensuring that the right computation is used for each superstep.

But let's look back to see what you have achieved with this separation of the functionality in different blocks. First, it is obvious that you did not have to encode the choice of the superstep in your application logic, leaving it clean and intact. From a code readability point of view, the two separated computations encode only what they intend to do and are much simpler and easier to understand.

Apart from this, it was previously said that it would be nice to be able to easily reuse code across the application. In particular, let's now revisit the scenario where you want to execute the connected components algorithm on an undirected graph. You already have code that converts a graph to an undirected one, and you already have code that implements the Connected Components algorithm. All you are going to do is use the composable API to glue them together in the MasterCompute with no modification in the main logic. The implementation of the MasterCompute class is shown in Listing 8-20.

Listing 8-20. Combining Graph Conversion with the Connected Components Algorithm

```
public class MasterCompute extends DefaultMasterCompute {

  @Override
  public final void compute() {
    long superstep = getSuperstep();
    if (superstep == 0) {
      setComputation(PropagateId.class);
    } else if (superstep == 1) {
      setComputation(ReverseEdges.class);
    } else {
      setComputation(ConnectedComponents.class);
    }
  }
}
```

As you may have observed, what you have done is moved the coordination and the selection of the logic to execute at each superstep from the main computation to a part of your code that is specifically intended for coordination—that is, the MasterCompute. Again, as a result, you did not have to modify the logic of the Connected Components algorithm at all. You just took three individual pieces of logic and put them together to form a new application. You only had to instruct Giraph to use the graph conversion logic for the first two supersteps, and then your main computation for all the subsequent supersteps. In a similar way, instead of the connected components algorithm, you could use these first two computations in combination with any other algorithm, simplifying code reuse and making it easy to build new applications.

■ **Note**　While inside MasterCompute, you can set the computation class at every superstep; if you do not set it explicitly, Giraph uses the value set during the last superstep.

Summary

Even though the basic Giraph API is flexible enough to allow you to express a wide range of algorithms, Giraph provides tools that give you more capabilities. These make it possible to express new programming patterns, such as centralized coordination, making your algorithms more efficient, and possibly making your life easier from a programming perspective.

- The graph mutation API allows an algorithm to modify the graph during execution. This is often a natural requirement by many applications that must change the graph, but may also come as a handy tool in algorithms that require the temporary creation of logical vertices.

- Giraph provides three ways to perform mutations: (i) direct mutations on a Vertex object, (ii) mutation requests, and (iii) mutations through messages.

- When using mutation requests, conflicts may occur. Use a `VertexResolver` to determine how you want the conflicts to be resolved.

- Aggregators allow you to compute global statistics, and Giraph already provides a rich set of common operations. At the same time, you can use the simple aggregator API to implement custom ones.

- Aggregators and the `MasterCompute` allow the expression of algorithms that require centralized coordination.

- Use the `MasterCompute compute()` method to implement your centralized coordination logic.

- Algorithms can quickly get complicated, but they are usually structured in phases that depend on the superstep executed. Using the composable API can simplify programming, enable code reuse, and result in code that is easier to maintain.

At this point, you have already started using the more advanced features of Giraph. This should allow you to write more sophisticated applications. In the following chapter, you continue to look at these advanced features of Giraph, including how to take control of the parallelization in Giraph to write more efficient applications.

PART III

■ ■ ■

Advanced Topics

Advanced Topics

CHAPTER 9

■ ■ ■

Exposing Parallelism in Giraph

This chapter covers

- Per-worker computations

- Thread safety in Giraph

- The importance of graph partitioning

- Implementing custom partitioners

In the previous chapters, you looked at the different programming primitives of Giraph and discovered how to use them to implement graph algorithms. In the process of implementing graph algorithms, you considered graphs in an abstract way, viewing them only as vertices and edges that can communicate each other through messages. Fortunately, you did not have to worry about what happens under the hood; that is, where these vertices and edges live when Giraph executes your algorithms on a cluster or how messages are actually implemented. Nor did you have to worry about how Giraph manages to parallelize the execution of your algorithm. In fact, this is one of the great benefits of Giraph; it hides from you the complexity that comes with executing a graph algorithm on large compute clusters.

Then in Chapter 6, you got more insight on the internal architecture of Giraph and discovered how it manages the parallel processing of very large graphs. You learned that Giraph groups vertices in partitions and distributes them to the actual machines, or workers, responsible for the processing of each partition, thus parallelizing the processing. You also learned that Giraph might even perform the processing on different threads within the same worker, achieving further parallelization.

In this chapter, you revisit how Giraph parallelizes the processing of a large graph, but with a different goal. You are shown that with a little knowledge about the parallel processing mechanism, you can build more efficient applications on top of Giraph. For instance, the chapter discusses how you can exploit the distribution of vertices to worker machines through the concept of worker computations. In particular, you learn how to augment your vertex-centric algorithms with per-worker computations that allow you to share data and computations among vertices that are physically collocated on the same worker. You see in several scenarios that sharing data and computation can help you add new functionality or significantly improve the performance of your application.

This chapter also discusses graph partitioning, the placement of vertices across the different worker machines. Simple examples describe the impact that different graph partitioning strategies can have on application performance. Then you learn how different graph mining scenarios may benefit from different partitioning strategies. Finally, you learn how to control graph partitioning to your benefit by implementing custom partitioners.

Worker Computations

As you learned in Chapter 6, Giraph splits the graph into partitions and assigns partitions to the cluster machines, or workers. This gives Giraph the ability to parallelize the processing of the graphs by assigning partitions to different worker machines and potentially by assigning different partitions to different CPU cores within the same worker machine. More specifically, each worker machine or core is responsible for calling the compute function for all the vertices resident on that worker or core. The nice thing is that so far, you as a programmer did not really have to worry about how the graph is partitioned under the hood and how Giraph processes each partition; you only needed to think about the graph and algorithm abstractly, as vertices with edges, and vertex computations.

In this section, however, you will discover that sometimes being aware of this concept of partition and worker distribution can help you build better applications. In particular, you will learn that some algorithms may benefit from doing computation not only on a per-vertex basis, but also on a per-worker basis. For the first time, you will perform computations that do not fall under the typical vertex-centric mentality of Giraph. Here, you review some common use cases where this occurs and you learn how to take advantage of the concept of partitioning, either to add functionality to your applications or to make them more efficient.

Use case: Sharing Data Across a Worker

One of the most common use cases where per-worker computation becomes very handy is when you want to share some data structure among all vertices of a worker. Often, the programming logic inside a vertex computation requires access to data, potentially read from some external service, like a distributed storage system, which is common to all vertices.

To make this more concrete, imagine you want to process our familiar Twitter graph, depicted in Figure 9-1, to count the number of male or many female followers each user has.

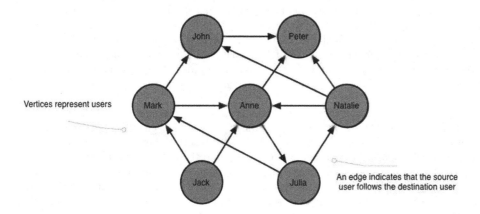

Figure 9-1. *An example Twitter graph*

In fact, in Chapter 3 you already saw pseudocode for a simpler version of this algorithm that simply counts Twitter followers. Before getting into the more complex algorithm, in Listing 9-1, let's see what this simpler application looks like in actual code.

Listing 9-1. Algorithm That Counts Twitter Followers

```
public static class CountFollowers
  extends AbstractComputation<Text, IntWritable, NullWritable, Text, Text> {

  @Override
  public void compute(Vertex<Text, Text, NullWritable> vertex,
    Iterable<Text> messages) throws IOException {

    if (getSuperstep()==0) {
      sendMessageToAllEdges(vertex, vertex.getId());          #1
      voteToHalt();                                           #1
      return;                                                 #1
    }

    int numFollowers = 0;
    for (Text msg : messages) {                               #2
      numFollowers++;                                         #2
    }                                                         #2

    vertex.setValue(new IntWritable(numFollowers));           #3
    voteToHalt();
  }
}
```

#1 In the 1st superstep, you send the ID to all neighbors and return

#2 In the 2nd superstep, a vertex counts messages, one from each follower

#3 Finally, a vertex sets its value as the number of followers, votes to halt, and returns implicitly

In this algorithm, during the first superstep, every vertex sends its ID to all the vertices that it has edges to; that is, to all users that it follows. In the second superstep, every vertex receives messages: one from every follower. At this point, all it has to do is to count the number of received messages and set the resulting count as its value.

Next, you modify this application to count followers based on their gender. First, notice that since the gender information might not be part of the profile of a user, you need a way to automatically decide whether a user is male or female and label them with a "female" or "male" tag. You assume that there is an external service that maintains this information; that is, the service can map a user to a gender. Such a service would essentially give you a classifier that can automatically detect the gender if you pass it a name.

You do not worry about the details of the service and the classifier here. You simply assume that calling getClassifierFromService() from within your applications contacts the external service and returns an object of type GenderClassifier. Once you have an instance of the GenderClassifier object, you can pass the name to the get(String) method of the instance, which returns a string indicating a "female" or "male" gender.

Now that you know how to access the external service and map a name to a gender, let's look at how you are going to implement this application in Giraph, as shown in Listing 9-2.

Listing 9-2. Algorithm That Counts Twitter Followers by Gender

```
public static class CountFollowersByGender
  extends AbstractComputation<Text, Text, NullWritable, Text, Text> {

  @Override
  public void compute(Vertex<Text, Text, NullWritable> vertex,
    Iterable<Text> messages) throws IOException {

    if (getSuperstep()==0) {
      GenderClassifier classifier = getClassifierFromService();    #1
      String gender = classifier.get(vertex.getId());             #2
      sendMessageToAllEdges(vertex, new Text(gender));            #3
      voteToHalt();                                              #4
      return;                                                    #4
    }

    int femaleFollowers = 0;
    int maleFollowers = 0;
    for (Text msg : messages) {                                  #5
      if (msg.toString().equals("female")) {                    #5
        femaleFollowers++;                                       #5
      } else {                                                   #5
        maleFollowers++;                                         #5
      }                                                          #5
    }                                                            #5
    String vertexValue =
     "female:"+femaleFollowers+", male:"+maleFollowers;         #6
    vertex.setValue(new Text(vertexValue));                     #6
    voteToHalt();
  }
}
```

#1 Connect to external service and get gender classifier

#2 Get gender based on name from classifier

#3 Send the gender to all the users this vertex follows

#4 If this is the very first superstep, vote to halt and returns

#5 In the second superstep, messages contain gender labels of followers

#6 Set the gender counters as the vertex value

This new algorithm is fairly straightforward as well. Compared to the simpler version, the difference is that during superstep 0, a vertex contacts the external service to get the gender classifier to decide whether the user it represents is male or female. After that, it sends a "female" or "male" label to all the users that it follows. In the second superstep, every vertex known maintains two separate counters: one for female counters and one for male counters. Notice that you also change the type of the value to Text, and you set it to a string representing the two counts. This is all you need to count followers by gender.

Now, this is a perfectly valid way to add this functionality to your algorithm, but it presents a problem. It forces you to access an external service for every vertex computation that occurs. This seemingly harmless feature can potentially add latency to the processing, significantly degrading the performance of your application.

However, you may have noticed that in this scenario, you repeatedly read exactly the same information for every vertex computation; that is, the gender classifier object. Ideally, you would like to avoid this and instead keep this information somewhere in memory and share it across all vertices. This would avoid all these unnecessary accesses and eventually speed up computation.

To achieve this, Giraph provides a facility called WorkerContext. A WorkerContext is an object that is shared across all vertices in a partition and lives throughout the execution of your algorithm; every worker has its own WorkerContext. You can use it in two ways. First, you can use it to perform per-worker computations at predefined points in the algorithm execution; for instance, before and after a superstep. Second, you can also use it to maintain shared data structures that you can access from within each vertex computation. Giraph provides an abstract class that you can extend to customize to your use case. Before looking at how you are going to use it in this particular scenario, let's look at the methods of the WorkerContext abstract class, shown in Listing 9-3.

Listing 9-3. The WorkerContext Abstract Class

```
public abstract class WorkerContext {
  public void preApplication();          #1
    throws InstantiationException, IllegalAccessException;
  public void postApplication();         #2
  public void preSuperstep();            #3
  public void postSuperstep();           #4
}
```

#1 Called on every worker before the computation begins

#2 Called on every worker after the computation finishes

#3 Called on every worker before the beginning of every superstep

#4 Called on every worker after the end of every superstep

To understand how to extend the WorkerContext abstract class, first you must know about the life cycle of a WorkerContext object. For every machine worker that the Giraph job runs on, Giraph creates a WorkerContext instance. Before the execution of your algorithm starts, Giraph calls the preApplication() method for each worker. You can use this method to set up the WorkerContext. After the execution starts, Giraph calls the preSuperstep() method of every WorkerContext right before a superstep begins. After a superstep finishes, Giraph calls the postSuperstep() method. Finally, at the end of your algorithm execution, Giraph calls the postApplication() method for every worker. During the entire execution of the algorithm, a vertex computation can access the WorkerContext object by calling the getWorkerContext() method.

Next, you see how to implement these methods in our particular scenario. You will create a WorkerContext that is used to contact the external service to get the gender classifier object and share it across all vertices on the same worker. Listing 9-4 shows the implementation.

Listing 9-4. Example Implementation of WorkerContext

```
public class MyWorkerContext extends WorkerContext {

  private GenderClassifier classifier;              #1

  @Override
  public void preApplication() throws InstantiationException,
    IllegalAccessException {

  classifier = getClassifierFromService();          #2
  }
```

```
  public GenderClassifier getClassifier() {
    return classifier;                              #3
  }
}
```

#1 Variable to hold the gender classifier

#2 Load data from external service before the application starts

#3 Gives access to the classifier variable

First, the WorkerContext implementation has a variable of type GenderClassifier that is the same as in the original algorithm and is holding the information that all vertices will share. Recall that in this case, you want to access an external service and read some data to be shared across the vertices. This is a step that you need to perform only once, since this data does not change during the execution of the algorithm. The preApplication() method is the perfect candidate to perform this operation. You use this method to get the classifier from the external service, using the same API call to connect to the service as before. This ensures that data is loaded before the computation of any vertex occurs and only once. In this example, you leave the rest of the methods of the WorkerContext class empty.

Let's now look at how you would modify the algorithm to access the shared data maintained in the partition context from within the vertex computation method. This is shown in Listing 9-5.

Listing 9-5. Modified Algorithm to Access Shared Data in WorkerContext

```
public static class CountFollowersByGender
    extends AbstractComputation< Text, Text, NullWritable, Text, Text> {

  @Override
  public void compute(Vertex<Text, Text, NullWritable> vertex,
    Iterable<Text> messages) throws IOException {

    if (getSuperstep()==0) {
      MyWorkerContext context = (MyWorkerContext)getWorkerContext();    #1
      GenderClassifier classifier = context.getClassifier();           #2
      String gender = classifier.get(vertex.getId());
      sendMessageToAllEdges(vertex, new Text(gender));
      voteToHalt();
      return;
    }

    int femaleFollowers = 0;
    int maleFollowers = 0;
    for (Text msg : messages) {
      if (msg.toString().equals("female")) {
        femaleFollowers++;
      } else {
        maleFollowers++;
      }
    }
    String vertexValue =
      "female:"+femaleFollowers+", male:"+maleFollowers;
    vertex.setValue(new Text(vertexValue));
    voteToHalt();
  }
}
```

#1 Gets a reference to the worker context and casts it to your implementaiotn type

#2 Gets access to the GenderClassifier contained in the WorkerContext

In this implementation, the only difference is how you get access to the GenderClassifier object. Instead of reading it every time from the external service, you get it from the WorkerContext object that has already loaded it for you. The Computation API gives you access to the getWorkerContext() method that returns the WorkerContext object. Remember that this method returns an object of the WorkerContext superclass, so you have to cast it to the type that just implemented before you start using it. At this point, you can access it as you wish. In this case, you simply get a reference to the GenderClassifier object and use it exactly as before.

By using this simple concept of per-partition computations, you have managed not only to simplify the application, but also to make it more efficient. This way you have avoided loading and maintaining a shared data structure once for every vertex. This saves computation overhead, communication overhead, and memory since you do not need to replicate the same data structure multiple times. Essentially, what you have done in this case is create a simple cache of objects that are filled once and then reused across all vertices. Once you start building more sophisticated applications, this technique may prove very useful.

One final detail that you must be aware of is how to tell Giraph to use a specific WorkerContext implementation when it starts your job. As with many customizable aspects of Giraph, you can pass this as a command-line argument when you start your job. Specifically, you can set the giraph.workerContextClass parameter to the full class name of your WorkerContext implementation. You can do this by adding the following in your command line:

```
-ca giraph.workerContextClass=MyWorkerContext.
```

Use Case: Per-Worker Performance Statistics

Another use case where per-worker computations are handy is monitoring and computing fine-grained statistics about the runtime execution of your algorithm. Note that Giraph already maintains some aggregate statistics for you, like total processing time, aggregate size of the graph, and others, and apart from these, you may always find useful performance statistics in the logs of a Giraph job. However, you may want to maintain a finer view of algorithm performance; for instance, to detect performance problems that are specific to your application. For example, you may want to compute the number of vertices each worker holds, or the number of computations that happen per worker to monitor how evenly your computation is distributed across the workers.

To do this, you are going to use the same functionality that the WorkerContext offers: the ability to share an object across all vertices on a worker. In this case, this object holds performance statistics and is accessed and modified by every vertex computation.

Before learning how to do this, let's first decide on the kind of information you want to maintain. In this scenario, you want to count the number of times that a vertex computation occurs on each worker. This gives you an idea of the processing load on each worker and can potentially help detect an uneven distribution across the workers. At the end of the application, you also want to somehow output this information so that it is presented to whoever is running the algorithm.

Let's now assume that you have constructed a class called Performance that holds information about runtime performance, such as the number of vertex computations. You will keep this class simple and assume that it only contains a method to increment the number of computations that occur. Obviously, you could enrich such a class with (and maintain) more performance metrics, as shown in Listing 9-6.

Listing 9-6. Class to Maintain Simple Performance Statistics

```
public class Performance {
  private int numComputations;        #1

  public void getNumComputations() {   #2
    return numComputations;
  }

  public void incNumComputations() {   #3
    numComputations++;
  }
}
```

#1 Holds the number of vertex computations on a worker

#2 Returns the current number of worker computations

#3 Increments the number of worker computations by 1

Next, you are going to implement a WorkerContext through which you manipulate these performance statistics. In this application, you need to update the performance statistics during every compute method; you also need to output the statistics. Let's look at the implementation of the WorkerContext in Listing 9-7.

Listing 9-7. Implementation of WorkerContext to Maintain Performance Statistics

```
public class PerfStatsWorkerContext extends WorkerContext {

  private Performance perf;

  @Override
  public void preApplication()
    throws InstantiationException, IllegalAccessException {
    perf = new Performance();                          #1
  }
  synchronized public void incComputations() {
    perf.incComputations();                            #2
  }

  @Override
  public void postApplication() {                      #3
    String id =
      getContext().getConfiguration().get("mapred.task.id");  #4
    Counter counter =                                  #5
      getContext().getCounter(                         #5
        "Worker Computations",                         #5
        "Worker_"+id);                                 #5
    counter.increment(perf.getNumComputations());      #5
  }

  @Override
  public void postSuperstep() {                        #6
    String id =
      getContext().getConfiguration().get("mapred.task.id");
```

```
    Counter counter =
      getContext().getCounter(                                    #7
        "Worker Computations",                                    #7
        "Worker_"+id+"_"+getSuperstep());                         #7
      counter.increment(perf.getNumComputations());
    }
}
```

#1 initialize the performance statistics class

#2 To be called upon each vertex computation

#3 Called at the end to output the statistics

#4 Obtain worker id

#5 Get Hadoop counter and set value

#6 Prints statistics after each superstep for fine-grained information

#7 You construct a counter based on worker id and superstep number

USING THE MAPPER.CONTEXT

Often times you may need to get some configuration information stored inside the Mapper.Context object, which is inherited from Hadoop. Every Giraph worker is implemented as a Hadoop mapper, and the Mapper.Context gives information about the mapper configuration. In this case, you used it to (i) get the unique index of the mapper, which is the value of the mapred.task.id property, and hence, the unique index of the Giraph worker, and (ii) to access to the Hadoop counters interface.

In this scenario, the WorkerContext implementation simply has a field of type Performance that holds the statistics. You use the preApplication() method to initialize this field. This WorkerContext implementation also provides a public method called incComputations() that increments the number of computations by calling the corresponding method in the Performance object. You look at how and when you call this method in a moment.

Furthermore, you implement the postApplication() method, which you use here to output the information you have collected during the execution of the algorithms; that is, the total number of computations per worker. Apart from this, you would like to calculate and output the total number of computations across all workers. In other words, you would like to aggregate the partial statistics maintained from the different workers. Giraph provides a number of different ways to output information to the user. For instance, Chapter 7 talked about output formats. Here, you make use of another mechanism, one that you borrow from Hadoop, called Counters. Hadoop Counters make it easy to aggregate and output statistics using a simple API.

Now, apart from the final computation count, you may want to measure this metric at a finer granularity, and observe performance during each superstep. You can easily perform the same operations inside the postSuperstep() method of the WorkerContext. Giraph calls this method after the end of each superstep, giving detailed information during the execution of the algorithm.

USING HADOOP COUNTERS

Counters are a mechanism that Hadoop provides to output useful statistics in a friendly way to the user. If you have run a Hadoop or a Giraph job, you have already seen what a counter looks like in the output. The Hadoop web interface or the output that you see in your terminal contains such counters.

Counters typically belong to a *counter group* with a specific name, such as "Giraph Stats". Each group may contain a number of counters, each of which has a name and associated value. For example, the Giraph counter group includes counters that show the time each superstep took.

You can create new counters and assign a value to them from within your code using the Hadoop API. In fact, counters resemble aggregators; you can increment the value of a counter, given its name, from anywhere in your code and Hadoop takes care of computing the final sum of the counter value.

Let's now look at the final missing part; that is, how you would update the performance statistics from within an application. Listing 9-8 shows the implementation of a computation function. You omit the details of the computation and focus on how to access the WorkerContext. As you can imagine, this would be any computation that you implement.

Listing 9-8. Modifying the Performance Statistics from Within the Application

```
public class MyComputation extends BasicComputation {

  @Override
  public void compute(Vertex<I, V, E> vertex,
      Iterable<M1> messages) {

    PerfStatsWorkerContext context =        #1
          (PerfStatsWorkerContext)getWorkerContext();
    context.incComputations()               #2

    ...                                     #3
    ...                                     #3
  }
}
```

#1 Get handle to worker context

#2 Increment number of computations

#3 Perform rest of computation as usual

From inside the compute() method, you simply get access to the worker context object. Recall that you first need to cast it to the type of your own implementation. After this, you can use any custom functionality you have implemented. In this case, you simply increment the number of computations. At this point, the implementation is done. Recall that to enable this WorkerContext implementation, you must add the following option in your command line:

```
-ca giraph.workerContextClass=PerfStatsWorkerContext
```

If you forget to do this, Giraph will use the default implementation and will obviously fail to cast the WorkerContext object to your own implementation at runtime.

Thread Safety in Giraph

The two use cases presented in the previous sections make use of the same functionality—sharing the WorkerContext across vertices, but have an important difference: while in the first use case, vertices only read the shared data structure, in the second use case, they also modify it. This difference requires your attention as Giraph users, as it impacts the thread safety of your programs.

You must have noticed that, even though this entire book is dedicated to designing and implementing parallel algorithms, this is the first time thread safety is discussed. This is normally an important concern and often a source of problems when writing parallel programs. But the core Giraph programming model that you have used up until now purposefully hid any complex synchronizations and thread-safety issues from you. When you write a vertex-centric program, you do not have to worry whether vertices are executed at the same time or in what order they compute. In the BSP model, you assume that every vertex computation is independent and synchronization is explicit through messages.

However, with the introduction of the WorkerContext, you are exposing a bit more of the internal mechanism of Giraph. This offers the benefits you saw, but requires you to be careful. In particular, you must now be aware that a WorkerContext instance may be accessed by multiple threads at the same time. Recall that Giraph splits the graph into partitions and a single worker machine may be holding multiple partitions. The processing of each partition may occur on a separate thread, leveraging any opportunities for parallelization within the same worker machine.

Therefore, you have to take care to ensure thread safety for your applications. In some cases, you may only be reading the data structures maintained by the WorkerContext, so it is easier to ensure thread safety. The preceding example—where you read data from an external service and made it available for reading to all vertices—falls under this category. In other cases, like the performance-monitoring example, you may also be modifying the WorkerContext object. In this example, you do this with the incComputations() method. You may have already noticed that you declared this method as synchronized since you expected this method to be called by different threads. This ensures consistent access and modification of the performance statistics. As your applications start maintaining more complex data structures inside the WorkerContext, thread safety is something to pay attention to.

■ **Note**   You do not need to synchronize the preApplication(), postApplication(), preSuperstep(), and postSuperstep() methods that you implement. Giraph calls these methods exactly once per superstep.

In this section, you revisited the concept of partitioning and distributing the graph and learned how you can leverage it in your benefit by doing per-working computations. You were presented with only a small sample of functionality that you can implement with the WorkerContext. You may become more creative and come up with advanced ways to use it. Often, doing this requires a bit more information and understanding about your algorithm. In all cases, you should remember that lifting some shared process or shared data from your main application logic and putting in the WorkerContext could improve the performance of your algorithm. Next, you go even further and discover how you can actually control the way Giraph partitions the graph across the workers.

Controlling Graph Partitioning

So far, you have not cared about how Giraph splits the graph into partitions, as this does not affect how you program our algorithm. If you think about it, you only specify per-vertex computation and message exchanges at a logical level. It does not really matter where the vertices are located or how these messages are exchanged. Even when you performed per-worker computations in the previous section, you did not care about which vertices a worker holds; Giraph abstracted these details.

In this section, you look a bit more into the details of how Giraph parallelizes the processing, and in particular, you look at how Giraph partitions a graph. You will see that even though the way Giraph partitions the graph across workers does not affect the correct execution of your algorithm, it may have implications with respect to the performance of the execution of a graph algorithm. You will briefly look at how intelligent partitioning can improve performance and then focus on the facilities that Giraph provides for users to control the partitioning of the graph at will.

The Importance of Partitioning

Before getting into the details of the Giraph API, let's look at why you should care about graph partitioning. Graph partitioning defines how the different vertices of the graph are assigned to partitions and affects the performance of the algorithm execution in different ways. First, it directly affects the amount of network traffic incurred on your cluster during the execution of your graph algorithm. To see this, consider the example graph shown in Figure 9-2; let's look at the impact of different partitioning strategies on network traffic.

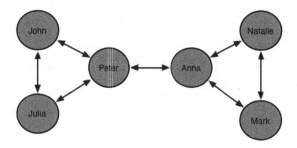

Figure 9-2. *An example graph representing a social network with six users*

Imagine that your cluster consists of only two worker machines, A and B, and that Giraph partitions the graph across the workers in a random way. In particular, Giraph computes a hash of the vertex ID, in this case the name of the user. If there is an even number, it places the vertex on worker A, and if it is odd, on worker B. This is pretty close to deciding randomly where to place vertices. It is like having Giraph flipping a coin and placing the vertex in one of the two workers. In fact, this is the default partitioning strategy of Giraph. Let's assume that the resulting placement is like the one shown in Figure 9-3.

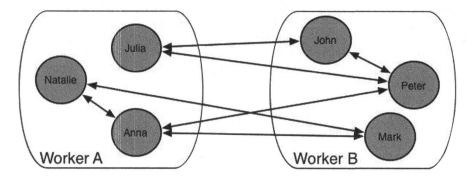

Figure 9-3. *An example partitioning of the six-node social graph across two worker machines*

HASH PARTITIONING IN GIRAPH

The default partitioning strategy in Giraph is *hash partitioning*. Assume that Giraph must partition the graph across P partitions. For every vertex, Giraph computes the hash code H of the vertex ID and the computes the number $H \bmod P$, the remainder of H when divided with P. The result of this operation is the partition to assign the vertex.

Hash partitioning is a very common data partitioning technique. It is easy to compute, and furthermore, it results in an even distribution of data items, in this case vertices, across the partitions.

Now, assume that during the execution of a superstep of your algorithm, every vertex sends a message to each of its neighbors. This is a very common pattern; for instance, the PageRank algorithm that you saw in Chapter 3 behaves like this. It is easy to see from this figure that 10 of the messages will cross the machines, incurring network traffic on your cluster. The rest of the messages, like messages sent from Natalie to Anna, will be exchanged through the main memory of each worker.

Next, let's see what happens if you do not leave the placement of vertices to fate by flipping a coin, but rather decide in a more intelligent and informed manner. In particular, you will try to ensure that each vertex has most of its friends on the same worker. You are shown an alternative partitioning in Figure 9-4. You will assume, for now, that you have a way to instruct Giraph to place vertices however you want.

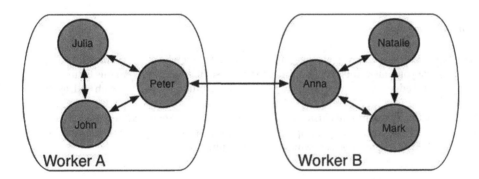

Figure 9-4. *An alternative partitioning across the two worker machines that causes less network traffic*

It is obvious that this partitioning is more favorable when it comes to the network traffic incurred. There is only one graph edge crossing the two machine workers; therefore, this results in only two messages traversing the network: one from Peter to Anna and one from Anna to Peter.

Even though for such a small graph the different partitionings may not make a visible difference in terms of performance, when you start using real, large graphs, partitioning will matter. A large number of messages across machine workers can place a big burden on your network, potentially impacting scalability and increasing the processing time of your Giraph jobs.

Note that finding a good partitioning is a whole area of research in graph management and may depend on several factors, such as the type of graph and algorithm at hand. Studying different partitioning techniques is beyond the scope of this chapter. Instead, here you focus on the Giraph API, which allows you to plug different partitionings. That is, once you have found an appropriate technique for mapping vertices to partitions, you are provided an easy way to plug it into Giraph and enforce the mapping during execution. You see this in the following subsections.

Implementing Custom Partitioners

As you saw, the default hash partitioning technique that Giraph uses does not take into consideration the incurred network traffic, so you may want to replace it with a more intelligent partitioning. In general, you will be doing this by implementing an interface called WorkerGraphPartitioner that is responsible for assigning vertices to partitions.

To illustrate the use of the interface, you will use the example of the web graph and describe an alternative partitioning technique that is commonly used when processing web graphs like the one shown in Figure 9-5.

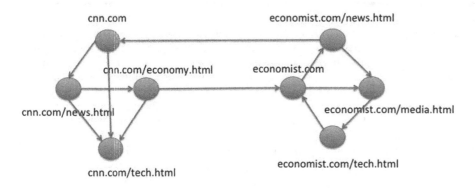

Figure 9-5. *An example web graph*

Recall that a web graph describes how web pages are connected through links. In Chapter 3, you saw examples of algorithms that you may want to run on the web graph, such as PageRank. In this figure, vertices represent the web pages and edges represent links from one page to another. You use the web URL as the ID of a vertex in Giraph.

Assume for simplicity that our compute cluster again consists of two worker machines, A and B. As before, if you use the default hash partitioning, Giraph places the vertices randomly, and the partitioning would look like Figure 9-6, with several edges crossing the two machines.

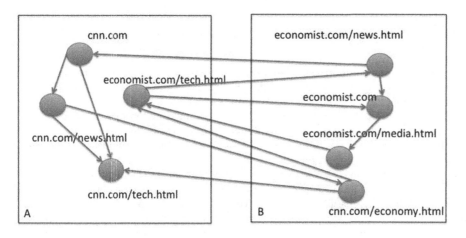

Figure 9-6. *A hash-based partitioning of the web graph across two workers*

In this web graph, you can use a better approach. Remember that in the example with the Twitter graph, the intelligent partitioning was based on the idea that you should do your best to place a vertex along with its neighbors on the same worker. In the small Twitter graph, this was simple, and you could easily do this assignment intuitively and just by looking at the graph. Here, you will use a more principled strategy of deciding where to place a vertex representing a web page. In particular, the strategy is based on the fact that most of the links on a web page are to other web pages that are usually located under the same domain name. For example, most of the links from `cnn.com/index.html` are to pages such as `cnn.com/technology.html` and `cnn.com/economy.html`, which are under the `cnn.com` domain. This must have already given you a hint as to how you are going to partition the graph. You are going to use a technique like hash partitioning, but instead of hashing the vertex ID, you are going to hash the domain prefix only; for instance the prefix `cnn.com`. This ensures that vertices with URLs that fall under the same domain are also placed on the same worker. This partitioning is illustrated in Figure 9-7.

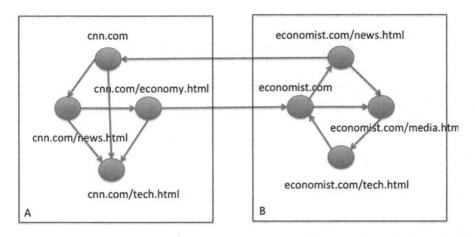

Figure 9-7. *A partitioning of the same web graph based on the domain prefix of each web page*

Next, you look at how you are going to implement the `WorkerGraphPartitioner` to apply this logic. But before getting to the core of the partitioner, let's describe a few other auxiliary but necessary interfaces that you have to implement first.

First, you have to implement the `GraphPartitionerFactory` interface. A `GraphPartitionerFactory` object is responsible for creating the instances of the actual partitioner implementations. In particular, Giraph uses the `GraphPartitionerFactory` to instantiate a `MasterGraphPartitioner` and `WorkerGraphPartitioner` object to perform all the main operations. In Listing 9-9, you can see the only two methods of the interface that you have to implement.

Listing 9-9. The GraphPartitionerFactory Interface

```
public interface GraphPartitionerFactory<I extends WritableComparable,
    V extends Writable, E extends Writable> extends
    ImmutableClassesGiraphConfigurable {

  MasterGraphPartitioner<I, V, E> createMasterGraphPartitioner();   #1
  WorkerGraphPartitioner<I, V, E> createWorkerGraphPartitioner();   #2
}
```

#1 Instantiates and returns a MasterGraphPartitioner object

#2 Instantiates and returns a WorkerGraphPartitioner object

Before going into the particular implementations of these interfaces, we will first describe their role and show the individual methods that you have to implement. After this, you implement them, one by one, to realize the actual hash partitioning technique.

The first interface, the `MasterGraphPartitioner`, is mainly responsible for setting up and maintaining the necessary data structures that hold all the partitioning information. Its methods are shown in Listing 9-10. They are explained in detail through an example in a bit.

Listing 9-10. The MasterGraphPartitioner interface

```
public interface MasterGraphPartitioner<I extends WritableComparable,
    V extends Writable, E extends Writable> {

  Collection<PartitionOwner> createInitialPartitionOwners(
      Collection<WorkerInfo> availableWorkerInfos, int maxWorkers); #1

  Collection<PartitionOwner> generateChangedPartitionOwners(
      Collection<PartitionStats> allPartitionStatsList,
      Collection<WorkerInfo> availableWorkers,
      int maxWorkers,
      long superstep);                                              #2

  Collection<PartitionOwner> getCurrentPartitionOwners();          #3

  PartitionStats createPartitionStats();                           #4
}
```

#1 Create PartitionOwner objects based on available workers

#2 Called at the beginning of each superstep, to potentially create new PartitionOwners

#3 Returns the current list of PartitionOwner objects

#4 Returns and object that will hold statistics about the partitions

The most important data structure here is that of the `PartitionOwner`. A `PartitionOwner` encapsulates information about the partition itself, but also about the worker machine that *owns* the partition; that is, the worker machine where the partition is located. As you see, a `MasterGraphPartitioner` is responsible for creating a collection of `PartitionOwner` objects that typically describe the current configuration of your scenario; that is, the set of workers that you are going to run a job on and the number of partitions that you want to split the graph into.

The `PartitionOwner` is actually an interface that can have different implementations, depending on the kind of metadata that you want to maintain about a partition owner. Here, you do not look into the details of the `PartitionOwner` interface; rather you use the `BasicPartitionOwner`, an implementation already provided by Giraph. The main information that a `BasicPartitionOwner` holds is the ID of the partition that it corresponds to and information about the worker where the partition is located. This implementation of the `PartitionOwner` should suffice for most scenarios.

But now let's look at how you are actually going to use these to implement a hash-based partitioner. First, you implement the `GraphPartitionerFactory` interface. The implementation is quite simple; it's shown in Listing 9-11.

Listing 9-11. An Example GraphPartitionerFactory Implementation

```
public class HashPartitionerFactory<I extends WritableComparable,
    V extends Writable, E extends Writable>
    implements GraphPartitionerFactory<I, V, E> {

  private ImmutableClassesGiraphConfiguration conf;                #1

  @Override
  public MasterGraphPartitioner<I, V, E> createMasterGraphPartitioner() {
    return new HashMasterPartitioner<I, V, E>(getConf());          #2
  }

  @Override
  public WorkerGraphPartitioner<I, V, E> createWorkerGraphPartitioner() {
    return new HashWorkerPartitioner<I, V, E>();                   #3
  }

  @Override
  public ImmutableClassesGiraphConfiguration getConf() {
    return conf;                                                   #4
  }

  @Override
  public void setConf(ImmutableClassesGiraphConfiguration conf) {
    this.conf = conf;                                             #5
  }
}
```

#1 Object holding the job configuration

#2 Returns an instance of the hash-based MasterGraphPartitioner implementation

#3 Returns an instance of the hash-based WorkerGraphPartitioner implementation

#4 Returns the configuration object

#5 Called upon initialization of the factory to set the configuration object

You see in this implementation that the factory simply instantiates two objects: of type MasterGraphPartitioner and WorkerGraphPartitioner. Next, let's look at the implementation of the HashMasterPartitioner in more detail to understand the typical way of setting up the PartitionOwner data structures (see Listing 9-12).

Listing 9-12. The Implementation of the HashMasterPartitioner

```
public class HashMasterPartitioner<I extends WritableComparable,
    V extends Writable, E extends Writable> implements
    MasterGraphPartitioner<I, V, E> {

  private ImmutableClassesGiraphConfiguration conf;                #1
  private List<PartitionOwner> partitionOwnerList;                 #2

  public HashMasterPartitioner(ImmutableClassesGiraphConfiguration conf) {
    this.conf = conf;                                             #3
  }
```

233

```
  @Override
  public Collection<PartitionOwner> createInitialPartitionOwners(        #4
      Collection<WorkerInfo> availableWorkerInfos, int maxWorkers) {
    int partitionCount = PartitionUtils.computePartitionCount(           #5
        availableWorkerInfos, maxWorkers, conf);
    List<PartitionOwner> ownerList = new ArrayList<PartitionOwner>();
    Iterator<WorkerInfo> workerIt = availableWorkerInfos.iterator();
    for (int i = 0; i < partitionCount; ++i) {
      PartitionOwner owner =                                            #6
        new BasicPartitionOwner(i, workerIt.next());
      if (!workerIt.hasNext()) {                                        #7
        workerIt = availableWorkerInfos.iterator();
      }
      ownerList.add(owner);                                             #8
    }
    this.partitionOwnerList = ownerList;
    return ownerList;
  }

  @Override
  public Collection<PartitionOwner> getCurrentPartitionOwners() {
    return partitionOwnerList;
  }

  @Override
  public Collection<PartitionOwner> generateChangedPartitionOwners(
      Collection<PartitionStats> allPartitionStatsList,
      Collection<WorkerInfo> availableWorkerInfos,
      int maxWorkers,
      long superstep) {
    return PartitionBalancer.balancePartitionsAcrossWorkers(
        conf,
        partitionOwnerList,
        allPartitionStatsList,
        availableWorkerInfos);
  }

  @Override
  public PartitionStats createPartitionStats() {
    return new PartitionStats();
  }
}
```

#1 Holds the configuration

#2 Keeps the list of PartitionOwner objects

#3 When you construct a HashMasterPartitioner, you pass it the job configuration

#4 Called upon before execution, to create the initial set of PartitionOwner objects

#5 Get the number of partitions, usually a configuration option

#6 For each partition, create a partition owner with the next available worker

#7 If you have assigned partitions to all workers, then start over

#8 Once created, add the PartitionOwner to the list you maintain

The most important method of the HashMasterPartitioner implementation is the createInitialPartitionOwners(). This method first calculates the number of partitions you want to have. This number is usually a configuration parameter and could depend on factors such as the total number of cores you have in your cluster. After this, it creates a PartitionOwner object for each partition and assigns to it an available worker in a round-robin fashion. Here, you could apply your own logic. For instance, if information about the load of each worker was available through some external service, you could incorporate it in your assignment of partitions to workers. At the end, the method simply returns the list of PartitionOwner objects created.

Then you need to implement the generateChangedPartitionOwners() method. Giraph calls this method after the execution of a superstep and gives you the opportunity to change the initial assignment of partitions to workers, if you wish. This could be, for instance, because you noticed that one of the workers crashed, in which case you need to assign its partitions to another one. Another possible scenario is that you notice that a particular worker ended up having more partitions than it can handle, in which case you may want to offload some of the processing to another one. In all of these cases, after the end of a superstep you have a chance to adjust the partitioning by choosing new locations for the partitions. Notice, though, that you are not allowed to change the assignment of vertices to partitions.

In this particular implementation, the decision to modify the partition owners is delegated to a special helper class that you call PartitionBalancer. In brief, the PartitionBalancer ensures that each worker is assigned pretty much the same amount of work, providing good cluster utilization. Discussion of the balancer and why it is important is deferred until the next section. There you will also see how to modify the partitioning through the generateChangedPartitionOwners() at runtime.

The createPartitionStats() method returns an object of type PartitionStats that holds useful statistics about the partition. Such statistics include the number of vertices and edges in a partition and the number of messages sent by vertices in this partition. Such statistics may be useful for applying sophisticated partitioning techniques that depend on runtime behavior. More specifically, you can use the information maintained in this class to decide how to update the partition owners when Giraph calls the generateChangedPartitionOwners() method. Giraph already provides a basic PartitionStats class that includes the most common statistics, and you will keep using this. You are free to extend this class to maintain more information if that suits your particular scenario.

Now that you have implemented all the auxiliary classes, you are ready to implement the main logic of the partitioner, which is defined by the WorkerGraphPartitioner interface. Once Giraph has used the partitioner factory and the HashMasterPartitioner to create the collection of PartitionOwner objects, it is ready to assign vertices to the constructed partition owners. This is the job of the WorkerGraphPartitioner. Its methods are illustrated in Listing 9-13.

Listing 9-13. The WorkerGraphPartitioner Interface

```
public interface WorkerGraphPartitioner<I extends WritableComparable,
    V extends Writable, E extends Writable> {

  PartitionOwner createPartitionOwner();                      #1

  PartitionOwner getPartitionOwner(I vertexId);               #2

  Collection<PartitionStats> finalizePartitionStats(          #3
      Collection<PartitionStats> workerPartitionStats,
      PartitionStore<I, V, E> partitionStore);
```

```
PartitionExchange updatePartitionOwners(                          #4
    WorkerInfo myWorkerInfo,
    Collection<? extends PartitionOwner> masterSetPartitionOwners,
    PartitionStore<I, V, E> partitionStore);

Collection<? extends PartitionOwner> getPartitionOwners();     #5
}
```

#1 Creates and empty PartitionOwner object

#2 Maps a vertex ID to a PartitionOwner

#3 Called at the end of a superstep

#4 Determines which partitions must be sent from a specific worker

#5 Returns the current list of ParttionOwner objects

The most important method is the getPartitionOwner() method that takes a vertex ID as an input and returns the PartitionOwner that this vertex corresponds to. Giraph uses this method to decide where to place vertices. Next, all the methods of the GraphWorkerPartition interface are explained through the example implementation of the hash-based partitioner. Listing 9-14 shows the implemented methods.

Listing 9-14. The HashWorkerPartitioner Implementation

```
public class HashWorkerPartitioner<I extends WritableComparable,
    V extends Writable, E extends Writable>
    implements WorkerGraphPartitioner<I, V, E> {

  protected List<PartitionOwner> partitionOwnerList =            #1
      Lists.newArrayList();

  @Override
  public PartitionOwner createPartitionOwner() {
    return new BasicPartitionOwner();                            #2
  }

  @Override
  public PartitionOwner getPartitionOwner(I vertexId) {
    URL url = new URL(vertexId.toString());                      #3
    String domain = url.getHost();                              #3
    return partitionOwnerList.get(                               #4
        Math.abs(domain.hashCode() % partitionOwnerList.size()));
  }

  @Override
  public Collection<PartitionStats> finalizePartitionStats(
      Collection<PartitionStats> workerPartitionStats,
      PartitionStore<I, V, E> partitionStore) {
    return workerPartitionStats;
  }
```

```
    @Override
    public PartitionExchange updatePartitionOwners(
        WorkerInfo myWorkerInfo,
        Collection<? extends PartitionOwner> masterSetPartitionOwners,
        PartitionStore<I, V, E> partitionStore) {
      return PartitionBalancer.updatePartitionOwners(partitionOwnerList,
          myWorkerInfo, masterSetPartitionOwners, partitionStore);
    }

    @Override
    public Collection<? extends PartitionOwner> getPartitionOwners() {
      return partitionOwnerList;
    }
}
```

#1 Keep a list of the current PartitionOwner objects

#2 Creates an empty BasicPartitionOwner object

#3 Extract domain name from URL

#4 Applied the hash partitioning logic on the domain

The core of the hash-based partitioning logic is inside the getPartitionOwner() method. Given a vertex ID—that is, a web page URL—this method extracts the domain name from the URL. Then it computes the hash code of the domain and calculates the remainder of the division with the number of partitions. The result indicates the partition where the specified vertex should go.

Note, in this case the partitioning logic is quite simple, but inside this method you could put arbitrary logic. You could even be connecting to some external service that is responsible for maintaining the partitioning of your graph.

Furthermore, you also need to implement the updatePartitionOwners() method. Giraph calls this method at the end of a superstep and uses it to determine whether and which partitions need to be sent from other workers. This decision is delegated to the PartitionBalancer as well, which you will see in a bit.

Finally, now that you have implemented all the necessary pieces of a custom partitioner, you may wonder how you actually instruct Giraph to use them. To enable the use of your custom partitioner, you only need to set the giraph.graphPartitionerFactoryClass parameter that takes the full class name of the partitioner factory as value. As usual, you can do this through the command line.

In this particular scenario, you need to add the -ca giraph.graphPartitionerFactoryClass=Hash PartitionerFactory option to the command line when you start your job. As always, Giraph takes care of the rest.

Partition Balancing

In the discussion on partitioning up until now, you only learned about how the partitioning affects the network traffic and eventually your application performance. There is another aspect of partitioning that can affect performance, called *partition balancing*. In general, a partitioning is balanced when all the workers contain pretty much the same number of vertices or edges. But first, let's look at why balancing is important.

Consider the graph you saw previously and imagine a third alternative partitioning across the workers, like the one shown in Figure 9-8.

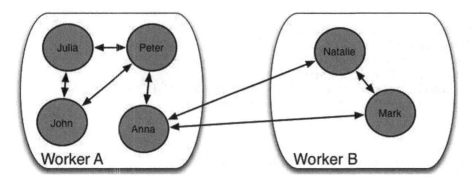

Figure 9-8. *An example of unbalanced partitions*

This is a perfectly valid way to partition your graph, but it presents an issue. Notice that worker A contains many more vertices than worker B; in this case, double the number. Because the larger the number of vertices on a machine is, the larger its CPU load is as well; such an uneven distribution of the graph across the workers also causes imbalance in terms of the CPU load. As a result, some of your workers—in this case, worker B—are underutilized and you do not get the maximum performance out of your cluster.

On the contrary, distributing the workload evenly makes efficient use of your cluster resources and reduces the total processing time. The PartitionBalancer is a helper class with methods that find the right placement of the partitions, so that the partitioning is balanced. Essentially, it makes use of the ability the partitioning API gives you to modify the location of the partitions at runtime through the generateChangedPartitionOwners() method of the GraphMasterPartitioner and the updatePartitionOwners() method of the WorkerGraphPartitioner as you saw previously.

Importantly, the PartitionBalancer supports different types of balancing. The first type of balancing, called *vertex balancing*, ensures that every worker holds approximately the same number of vertices. Apart from this, it also supports *edge balancing*, which tries to ensure that all workers contain approximately the same number of edges. Which one is more appropriate depends on the type of algorithm that you want to run and whether the CPU load of each worker depends on the number of vertices or the number of edges it holds.

As an example, consider the PageRank algorithm that you saw in Chapter 4, used to rank web pages. In PageRank, every vertex receives messages from its neighbors with each message carrying a number. For every message it receives, the vertex performs an operation, adding the message to its local rank value. Therefore, the total amount of work that a vertex does depends mainly on how many neighbors, or edges, it has. Therefore, to distribute workload even across the workers, you would like that every worker has approximately the same number of edges.

In general, choosing between vertex and edge balancing requires some insight about the computation and communication pattern of your algorithm. Often, simply by comparing the computation and communication pattern of your algorithm to some other, say PageRank, you may be able to understand which type of balancing fits better. In any case, balancing your partition correctly can impact the performance of your algorithm and the utilization of your cluster resources positively.

Summary

The core programming API of Giraph hides the details of how it parallelizes the processing of a graph. At the same time, by exposing this process to the user, you get a chance to write more efficient applications.

- The need to share data across vertices may arise in several application scenarios, and at first glance, breaks the vertex-centric computation model.

- The WorkerContext becomes very handy in these cases. It allows you to do computations on a per-worker basis and share a data structure across all vertices on the same worker. This can simplify your application but also improve performance.

- Do not forget that using the WorkerContext may raise thread-safety issues. Some application may be modifying the data held by the WorkerContext and this can happen concurrently by different threads. Be sure to declare methods as synchronized to avoid race conditions.

- Graph partitioning can have a significant impact on the scalability and performance of your application.

- By default, Giraph uses hash partitioning, a common approach that achieves even distribution of the graph across machine, but can result in high network traffic.

- Partitioning the graph in an intelligent manner is often dependent on the graph and application at hand.

- Once you have settled on the right partitioning scheme for you scenario, use the WorkerGraphPartitioner and all the related interfaces to implement a custom partitioner.

The last few chapters focused on the internals of Giraph and on the more advanced features. In the next chapter, you continue in the same spirit. You revisit reading and writing graphs from and to various storage systems, but this time you go over scenarios of more advanced storage systems, such as HBase that enable you to connect Giraph to any data source.

CHAPTER 10

■ ■ ■

Advanced IO

This chapter covers

- Using Hive tables for IO

- Accessing data through Gora

In this chapter, you continue with the more advanced features of Giraph, focusing on ways to read input data and write output data. Recall that as your input graph is usually stored in some special format on a storage system, Giraph must be instructed on how to read data from your storage system and how to convert it to its own internal representation; that is, vertices and edges using the `VertexInputFormat` and `EdgeInputFormat` implementations. For example, in Chapter 7, you described the basics of reading and writing data from the Hadoop Distributed File System (HDFS) that are typically in text format, a very common scenario for storing graph data.

However, HDFS is not the only option. There is a variety of storage systems and different formats in which you might store your data. For instance, you might use a NoSQL type of storage systems, like Hive and HBase, to store your data in the form of semistructured tables. Accumulo is another distributed key-value store that allows you to store data in a table format and provides access control mechanisms to enable security-related policies. In-memory data stores, such as Redis, have also become popular because they allow low-latency data access.

To facilitate all of these different storage options, Giraph provides an API that makes it easy to extend to different formats and storage systems. Aside from this, it also provides implementations of these APIs for a variety of storage systems.

In this chapter, you look at two specific cases of storage systems: Hive and Gora. Note that the goal of this chapter is not to present an exhaustive list of systems or data formats. Rather, by the end of this chapter, you will know how to customize the existing API implementation or extend it to cover your particular application scenario.

Accessing Data in Hive

First, you will look at table-based formats— Hive, in particular. In general, table-based formats for storing data are quite common. Hive allows you to view your data as a semistructured table and to query data using an SQL-like query language. Hive runs on top of the Hadoop clusters. Hive tables are stored on HDFS, and Hive queries are executed as Hadoop jobs, providing a scalable platform for storing and querying semistructured table data. For these reasons, it is a popular storage system.

A Hive table stores data in rows, with each row consisting of a number of columns. In, Hive the number and the type of the data contained in the columns of a row is agreed in advance; this is the schema of the table.

Here, you see how a Hive table may store information representing an input graph. Now, just like with the text-based input formats that you saw in the early chapters, a Hive table might store data about the input graph in an edge-based format or a vertex-based format. In an edge-based format, a Hive row contains all the information required to construct an edge; that is, the identities of the two endpoints of the edge and its edge value. In a vertex-based format, a Hive row may contain all the information required to construct a vertex; that is, the vertex ID, its value, and information about all the edges of the vertex. For instance, in a vertex-based format, a single Hive column may contain a list of all the vertex edges.

Reading Input Data

Let's assume that you are running a microblogging service where users can post articles and share them with other users. Typically, whenever somebody makes such a post, you want store it for further analysis; for instance, find which users post the most articles, which people interacted the most over the last week, what are the most common topics users post about, and other useful metrics. Table-based stores are a good choice for storing these data. Table 10-1 shows an example of how you might store data there.

Table 10-1. *A Table Storing Article Share Events from Our Example Service*

user	shared_with	date	article
George	John	2015-03-01	www.cnn.com/big-data-economy.html
Mary	Nick	2015-02-14	www.howto.com/analyze-graphs.html
Mark	George	2015-04-29	www.techcrunch.com/monetize-data.html
John	Maria	2015-03-02	www.cnn.com/big-data-economy.html
Maria	Mark	2015-03-03	www.cnn.com/big-data-economy.html
...	...	...	...
...	...	...	...

Even though it might not be immediately obvious, the users that are active on your service form a type of network. Whenever user A shares an article with another user, this makes for a connection between two users. Now let's assume that George is a popular and frequent user of your service and that you are interested in finding out how fast information that George posts can travel on your service. In other words, once George posts an article, how many times does it have to be shared between users before it reaches everybody?

This type of analysis may sound familiar; you essentially want to compute distances between users. In Chapter 4, you saw how to write a Giraph application that computes the shortest paths between users. This is exactly what you want here too. Since you have already seen how to compute this metric, in this chapter you will only look into how to form the input graph from the input table data.

Now, recall that Giraph provides two basic kinds of input formats: edge-based and vertex-based. In edge-based formats, your data contains information about the connections of vertices in the graph; whereas in vertex-based formats, the data contains information about the vertices themselves. If you look at Table 10-1, you see that it stores information about which user shares articles with a particular other user; that is, the connection among users.

As with text-based formats and HDFS, Giraph already abstracts most of the details of reading data from Hive tables and provides a simple interface that you need to implement. Implementing this interface essentially requires you to specify how to create an edge from a table row, like the ones you saw in Table 10-1. Recall that the constituents of an edge definition are its source vertex ID, its destination vertex ID, and its edge value. These are the pieces of information that you have to extract from a Hive table row.

But let's look at the exact methods that you need to implement. In a moment, you will see them in action with an example implementation for your particular scenario. Listing 10-1 shows the SimpleHiveToEdge abstract class that Giraph provides.

Listing 10-1. The SimpleHiveToEdge Class for Converting Hive Rows to Edge Objects

```
public abstract class SimpleHiveToEdge<I extends WritableComparable,
    E extends Writable> extends AbstractHiveToEdge<I, E> {

  public abstract I getSourceVertexId(HiveReadableRecord hiveRecord);

  public abstract I getTargetVertexId(HiveReadableRecord hiveRecord);

  public abstract E getEdgeValue(HiveReadableRecord hiveRecord);
}
```

This is the simplest class that you may have to implement to read data from Hive. The first method that you have to implement, getSourceVertexId(), receives as input an object of type HiveReadableRecord, which is an abstraction of a Hive row. The role of the method is to construct the source vertex ID from the HiveReadableRecord. You will see how to do this in a moment. The second method, getTargetVertexId(), has a similar role, to extract the ID from the hive record. Finally, through the getEdgeValue() method, you are defining what the edge value is given an input Hive record. Notice that as with the input formats you saw in earlier chapters, this input format has parameters too: the type of the vertex ID (I), the type of the edge value (E).

Now, let's see all of these methods in action. In this particular scenario, there are two columns in the Hive table that you are interested in: the "user" column that is going to be the source of an edge and the "shared_with" column that is going to be the target of an edge. The vertex IDs represent the name of the users, so in your implementation, they are of type Text. Listing 10-2 shows the implementation.

Listing 10-2. An Example Implementation for Edges with IntWritable Vertex IDs and a NullWritable Edge Value

```
public class TextNullHiveRowToEdge
    extends SimpleHiveToEdge<Text, NullWritable> {

  @Override
  public NullWritable getEdgeValue(HiveReadableRecord hiveRecord) {
    return NullWritable.get();
  }

  @Override
  public Text getSourceVertexId(HiveReadableRecord hiveRecord) {
    return new Text(hiveRecord.getString(0));
  }

  @Override
  public Text getTargetVertexId(HiveReadableRecord hiveRecord) {
    return new Text(hiveRecord.getString(1));
  }
}
```

First, notice that all the methods are passed as input an object of type HiveReadableRecord, which represents a Hive record, essentially a row in the table. In your implementation, you will use the methods of the HiveReadableRecord class to get access to the different columns of a Hive row.

Let's look at the implementation of the methods one by one. The first method, getEdgeValue(), must return an object representing the edge value. In this particular scenario, you do not want to assign a particular value to an edge; rather it is of type NullWritable, so the method simply returns a NullWritable object through the static NullWritable.get() method.

Next, you must read the source and target IDs. You already know that the source ID of an edge is placed in the first column of a row and the target ID is placed in the second column. A HiveReadableRecord allows you to retrieve the data stored in a row by using the index of the column, with indices starting at column 0. Listing 10-3 shows all the methods of the HiveReadableRecord interface.

Listing 10-3. Use the Methods of the HiveReadableRecord Interface to Access Data of Various Types from the Columns of a Hive Table Row

```
public interface HiveReadableRecord {
  int numColumns();
  int numPartitionValues();
  HiveType columnType(int index);
  Object get(int index, HiveType type);
  boolean getBoolean(int index);
  byte getByte(int index);
  short getShort(int index);
  int getInt(int index);
  long getLong(int index);
  float getFloat(int index);
  double getDouble(int index);
  String getString(int index);
  <K, V> Map<K, V> getMap(int index);
  <T> List<T> getList(int index);
  boolean isNull(int index);
}
```

Another aspect you must be aware of is the type of data that each column stores. Each column may store data of different types. For instance, usernames may be stored in a column as strings; whereas a date may be stored in a column in the form of a timestamp of type long. Depending on the data type of each column, you are going to use the right method of the HiveReadableRecord interface. In the previous listing, you see that the interface provides methods for accessing data of all basic types, such as integers, floats, strings, and structures such as lists and maps.

Going back to your implementation, you now see how you can implement the getSourceVertexId() and getTargetVertexId() methods. You use the getString() method of the HiveReadableRecord interface to retrieve the columns with indices 0 and 1 accordingly, as these are of type string and then put them inside a Text object. And you are done.

The only thing that is missing now is instructing Giraph to use this particular implementation of an input format, but also to tell it which Hive table to read data from. You are going to do this using the familiar -eif command line parameter when you start the Giraph job.

Note that the implementation of the actual computation was not discussed here. The important thing is that Giraph allows you to run the same computation on data stored in different storage systems with no changes. All you have to do is write an input format; the rest remains the same. In fact, in many cases, you may not even have to write your own implementation. The Giraph code base already includes a variety of input format implementations to read data from Hive tables. One of these may already suit your particular scenario.

In this example, the input graph was stored in an edge-based format. As mentioned, it is possible to have vertex-based input formats in Hive too. Even though we do not go into the details, the concept of a vertex-based input format in Hive does not differ much from the text-based format that you saw in previous chapters. Giraph provides an interface to convert a single Hive row, just like with text lines in text-based formats, to a vertex. Listing 10-4 shows the interface that you would have to implement.

Listing 10-4. The SimpleHiveToVertex Class for Converting Hive Rows to Vertex Objects

```
public abstract class SimpleHiveToVertex<I extends WritableComparable,
    V extends Writable, E extends Writable>
    extends AbstractHiveToVertex<I, V, E> {

  public abstract I getVertexId(Row record);

  public abstract V getVertexValue(Row record);

  public abstract Iterable<Edge<I, E>> getEdges(Row record);
}
```

Similar to vertex-based formats, in this case you have to specify how to extract the following information from a row: the ID of a vertex, the value of the vertex, and the edges of the vertex. Just like the edge-based format, you can access the columns of the Hive table using the methods provided by the HiveReadableRecord interface. As an exercise, we suggest that you implement a vertex-based input format that builds a graph from Table 10-2.

Table 10-2. *Example Input Hive Table in Vertex-Based Format*

user	shared_with
George	{John, Mark}
Mary	{Nick, Maria, Mark, George}
Mark	{George}
John	{Maria, Helen, George}
Maria	{Mark, John, Nick, Helen}
...	...
...	...

This table contains similar information as the previous one; only in this case, all the information necessary to create a vertex: the vertex ID representing a user and the people who he or she shares with exist in the same row.

Writing Output Data

So far we talked about how to read input data from Hive, but naturally you may want the output of your application to be stored in a Hive table. This way you can query the results of your analysis in an easy way using the Hive Query Language. In this section, you see how to achieve this.

First, let's decide what the output Hive table will look like. In this particular scenario, you compute every user's distance from George. In the simplest form, the output table contains the computed distance for each user (see Table 10-3).

Table 10-3. *Exampke Hive Output Table*

user	value
George	0
Mary	4
Mark	3
John	1
Maria	2
...	...
...	...

The output table contains one row per user. Each row has one column for the username, called "user", that contains data of type string, and one column to store the computed distance, called "value", that contains data of type double. Here, you assume that this table already exists and has this particular schema. If you want to learn more about how to create Hive tables, please refer to the Hive documentation.

Similar to input formats, Giraph abstracts most of the details and makes it easy for you to fill this table with output. In practice, you are performing the inverse actions; you are telling Giraph how to extract information from a vertex table and where to place it in a table row. Let's go over the implementation shown in Listing 10-5.

Listing 10-5. The SimpleVertexToHive Class for Converting Vertex Objects to Hive Rows

```
public abstract class SimpleVertexToHive<I extends WritableComparable,
    V extends Writable, E extends Writable> extends
    AbstractVertexToHive<I, V, E> {

  public abstract void fillRecord(Vertex<I, V, E> vertex,
      HiveWritableRecord record);
}
```

The only method that you need to implement here is the fillRecord() method, which passes as input a Vertex object and a HiveWritableRecord object. Giraph calls this method for every Vertex in your graph and passes the Vertex object as input to this method. It also passes as input a HiveWritableRecord object that represents an output Hive table row. Your responsibility is to specify how to fill that row with information extracted from the Vertex object.

Like the HiveReadableRecord, the HiveritableRecord interface provides a set of methods that make it easy for you to store information of different types in the columns of a Hive row. These are shown in Listing 10-6.

Listing 10-6. The Methods for Modifying a Hive record

```
public interface HiveWritableRecord {
  void set(int index, Object value, HiveType type);
  void setBoolean(int index, boolean value);
  void setByte(int index, byte value);
  void setShort(int index, short value);
  void setInt(int index, int value);
  void setLong(int index, long value);
  void setFloat(int index, float value);
  void setDouble(int index, double value);
```

```
    void setString(int index, String value);
    void setMap(int index, Map value);
    void setList(int index, List value);
    List<Object> getAllColumns();
}
```

Typically, you just need to specify the index of the column that you want to store information and the value that you want to store. Aside from this, you need to ensure that you use the right method, depending on the schema of the table; that is, the type of data stored in each column.

Now, let's consider your particular scenario. The output table has two columns: one of type string, where you store the name of the user, and one of type double, where you store the computed distance for that user. Listing 10-7 shows the implementation.

Listing 10-7. The Example Implementation That Writes a Vertex Object to a Hive Row

```
public class MyVertexToHiveRow extends
    SimpleVertexToHive<Text, DoubleWritable, NullWritable> {

    public void fillRecord(Vertex<Text, DoubleWritable, NullWritable> vertex,
        HiveWritableRecord record) {
      record.setString(0, vertex.getId().toString());
      record.setDouble(1, vertex.getValue().get());
    }
}
```

As with input tables, you use an index to identify which column you are referring to with column indices starting at 0. Keep in mind that you must be aware of the schema of the output table and write the correct values to the correct columns. In cases where you actually attempt to write a value to a column of a different type, this will raise an exception.

And this concludes the implementation of your output format. Now, similar to the input format, you need to instruct Giraph to use this particular implementation of the output format, and tell it what Hive table to write the data to. To do this, you are going to use the –of command-line parameter when you start the Giraph job.

Accessing Data in Gora

Next, let's look at another type of storage system, specifically in-memory storage systems. In-memory storage systems are another popular class of systems because they provide fast access to data. This section discusses Apache Gora, a framework that provides an in-memory data model that also supports data persistence to different underlying storage systems: databases (like MySQL), column stores (like HBase and Cassandra), key-value stores (like Redis), and even simple files on HDFS. For a complete list of the supported storage systems, visit the Apache Gora site.

While Apache Gora provides persistence to such systems, it abstracts the details of how to persist objects, allowing the user to work with the in-memory representation of objects, making access and manipulation of objects much easier from a programming perspective. Typically, the Gora system is set up to map data in the underlying storage system; for instance, an HBase row to an in-memory Java object. From then on, a user may access data from Gora based on an object key, like an index.

In the remainder of this section, you will go over using the Giraph input formats to read data from Gora. Setting up Gora with an underlying persistent storage system is not covered, nor is how to define the mapping between the underlying storage system and the in-memory objects. Instead, you see how to access in-memory Gora objects and convert them to Vertex and Edge objects that represent your graph inside Giraph.

You will use the same application scenario that you used in the previous section. The input data represents a set of users that share articles with each other, forming a social network, and you want to compute the distance of all users from user George in this social network.

Reading Input Data

Let's look at how you can read input data from the Gora in-memory storage system. First, recall that Gora provides an in-memory data model. This means that it presents data to applications in the form of Java objects. While underneath the hood data may persist in different storage systems, as a user you are manipulating the in-memory version of it. The conversion is left to the Gora system itself.

Let's assume for a moment that the in-memory representation of a user in Gora is an object called GoraUser that has the form shown in Listing 10-8.

Listing 10-8. The Example Representation of a User in the Gora in Memory Data Model

```
public class GoraUser {
  String name;
  Map<String, String> sharedArticles;
}
```

This object represents a user and contains information similar to what you saw in the previous section with data stored in Hive. Each user has a name that he or she is identified by and also a set of users with whom they share articles with. The sharedArticles map contains a mapping from a username to an article that was shared with that user. For simplicity, here you assume that a user can share only one article with another user.

As with input format in general, you have to decide whether you need to implement an edge-based or a vertex-based input format. In fact, the Giraph code contains both edge-based and vertex-based implementations for reading data from Gora. Now, you must have noticed that the representation of a user in memory contains all the necessary information to construct a vertex; therefore, a vertex-based input format is the natural choice here.

So let's look at the interface that you have to implement to read vertex-based data from Gora. As expected, the Gora interface is implemented by extending the familiar VertexInputFormat class. Several of the details for setting up a GoraVertexInputFormat are already implemented. The remaining function, shown in Listing 10-9, requires you to essentially create a GoraVertexReader.

Listing 10-9. The GoraVertexInputFormat for Converting In-Memory Objects to Vertex Objects

```
public abstract class GoraVertexInputFormat
    <I extends WritableComparable, E extends Writable>
    extends VertexInputFormat<I, E> {

  @Override
  public abstract GoraVertexReader createVertexReader(InputSplit split,
      TaskAttemptContext context) throws IOException;
}
```

As you saw in Chapter 7, a VertexReader is responsible for performing the bulk of the work when reading input data; that is, for converting input data to the Vertex objects that comprise the graph. A GoraVertexReader, in particular, is responsible for converting an in-memory Gora object to a Vertex. Listing 10-10 shows the single method that you have to implement to be able to read data from Gora.

Listing 10-10. The GoraVertexReader API

```
protected abstract class GoraVertexReader extends VertexReader<I, V> {
    protected abstract Vertex<I, V> transformVertex(Object goraObject);
}
```

Notice that the API provided is quite flexible. It passes a generic Java object to you and expects you to transform it to a vertex. This object is nothing more than the in-memory representation of a user, which you saw earlier, and your implementation of the input format specifies how to translate this to a Giraph Vertex object. Let's look at the implementation in Listing 10-11.

Before getting into the details, notice how this class takes the types of the vertex ID, vertex value, and the edge value as parameters. Similar to the application scenario in the previous section, you represent the ID as an object of type Text, the value of the vertex (that is, the computed distance) as an object of type DoubleWritable, and finally, since you do not care about the value of edges, you use objects of type NullWritable. The implementation is shown in Listing 10-11.

Listing 10-11. Our Example Implementation for Transforming Gora Objects to Vertex Objects

```
public class GoraUserVertexInputFormat
    extends GoraVertexInputFormat<Text, DoubleWritable, NullWritable> {

    public GoraVertexReader createVertexReader(
        InputSplit split, TaskAttemptContext context) throws IOException {
        return new GoraGVertexVertexReader();
    }

    protected class GoraGVertexVertexReader extends GoraVertexReader {

        @Override
        protected Vertex<LongWritable, DoubleWritable, FloatWritable>
        transformVertex(Object goraObject) {
            Vertex<LongWritable, DoubleWritable, FloatWritable> vertex;
            vertex = getConf().createVertex();                              #1
            GoraUser user = (GoraUser) goraObject;                         #2
            Text vrtxId = new Text(user.getName())                        #3
            DoubleWritable vrtxValue = new DoubleWritable(0);             #4
            vertex.initialize(vrtxId, vrtxValue);                        #5
            for (Map.Entry<String, String> entry : user.getSharedArticles()) {
                String userName = entry.getKey();                        #6
                String article = entry.getValue();                       #6
                Edge<Text, NullWritable> edge = EdgeFactory.create(
                        new Text(userName), NullWritable.get());         #7
                vertex.addEdge(edge);                                    #8
            }
        }
        return vertex;                                                   #9
    }
}
}
```

#1 Create an empty vertex object

#2 Cast the input generic object to the GoraUser specific type

#3 Extract the name and create a new vertex Id of type Text out of it

#4 The initial value of the vertex should be 0

#5 You initialize the empty vertex object with its id and value

#6 For every entry in the shared articles map for this user

#7 Create an edge with the key of the entry as the target id

#8 Each newly created edge, you add it to the vertex

#9 At this point you have created our vertex object and can return it

Now, let's dive into the implementation of the API. The first action is to create an empty Vertex object that you fill with information from the Gora object. Recall that Giraph already provides some utility methods to instantiate objects, like edges and vertices. In this case, you can simply call getConf().createVertex() to create an empty, uninitialized Vertex object. Next, you will see how to fill it with the necessary information: its value and its edges.

Next, you need to extract the information for the in-memory object. Since the API input is a generic object, you first need to cast it to the internal representation that you expect it to have; in this case, the GoraUser class. If there is a disagreement between the type that you are trying to cast it to and the actual representation of the objects in memory, this naturally throws an exception at runtime.

First, you extract the ID of the vertex. For this, you are going to use the "name" field of the GoraUser object, assuming that the username uniquely identifies a user. As you have already seen, in Giraph you typically represent IDs of type string as Text objects. Similarly, out the string field, you create a new Text object to represent the vertex ID.

Then, you are going to set the value of the vertex. Recall that in the algorithm that computes distances, the initial value of a vertex is zero. All you have to do is create a DoubleWritable object set to 0. Now, that you have created the ID and the value of the vertex, you can call the Vertex.initialize() method to set these values in your newly created Vertex object.

Finally, you are going to extract the edges of the vertex. Here, you assume that each user object stores the articles that the user shares with others in a map. The vertex reader implementation iterates over all the map entries containing user-article pairs and creates an edge using the EdgeFactory utility class. Each edge created has the corresponding username as the target ID, which is the key of the map entry. In this scenario, you do not need an edge value, so you set it to NullWritable. Once you have created an Edge object, you can add it to the vertex using the familiar Vertex.addEdge() API.

Writing Output Data

Next, you learn how to output. In this case, you need to convert the Giraph representation of vertices and objects to the Gora in-memory model. Again, you are not going to worry about how Gora persists objects— this is left to the underlying system, simplifying your job here. Giraph abstracts the details of this process by providing output format implementations—the GoraVertexOutputFormat and the GoraEdgeOutputFormat— and requires you to implement a couple of methods, making the process easy.

Before getting into the details of the output format implementations, as always, you need to make a choice about whether you need a vertex-based or an edge-based output format. In this case, you want to output the per-vertex distances that the job calculates, so a vertex-based output format is the natural choice. Next, let's look at the API that you have to implement.

Recall that VertexWriters are the way Giraph outputs a Vertex object to any storage system. This is not different here. Giraph abstracts the details into the GoraVertexWriter class, shown in Listing 10-12.

Listing 10-12. The GoraVertexWriter API

```
protected abstract class GoraVertexWriter  extends VertexWriter<I, V, E> {
  protected abstract Object getGoraKey(Vertex<I, V, E> vertex);
  protected abstract Persistent getGoraVertex(Vertex<I, V, E> vertex);
}
```

The interface is quite simple and requires you to implement only two methods. The first method, getGoraKey(), takes the Vertex object as input and returns a key. This is going to be used to uniquely identify this object in the in-memory storage. Normally, the Gora key coincides with the vertex ID, used to uniquely identify the vertex. You will see this in action in a bit.

The second method that you have to implement, getGoraVertex(), takes the Vertex object as input and returns an object that implements the Persistent interface of Gora. Even though you can have your own implementation of the Persistent interface for storing the output, Giraph already provides a generic implementation—called GVertexResult—that allows you to store any information you want about a vertex, such as its ID and value. In the following examples, you will see this put in use.

First, let's take a look at the implementation of the input format shown in Listing 10-13. As mentioned, Giraph abstracts many of the details through the GoraVertexOutputFormat, an abstract class that you have to extend. Call your implementation the GoraUserVertexOutputFormat. As usual, you need to implement the createVertexWriter() method that returns an object implementing the VertexWriter interface that does the bulk of the work. Let's call this GoraUserVertexWriter. Now let's look at the VertexWriter implementation in more detail.

Listing 10-13. Example Implementation for Saving Vertex Objects As Gora Objects

```
public class GoraUserVertexOutputFormat
  extends GoraVertexOutputFormat<Text, DoubleWritable,  NullWritable> {

  @Override
  public VertexWriter<Text, DoubleWritable, NullWritable>
  createVertexWriter(TaskAttemptContext context)
    throws IOException, InterruptedException {
    return new GoraUserVertexWriter();
  }

  protected class GoraUserVertexWriter extends GoraVertexWriter {

    @Override
    protected Object getGoraKey(
        Vertex<Text, DoubleWritable, NullWritable> vertex) {
      String goraKey = vertex.getId().toString();                    #1
      return goraKey;
    }

    @Override
    protected Persistent getGoraVertex(Vertex<Text, DoubleWritable, NullWritable> vertex) {
      GVertexResult userVertex = new GVertexResult();                #2
      userVertex.setVertexId(vertex.getId().toString());             #3
      userVertex.setVertexValue(vertex.getValue().get());            #4
      return userVertex;
    }
  }
}
```

251

#1 Exctract the vertex ID and return its string representation as the Gora key

#2 Create an empty GVertexResult object to hold the output for this vertex

#3 Set its vertex ID using the provided API

#4 Similarly, set its vertex value

First, let's take a look at the implementation of the getGoraKey() method. This method should return an object used to uniquely identify the object in memory. Even though this is not necessary, using the vertex ID is a perfect candidate for this, as you already know that it uniquely identifies a vertex. For the Gora key, you do not need a Writable object, such as Text, so you can just extract the String representation of the key and return it.

Next, let's look into the implementation of the getGoraVertex() method, which performs most of the work. In this method, you need to extract all the necessary information that you want to store in a Vertex object and encapsulate it in an object of a type that Gora understands; that is, an object that implements the Persistent interface. The Giraph source code contains a generic object—called GVertexResult— that implements this interface and allows you to store information in it. We are not going to describe its implementation in detail here; the important aspect is that it abstracts the details of the Persistent interface and provides simple API calls, such as getting and setting the ID, value, and edges of a vertex. The implementation handles the details of converting those fields to the correct underlying representation.

Let's now look at the body of the method implementation. The first action is to create a new object of type GVertexResult that is going to hold the output. Recall that in this particular scenario, the output you care about is the vertex ID and its value; that is, the computed distance. For this, you use the setVertexId() and setVertexValue() methods of the GVertexResult class. Even though you do not show it here, the internal implementations of these methods convert the string and double values into the expected formats of the GVertexResult.

Further, although this is not necessary in your particular scenario, the GVertexResult allows you to set the edges of the vertex as well. The data structure used to maintain information about the edges is essentially a map. The code fragment in Listing 10-14 shows how to use it.

Listing 10-14. Setting the Edges of a GVertexResult

```
Iterator<Edge<Text, DoubleWritable>> it = vertex.getEdges().iterator();  #1
  while (it.hasNext()) {
      Edge<Text, DoubleWritable> edge = it.next();
      userVertex.getEdges().put(                                          #2
          edge.getTargetVertexId().toString(),                            #3
          edge.getValue().toString()
  }
```

#1 Iterate of the edges of a vertex

#2 Access the edges data structure throught the getEdges() method

#3 Use the put() API to insert the target ID and edge value

This concludes accessing data from Gora. Although you didn't learn it here, Giraph also provides an implementation of an edge-based output format for writing to Gora. Note also that even though the Giraph code base implementations of the output formats and interfaces, such as the GVertexResult, would cover most of your needs, you may still need to customize them. The existing implementations, though, serve as perfect guides to extending your own implementations.

Summary

In previous chapters, you saw how to use HDFS as the basic storage for data input and output. Giraph, however, provides a flexible API that allows the extension to a variety of storage systems. In this chapter, you explored table-based and in-memory storage systems.

- Table-based storage systems have become quite popular because they allow the storage of and the query of semistructured data.

- Among other table-based storage systems, Giraph provides vertex-based and edge-based APIs for accessing data in Hive tables. Typically, you are transforming a single Hive table row to a `Vertex` or an `Edge` object.

- In-memory storage systems are another popular class of systems that allow fast access to data. Apache Gora provides an in-memory data model that abstracts the details of the underlying storage and allows you to plug it into different storage systems for persistence.

- Giraph further hides many of the complexities of reading data from and writing data to the Gora in-memory model. You can use the `GoraVertexInputFormat` and `GoreEdgeInputFormat` APIs to convert generic Gora objects to and from Giraph `Vertex` and `Edge` objects.

By now you should be familiar with accessing data from a variety of data stores. Giraph provides interfaces to more systems than covered in this chapter. You should take a look at the code base and documentation for more information. You continue in the next chapter with more advanced features—in particular, tuning and performance.

CHAPTER 11

■ ■ ■

Tuning Giraph

This chapter covers

- Key performance factors and bottlenecks in Giraph

- Optimal setups of the Hadoop cluster hosting Giraph

- Designing and implementing ad hoc data structures optimized for your algorithms

- Spilling excessing data to disk when necessary

- The different Giraph parameters and knobs

So far, you have seen how you can use Giraph to compute graph analytics on large graphs across hundreds of machines. You have been presented with Giraph's architecture and the programming model that allows you to write programs that scale. It is neat to write programs with the vertex-centric programming model, without worrying about the headaches of programming a parallel and distributed system. You write a compute function, and Giraph takes care of executing it on vertices, exchanging messages, and performing all the functioning of the system across your machines. How all this machinery works is hidden from you so that you can focus on the semantics of your application.

There will be times, however, when you will want to obtain all the performance you can from your Giraph application that is running perhaps slower than expected. Or worse, there will be times when your application is too eager of resources to be able to run until completion; for example, running out of memory. In either case, you need an understanding of how Giraph works under the hood, and how to leverage the hooks and knobs that are available to you to pass these limits. You have already seen some of the pieces in the previous chapters. In this chapter, you will look back at them, this time with a particular focus on performance tuning, and you are introduced to a new set of knobs and classes that you can play with.

Performance tuning is a bit of "black magic." It is a mixture of intuition, experience, and gut feeling. Still, it is not about luck. There are two things you need to master to tune the performance of a large system. First, you need a good knowledge of the principles behind the functioning of the system. That is, you have to do your homework learning the architecture and understanding how data and computation flow through all the different components. Second, you need the patience and the perseverance of trying out different solutions, and *measure* them.

A big misconception in computer systems, or even computer science in general, is that if you have an idea of the functioning of a system then you can decide the best tuning of the system on paper. Sure, you can make some assumptions and estimations, perform complexity analysis, but in the end the devil is in the details (or in the constants) and you have to test your assumptions and calculations against the clock. This means that in this chapter, you are given some ideas, and the principles behind the tuning knobs of Giraph are explained, but we cannot tell you how to improve your very own application. Only you can, with "educated" trial and error. Hopefully, it will be easier with the help of this chapter.

Key Giraph Performance Factors

There are many ways of characterizing the performance of a computation (or a system, an application, etc.). One way of doing it is by putting the computation in one of the two following classes.

- **Compute-intensive**: A computation that devotes most of its time to computational effort, due to high complexity. In other words, a computation that spends most of its time executing instructions on a relatively small piece of data. Examples of compute-intensive computations are the calculation of prime numbers, satisfiability solvers (SAT), simulations, and so forth.

- **Data-intensive**: A computation that processes large amounts of data, on which it does not perform particularly complex calculations, hence devoting most of its time to IO. In other words, a computation that spends most of its time going through data and moving it around. Examples of data-intensive computations are most of Hadoop computations, data filtering and aggregation, indexing, and so forth.

Intuitively, one way of deciding whether a computation is compute-intensive or data-intensive is to identify whether it would take advantage more of faster CPUs or of faster disks / network.

Many computations in the Hadoop world are data-intensive. For example, the majority of MapReduce applications spend most of their time reading through data files, filtering, chopping and sorting records, passing them between mappers and reducers, aggregating them, and writing them back to files. Many of these operations require reading and writing from and to disks and network. In fact, to speed up a MapReduce cluster you often want multiple disks on each machine, so that when a core performs IO on a disk, the other cores do not have to idle waiting for their turn to access that same disk. Ideally each core would read from a different "dedicated" disk. This is also why MapReduce jobs can take advantage of fast network, as a lot of data is just passed around in the shuffle and sort phase between the map and reduce phases.

After all, this is why it is called Big Data. Big Data is about leveraging small computations performed on many little pieces of information, and putting the result all together. This is pretty much the definition of a data-intensive computation. Graph algorithms are not different; in fact they are perfect examples of data-intensive computations. As you have seen, information in a graph is contained in the connections across the vertices, and that is pretty much it. Graph algorithms navigate through these connections and vertices multiple times, and perform small computations on each of them. Think of PageRank. At every iteration, each vertex is computed by considering the incoming messages and outgoing edges. But what it does on each vertex is really just sum up the incoming PageRank values, dividing the sum by the number of outgoing edges, and pass the result along through the outgoing edges. These are not complex operations. The more data you have, hence edges and vertices, the longer the computation will last. And the same holds for the SSSP algorithm. Each vertex goes through its incoming messages and identifies the smallest one. If it finds one that is smaller than its current value, it passes it along. This is also pretty cheap computation and its cost depends pretty much only on the size of the data it works on.

■ **Note**   Graph computations tend to be *data-intensive*, as they require visiting the data graph multiple times and performing few computations on each vertex or edge. Giraph computations are data-intensive, as most of the runtime is spent processing graph data. That is, executing the compute function on each vertex, its associated edges and values, and the messages sent to it, and to exchange the messages between workers.

Now, in a Giraph computation there are two classes of data items that fill-up the workers memory:

- **The graph**: The vertices, each with their edges, are what fills up a large portions of the heap. By default Giraph stores the graph in memory during the whole computation, so the larger is the graph, the more memory you need.

- **Computation state**: The state in a Giraph computation comprises the vertex values (sometimes also the edge values if they change during the computation) and the messages that are transitioning. Often, the number of messages is proportional to the number of edges, which are the most frequent items in any reasonable graph.

In other words, graph and messages together occupy more or less all the used heap. Moreover, during each superstep, workers spend a lot of time "just" transmitting the messages over the network. That is why combiners and good partitioning can speed up the computation, sometimes even halve it, and drastically reduce memory usage: simply because they can drastically reduce the amount of data being stored and exchanged overt the network.

The bottom line of the this chapter is that to reduce the amount of memory and speed up your computation you'll have to reduce the amount and size of the data stored in memory; that is, the vertex and edge values, and the amount and size of the data sent over the network, i.e. the messages. In the remainder of this chapter you'll have a look at how to tackle these problems through "memory-tight" implementations of some of the data-structures used by Giraph. Unfortunately, there are other things that cannot be discussed in this chapter; for instance, there are a number of ways to implement each algorithm in the Giraph programming model. However, we are not able to discuss algorithmic-specific implementation problems, as they really depend on each specific algorithm.

Giraph's Requirements for Hadoop

Giraph runs as a Hadoop application. As such, it can take advantage, or be penalized, by some decisions that regard the setup of the hosting Hadoop cluster, and the cluster-related job configuration parameters.

Hardware-related Choices

Often, you'll be running Giraph on a cluster over which you do not have much control. You'll be just a user without administrative power on the cluster. But, if you can discuss with your Hadoop administrators and ask for some changes, in this section you look at what you should be asking. Moreover, some organizations have specific clusters, or subsets of nodes, that they use for Giraph. If the number of workloads that you'll be able to solve with Giraph will increase over time, it could make sense to make some Giraph-oriented decisions when your cluster is expanded.

The profile of machine that is more appropriate for Giraph is a machine with a lot of main memory; that is, RAM. Giraph does not need large amounts of disk storage, so disk are not something you want to focus on when you think about Giraph. As you will see later in this chapter, Giraph has the ability to store portions of graph and messages on local disk, but you should consider it as a plan B and not design for it. Traditionally, MapReduce jobs have not asked for much main memory, as data was directly streamed from and to disk and network. However, recently main memory has played a more important role even for MapReduce jobs, for example, to speed up the disk-based shuffle-and-sort phase and to execute some of the more memory-eager operators of computations like Hive and Pig. This means it's likely that your machines already have a lot of memory nowadays, but if you plan to extend your cluster buying machines for Giraph, focus on memory.

Second, as said, an important fraction of the time of a Giraph computation is spent exchanging messages between workers. The way to speed up this phase (apart from algorithmically reducing the size and number of messages) is by using a fast network. Compared to memory, you will probably have even less control on the network available to you. Still, it is worth considering bonding interfaces together, or utilizing machines on the same rack. Giraph is more traditional when it comes to scalability, and while it can scale horizontally, it prefers vertical scalability; that is, few beefy machines are better than many small ones. Hence, a solution could be to fit some beefy machines in a rack, and just use those with Giraph.

■ **Tip**   Giraph prefers few beefy machines than many small machines. When you consider buying machines to extend your Hadoop cluster to accommodate Giraph, focus on memory rather than storage, and try to put the machines on the same rack.

This consideration about vertical scalability is important, so you will spend the rest of this section on it. Before you start, let's be clear about what is considered vertical scalability here. In the Hadoop world, vertical scalability is often considered an expensive "old-fashioned" way of deploying infrastructure, which may be less reliable as it depends on few machines (and hence less redundancy). Here, you refer to the choice of using few machines but with more cores, to achieve the desired amount of cores in the cluster, while still remaining in the domain of "commodity machines." The reason is explained next.

In Giraph, given the same algorithm and the same graph, the amount of data that is sent over the network increases as you add more machines/workers. Intuitively, when you add one machine/worker to your cluster, more messages than before need to be transmitted over the network, for the simple reason that vertices that were *local* before (as-in stored on the same machine) are now *remote* (some of those vertices is on the new machine/worker). A message between two vertices that used to be exchanged within the same worker (and hence not transmitted) now needs to be transmitted over the network. Needless to say, this is a slower operation than simply putting the message in the vertex inbox within the same worker (which, inside of the same JVM, is just putting an object in a map).

Now, increasing the number of workers also increases parallelism, hence adding more computing units to your computation *is* something you want to do in order to increase performance. However, you want to do it by increasing the number of cores instead of the number of machines, if possible. Increasing the number of cores without increasing the number of workers increases parallelism without increasing network usage. Concretely, it is preferable to have 10 machines with 8 cores each, than 80 machines with one core each. The total number of cores is still 80, but with fewer machines you obtain much more local vertices and hence less network usage. Note that while increasing the number of machines increases parallelism, it also increases the amount of data sent over the network. The is only so-much runtime improvement you can get by adding machines, and that is a trade-off you have to discover by running tests.

Job-related Choices

Per-worker parallelism with Giraph is achieved through *compute threads*. In Giraph, every worker has a number of partitions assigned to it, each partition with a number of vertices in it. The granularity of parallelism of Giraph is at partition level, meaning that each compute thread is assigned a number of partitions, and the thread is responsible to execute the compute function on each vertex belonging to those assigned partitions. Through compute threads, a worker can make use of all the cores available. Note that Giraph has also a number of other threads running during a computation, but that is discussed later. The relevant parameters are described later in this chapter.

There are two ways of exploiting the cores available to a cluster. Imagine you have 10 machines with 10 cores each. One way is to run 100 workers, assigning 10 workers to each machine, and using 1 compute thread in each worker. The other way is to run 10 workers, assigning 1 worker to each machine, and using 10 compute threads in each worker. Although theoretically the end result is full usage of the 100 cores, the two main differences are as follows:

- By using 100 workers, you obtain less locality when compared to using 10 workers, hence more messages are transmitted over the network. You may say that as 10 workers are on each machine, less of the network is used anyway; that is, between workers on the same machine. This is true. However, whereas two workers assigned to the same machine can communicate over the loopback, Giraph still has to treat them internally as remote communications (Giraph is not aware of this worker-worker per-machine locality); for example, serializing and deserializing messages via the Netty-based components, writing and reading data to and from sockets, and so forth.

- Memory is wasted on overhead. Having 100 workers means 100 Hadoop tasks and 100 JVMs. All this redundancy results in wasted memory, as you can obtain the same results with 10 times fewer tasks and JVMs. Plus, many data structures within a Giraph worker occupy space proportionally to the number of workers used, hence utilizing more heap memory.

You may think this is obvious and even unnecessary to point out. Why would you use multiple workers per machine? As it turns out, it is not trivial to request Hadoop to run one task per machine, particularly on clusters that you have to share; for example, MapReduce jobs. The setting is often not decided by the cluster administrators (i.e., mapred.tasktracker.map.tasks.maximum) and cannot be overridden by the user submitting the job.

■ **Tip** Avoid using multiple workers per machine and try to run as many workers as the number of machines available to you. Achieve parallelism within each worker on each machine by using multiple compute threads instead.

On pre-YARN clusters, a Giraph computation runs as a single map-only MapReduce job, where each worker would be executed as a mapper task. Traditionally, each machine, or its TaskTracker, would be set up to accommodate multiple concurrent mapper and reducer tasks, usually proportionally to the number of cores or disks. In addition, the maximum heap size of each task would be proportional to the total memory size divided by the number of tasks accepted, to avoid overcommitting. This means that to obtain all the available resources in a machine, you have to request the total number of available mapper tasks to use as workers, resulting in the suboptimal scenario described earlier with multiple workers per machine. On older versions of Hadoop, this could be overcome by setting the maximum number of mapper tasks per TaskTracker to 1 on the client-side. However, this option was discontinued in the more recent versions.

With YARN, things have changed. YARN is designed around the concept of *containers*. Containers represent a collection of physical resources, such as memory, CPU cores, and disks. YARN does not have a fixed number of mapper and reducer tasks, but instead depends on the available resources and the requested resources. A container can be defined to be of minimum and maximum size, relatively to the available resources to the machine; for example, 80% of the total memory. When a container is assigned 80% or 90% of the available memory in a machine, the ResourceManager will not claim more containers on that machine. When Giraph is running in YARN, each worker utilizes its own container. This means that you can obtain a setup of 1 worker per machine by requesting all the resources available to each machine for each worker/container; for example, memory. Here, we are assuming that all that memory is necessary and is used, and hence discussing a way to claim it to minimize redundancy. In practical terms, this can be obtained by submitting Giraph jobs with a heap size close to the size of the memory available to the machine (the smallest in the cluster, in case of heterogeneous clusters).

■ **Tip**   When using YARN, request for each container the amount of memory and CPUs available to a single machine. This allows a single worker to run on each machine.

This section concludes with a short discussion about homogenous and heterogeneous clusters. This discussion relates to both hardware- and job-related aspects of performance. Because of the synchronization barrier, a Giraph superstep last as long as it takes to the slowest or most loaded machine to compute its share. For this reason, you want to be able to use comparable machines, with similar computational resources, so that your faster machines do not have to idle at the synchronization barrier waiting for the slowest (or most loaded) to finish. Also, you want to make sure that each machine has the same load as the others. This is usually achieved through partitioning, by ensuring that vertices are partitioned in similar number across workers. Default hash-partitioning scheme usually guarantees that. Note that similar computational resources means similar CPU, memory, and network. All of these resources are as important. If a machine has a slower CPU, it takes more time to compute its part of the work. If it has less memory, it could fill-up its heap before the others, and run out of memory (note that your request for heap is global to all the workers). If it has a slower network, it takes more time to transmit its messages. In any case, it causes the other workers to wait idling at the barrier. At the same time, if a machine is assigned more workers, it is more loaded, and hence the workers assigned to it take more time to conclude their portion of the superstep. This has the same effects as having a machine with less computational resources.

■ **Tip**   Giraph prefers homogenous clusters, composed of machines with similar computational resources such as CPU, main memory size, and network. The slowest machine defines the duration of each superstep, causing faster machines to idle waiting for it at the synchronization barrier.

Tuning Your Data Structures

Giraph is designed internally to minimize use of memory. Initial versions of Giraph used pure Java object for all edges, vertices and messages. With graphs comprising billions of edges and vertices, and more billions of messages created and consumed at each superstep, the JVM would experience substantial pressure on its memory-management components. For all objects created by an application, the JVM has to allocate internally additional memory for the accounting of those objects. Moreover, the JVM garbage collector constantly tracks objects being created and destroyed, and moves them around to avoid fragmentation of the heap. These are all CPU cycles that the JVM could be using to compute application code, namely Giraph and your code. For this reason, Giraph uses a number of non-orthodox tricks to minimize memory footprint and time spent performing garbage collection.

In a nutshell, Giraph does some memory management on its own. For instance, it stores some of the data, like edges, messages and vertex values, serialized inside of byte arrays that it allocates at the beginning of the computation. Still, it offers a pure Java object-oriented API. To achieve this abstraction, Giraph keeps a number of objects around, internally called *representative* objects, which it reinitializes with data coming from these binary arrays before passing them to the user, and which it serializes back to the arrays after the user is done with them. This mechanism is used, for example, for the Iterable containing the messages passed to the compute function, or the default implementation of the OutEdges interface where edges are stored for each vertex. To understand how this works, let's have a look at the default implementation for these two classes.

The OutEdges Interface

The edges of a vertex are contained in a class implementing the OutEdges interface. The OutEdges interface extends the Iterable<Edge<I, E>> interface with the ability to add and remove edges, and to count the number of edges stored. The interface is used to access the edges, each contained in an Edge<I, E> object. Intuitively, you would expect that the outgoing edges of a vertex would be stored, for example, either in a Map<I, E>, to allow efficient *random* access, or in List<Edge<I, E>>, to allow efficient *iteration*. Before getting into the details about the implementation, it is important to understand these two access patterns to the edges in Giraph computations.

For example, think of PageRank and SSSP. In both, the compute method iterates through the outgoing edges to send messages to the other endpoints and take into account the edge weights as needed during the iteration over the edges. As you never look up a specific edge by its vertex ID (or by the other endpoint ID), it is sufficient to store edges in a data structure that supports efficient iteration, like an array or a list. Note that storing edges in an array is faster to iterate, as data is stored in a continuous chunk of memory, and more space-efficient, as you do not need anything else than the array of pointers to the edges. If you consider Label Propagation Algorithm described in Chapter 4, on the other hand, you access edges based on the senders of the messages, to update the edge values with new labels. For this kind of access, storing the edges in a map allows faster lookup (the cost of searching for an element in an unsorted array is proportional to the size of the array), but incurs additional overhead due to the map implementation itself, like a tree or a hashmap. Moreover, a tree is typically slower to iterate than an array. Hence, depending on the access pattern to the edges dictated by your algorithm, you want to use the right implementation of the OutEdges interface.

In additional to the iteration-friendly API of the OutEdges interface, Giraph comes with an additional interface, the StrictRandomAccessOutEdges<I, E>. This interface extends the OutEdges interface with methods to get and set edge values through lookups of the edge target vertex ID. Giraph provides an implementation for such interface called HashMapEdges, which as the name suggests, is backed by a Java hashmap for random access to the edges. By default, Giraph uses a class called ByteArrayEdges, which stores the edges serialized in a byte array, and it does not support random access to the edges. You can choose which class to use to store outgoing edges through the giraph.outEdgesClass Giraph parameter, including your own implementation. Let's have a look at why and how ByteArrayEdges makes use of byte arrays to minimize memory-footprint.

Iteration-Friendly OutEdges

As OutEdges extends the Iterable interface, ByteArrayEdges has to provide an implementation of Iterator<Edge<I, E>>. In other words, it has to be able to return an Edge<I, E> object at each call to the next() method of the iterator. This protocol is important as it allows the user to ignore of how things are functioning inside of any class implementing OutEdges. As mentioned earlier, a simple implementation would use a number of native Java objects—for example, one for each edge—and in turn, one for each edge value and target vertex ID associated to the edge. Having all of these objects around introduces additional memory overhead and pressure on the Garbage Collector(GC). Instead of creating and keeping these objects around, ByteArrayEdges stores them inside of a byte array via the methods provided by the Writable interface that these objects have to support for check-pointing. Storing the edge data serialized in byte arrays drastically reduces the number of objects, but you still need to provide Java objects to comply with the Giraph API. For this reason, ByteArrayEdges keeps a number of objects that it reinitializes based on the data in its byte array under the hood before returning them to the caller. Because internally these representative objects are constantly overwritten, the caller cannot keep a reference to them, but must make a copy if needed. This is fine, because the typical access pattern toward outgoing edges, like in the algorithms mentioned earlier, is just to iterate the edges and use their values locally at each step of the iteration. To ensure that the user knows this caveat, ByteArrayEdges implements the ReusableObjectsOutEdges interface, which precisely states this behavior. To have a better understanding of the mechanism, let's have a look at the implementation of the Iterator<Edge<I, E>> implemented as an inner class of ByteArrayEdges in Listing 11-1.

Listing 11-1. ByteArrayEdgeIterator

```
private class ByteArrayEdgeIterator
    extends UnmodifiableIterator<Edge<I, E>> {
  /** Input for processing the bytes */
  private ExtendedDataInput extendedDataInput =
      getConf().createExtendedDataInput(
          serializedEdges, 0, serializedEdgesBytesUsed); #1
  /** Representative edge object. */
  private ReusableEdge<I, E> representativeEdge =
      getConf().createReusableEdge(); #2

  @Override
  public boolean hasNext() {
    return serializedEdges != null && !extendedDataInput.endOfInput(); #3
  }

  @Override
  public Edge<I, E> next() {
    try {
      WritableUtils.readEdge(extendedDataInput, representativeEdge); #4
    } catch (IOException e) {
      throw new IllegalStateException("next: Failed on pos " +
          extendedDataInput.getPos() + " edge " + representativeEdge);
    }
    return representativeEdge; #4
  }
}
```

#1 Internally we encapsulate the byte array coming from the outer class into an ExtendedDataInput that we need to feed the Writable interface methods

#2 Create a reusable edge that we will return to the user

#3 Check whether edges exist and we haven't consumed all of them

#4 Rewrite the representative edge with the next data available in the array and return it to the user

First, note that serializedEdges is a private field of type byte[] used to store the edges and it belongs to the outer class ByteArrayEdges. In the same way, serializedEdgesBytesUsed is a private field of type int that counts the number of bytes used in that array. At construction, the iterator wraps the byte array into a DataInput object, used by Writable to read data. Furthermore, you construct a representativeEdge object of type ReusableEdge. ReusableEdge is an interface used internally to Giraph and not exported to the user, which allows overwriting a target vertex ID in an edge when the representativeEdge is reused. Second, note how at every call of the next() method the representativeEdge object is reinitialized consuming the wrapped array. This technique allows you to save memory and keep a generic interface, as you rely on the Writable interface that edges have to implement. Keep in mind that this mechanism puts more pressure on the CPU, which has to constantly serialized and deserialize objects, but usually the advantage is worth the cost.

Random Access-Friendly OutEdges

As mentioned earlier, HashMapEdges supports efficient lookup of edge values but to do so, it generates a number of objects for each edge. Supporting random lookups in a byte array would require additional complexity, such as sorting the edges and performing a binary search, or performing a linear search at each lookup. Moreover, the former approach would be feasible only assuming fixed-sized edge values, which is not always the case in a Giraph application. Another approach that is addressed in this section is using a hashmap of primitive types. Open source project *fastutil*[1] provides fast and compact type-specific implementations for common Java Collections data structures, such as maps, sets, lists, etc. Differently from using native Java objects, these data structures are automatically generated through a preprocessor, and work for different combinations of Java primitive types. For example, it provides an Int2ByteMap interface that extends the Java Map<Integer, Byte> interface, but the provided Int2ByteOpenHashMap implementation underneath stores a key as an int and a value as a byte. While the interface is complaint to the Collections API from Java, namely the methods of the map accept and return objects of type Integer and Byte, underneath these objects are not used in favor of their primitive-type counterpart. It has been shown that the implementations in this package are often the fastest out there, and with smallest memory footprint. Now, with the help of these classes, you can play a similar trick to the one in ByteArrayEdges. Imagine you want to optimize the implementation of LPA. Imagine that the vertex IDs are integers, and so are the edge values. The StrictRandomAccessOutEdges<I, E> interface extends OutEdges<I, E> with two methods in which you are interested, namely E getEdgeValue(I targetVertexId) and void setEdgeValue(I targetVertexId, E edgeValue) that get and set an edge value respectively. You need these methods when you iterate over the labels contained in the messages to update the corresponding labels that you store in the edge values. Listing 11-2 shows the implementation of a class OpenHasMapEdges<IntWritable, IntWritable> based on fastutil. In the interest of space, the implementation of some unrelated methods have been omitted.

Listing 11-2. OpenHashMapOutedges

```
public class OpenHashMapEdges extends
    ConfigurableOutEdges<IntWritable, IntWritable> implements
    StrictRandomAccessOutEdges<IntWritable, IntWritable>,
    ReuseObjectsOutEdges<IntWritable, IntWritable> {
  private Int2IntMap map; #1
  private IntWritable repValue = new IntWritable(); #2

  @Override
  public void initialize(Iterable<Edge<IntWritable, IntWritable>> edges) {
    EdgeIterables.initialize(this, edges);
  }

  @Override
  public void add(Edge<IntWritable, IntWritable> edge) { #3
    map.put(edge.getTargetVertexId().get(), edge.getValue().get()); #3
  }

  @Override
  public void remove(IntWritable targetVertexId) { #4
    map.remove(targetVertexId.get()); #4
  }
```

---

[1]http://fastutil.di.unimi.it/

```java
@Override
public int size() {
  return map.size();
}

@Override
public Iterator<Edge<IntWritable, IntWritable>> iterator() {
  return new PrimitiveTypedIterator();
}

private class PrimitiveTypedIterator
    extends UnmodifiableIterator<Edge<IntWritable, IntWritable>>{
  private Iterator<Entry<Integer, Integer>> it = #5
    map.entrySet().iterator(); #5
  private ReusableEdge<IntWritable, IntWritable> repEdge = EdgeFactory #6
    .createReusable(new IntWritable(), new IntWritable()); #6

  @Override
  public boolean hasNext() { #7
    return it.hasNext(); #7
  }

  @Override
  public Edge<IntWritable, IntWritable> next() {
    Entry<Long, Short> entry = it.next(); #8
    repEdge.getTargetVertexId().set(entry.getKey()); #8
    repEdge.getValue().set(entry.getValue()); #8
    return repEdge; #8
  }
}
@Override
public void readFields(DataInput in) throws IOException {
  int numEdges = in.readInt();
  initialize(numEdges);
  for (int i = 0; i < numEdges; i++) {
    int id = in.readInt();
    int v = in.readInt();
    map.put(id, v);
  }
}

@Override
public void write(final DataOutput out) throws IOException {
  out.writeInt(map.size());
  for (Entry<Long, Short> e : map.entrySet()) {
    out.writeInt(e.getKey());
    out.writeInt(e.getValue());
  }
}
```

```
@Override
public IntWritable getEdgeValue(IntWritable targetVertexId) {
  int v = map.get(targetVertexId.get()); #9
  repValue.set(v); #9
  return repValue; #9
}

@Override
public void setEdgeValue(IntWritable targetVertexId,
    IntWritable edgeValue) {
  map.put(targetVertexId.get(), edgeValue.get()); #10
}
}
```

#1 We keep the values in a map based on primitive types

#2 Also we keep a representative object for edge values of the actual type

#3 Proxy a request to add an edge as a put in the map

#4 Proxy a request to remove an edge as a remove from the map

#5 We keep an iterator of the map that we will proxy

#6 We keep a representative Edge in the iterator to reinitialize with data coming from the map iterator

#7 Proxy the request to the internal iterator

#8 Reinitialize the representative object with data coming from the map

#9 Fetch the data from the map and return the update representative object

#10 Update the backing map with the new value

The class is very simple. You store internally a fastutil map of primitive types that correspond to the specific types the algorithm is expecting. In this sense, the approach is less general than ByteArrayEdges, which accepts Generic types. In fact, if you want to go through this road, you have to modify the given class with specific primitive types for each different algorithm. It may take some copy-pasting, but it is worth the effort. For the rest, the class proxies the calls to the internal map to get and put edge values coming from the user. Note that you can also proxy the calls for the iterator, by relying on the internal iterator that you proxy under the hood. This is pretty much all there is to know about tailoring OutEdges to your algorithms to save memory and increase speed.

The MessageStore Interface

Messages are the other data that fills the heap of the workers. Some algorithms produce a number of messages proportional to the number of edges, such as PageRank, other produce less and the number at each superstep depends on many factors as, for example, the topology of the graph. Moreover, certain algorithms generate very large messages, as, for example, the algorithm to compute clustering coefficients, which sends lists of neighbor IDs as messages to detect triangles. To minimize the impact of messages, Giraph stores messages through a technique similar to that presented in the previous section regarding edges. Similarly to the ReusableObjectsOutEdges interface, also the references to messages are valid only before the next call to the next() method of the iterator passed to the compute() method for each vertex. The reason is the following. Messages are stored in a serialized format inside of byte arrays, and what the iterator passes to the user is a representative message that is reinitialized with message data at each call. This trick is implemented by the ByteArrayMessagesPerVertexStore class, which is the default implementation of the MessageStore interface. A MessageStore is the inbox where Giraph stores all messages sent to the vertices. Each worker has one of such stores, and within the store messages are grouped by destination vertex, which are themselves organized per partition. In other words, you can imagine a MessageStore as a nested map, with partitions as keys of the outer map, and vertex IDs as keys of the inner maps. One thing to notice is that combiners drastically reduce the number of message transiting the system. This can have particular impact on the runtime management of a store. In fact, Giraph uses a different default implementation when combiners are involved, as Giraph combines all incoming messages at the store-level as they arrive to the destination worker. In other words, Giraph only stores a single message, if any, in the store when a combiner is involved.

The question is then, how can you save more memory given that the default message stores provided by Giraph are already efficient? The idea is again similar to what you have seen in the previous section. You can use data structures tailored to the data types that you use in your application, through primitive types-based classes provided by fastutil. While compact and minimizing the number of objects, the original stores are still based on the original Java HashMap class, in order to be as general as possible to the vertex ID type and message value. But as you want to minimize the memory footprint of the store as much as you can, you can drop this requirement for generality and implement a store that is defined precisely around the data types. Let's first have a look at how the default implementation in Giraph works, and then let's have a look at how to write a message store for a vertex ID of type int and message value of type float. So, first you will look at the SimpleMessageStore abstract class, which implements some of the routines that are used by, for example, the OneMessagePerVertexStore class. Listing 11-3 presents a selection of methods from the class.

Listing 11-3. SimpleMessageStore

```
public abstract class SimpleMessageStore<I extends WritableComparable,
    M extends Writable, T> implements MessageStore<I, M>  {
  /** Message class */
  protected final MessageValueFactory<M> messageValueFactory;
  /** Service worker */
  protected final CentralizedServiceWorker<I, ?, ?> service;
  /** Map from partition id to map from vertex id to messages for that vertex */
  protected final ConcurrentMap<Integer, ConcurrentMap<I, T>> map; #1
  /** Giraph configuration */
  protected final ImmutableClassesGiraphConfiguration<I, ?, ?> config;

  /**
   * Constructor
   *
   * @param messageValueFactory Message class held in the store
   * @param service Service worker
   * @param config Giraph configuration
   */
  public SimpleMessageStore(
      MessageValueFactory<M> messageValueFactory,
      CentralizedServiceWorker<I, ?, ?> service,
      ImmutableClassesGiraphConfiguration<I, ?, ?> config) {
    this.messageValueFactory = messageValueFactory;
    this.service = service;
    this.config = config;
    map = new MapMaker().concurrencyLevel(
        config.getNettyServerExecutionConcurrency()).makeMap(); #1
  }
```

```
  /**
   * Get messages as an iterable from message storage
   *
   * @param messages Message storage
   * @return Messages as an iterable
   */
  protected abstract Iterable<M> getMessagesAsIterable(T messages); #2

  /**
   * If there is already a map of messages related to the partition id
   * return that map, otherwise create a new one, put it in global map and
   * return it.
   *
   * @param partitionId Id of partition
   * @return Message map for this partition
   */
  protected ConcurrentMap<I, T> getOrCreatePartitionMap(int partitionId) {
    ConcurrentMap<I, T> partitionMap = map.get(partitionId);
    if (partitionMap == null) {
      ConcurrentMap<I, T> tmpMap = new MapMaker().concurrencyLevel(
          config.getNettyServerExecutionConcurrency()).makeMap();
      partitionMap = map.putIfAbsent(partitionId, tmpMap); #3
      if (partitionMap == null) { #3
        partitionMap = tmpMap; #3
      }
    }
    return partitionMap;
  }

@Override
  public Iterable<M> getVertexMessages(I vertexId) throws IOException {
    ConcurrentMap<I, T> partitionMap = map.get(getPartitionId(vertexId));
    if (partitionMap == null) {
      return Collections.<M>emptyList();
    }
    T messages = partitionMap.get(vertexId);
    return (messages == null) ? Collections.<M>emptyList() : #4
        getMessagesAsIterable(messages); #4
  }
}
```

#1 Messages are organized per-partition, inside of 2-levels of nested maps

#2 The method to pack messages in an iterable is left abstract

#3 Partitions messages are returned or an empty map is created instead

#4 Vertex messages are fetched from the map and packed in an iterable object

There are a couple of things to notice here. First, being abstract the class cannot be instanced, but it provides some basic functionality to manage the nested map mentioned before. In particular, note how the getVertexMessages() method first gets the inner per-partition map, then it fetches the vertex messages, and then it relies on the getMesssagesAsIterable() method to pack the messages in an iterable object that is passed to the compute method together with the vertex. Second, as you may have noticed, the implementation of getMesssagesAsIterable() is not provided. This is exactly because, depending on the store, for a given vertex, you may have either a list of messages serialized in a byte array, or a single message when a combiner is involved. You need to let the implementer handle this aspect further. Let's have a look at this latter case. Listing 11-4 presents some of the implementation of OneMessagePerVertexStore, which, as mentioned, stores only one message per vertex through a combiner.

Listing 11-4. OneMessagePerVertexStore

```
public class OneMessagePerVertexStore<I extends WritableComparable,
    M extends Writable> extends SimpleMessageStore<I, M, M> {
  /** MessageCombiner for messages */
  private final MessageCombiner<? super I, M> messageCombiner; #1

  /**
   * @param messageValueFactory Message class held in the store
   * @param service Service worker
   * @param messageCombiner MessageCombiner for messages
   * @param config Hadoop configuration
   */
  public OneMessagePerVertexStore(
      MessageValueFactory<M> messageValueFactory,
      CentralizedServiceWorker<I, ?, ?> service,
      MessageCombiner<? super I, M> messageCombiner,
      ImmutableClassesGiraphConfiguration<I, ?, ?> config) {
    super(messageValueFactory, service, config);
    this.messageCombiner = messageCombiner; #1
  }

  @Override
  public void addPartitionMessages(
      int partitionId,
      VertexIdMessages<I, M> messages) throws IOException {
    ConcurrentMap<I, M> partitionMap =
        getOrCreatePartitionMap(partitionId);
    VertexIdMessageIterator<I, M> vertexIdMessageIterator =
      messages.getVertexIdMessageIterator();
    // This loop is a little complicated as it is optimized to only create
    // the minimal amount of vertex id and message objects as possible.
    while (vertexIdMessageIterator.hasNext()) {
      vertexIdMessageIterator.next();
      I vertexId = vertexIdMessageIterator.getCurrentVertexId();
      M currentMessage =
          partitionMap.get(vertexIdMessageIterator.getCurrentVertexId());
      if (currentMessage == null) { #1
        M newMessage = messageCombiner.createInitialMessage(); #1
        currentMessage = partitionMap.putIfAbsent( #1
```

```
          vertexIdMessageIterator.releaseCurrentVertexId(), newMessage); #1
      if (currentMessage == null) { #1
        currentMessage = newMessage; #1
      }
    }
    synchronized (currentMessage) {
      messageCombiner.combine(vertexId, currentMessage, #2
          vertexIdMessageIterator.getCurrentMessage()); #2
    }
  }
}

@Override
protected Iterable<M> getMessagesAsIterable(M message) { #3
  return Collections.singleton(message); #3
}
}
```

#1 We make sure we have a message for each vertex

#2 We update the message by combining the previous one with the new one

#3 We return an iterable implemented by a singleton

Here you can note how getMesssagesAsIterable() packs the only message in a singleton iterator. You can also notice the way addPartitionMessages() inserts messages in the store. Also, messages are put into the store in groups, as they are buffered upon reception. In particular, note how messages are combined one after the other with the message currently stored, if any, in the store.

You are now ready to look at a primitive type-based implementation of such a class. Listing 11-5 presents the IntFloatMessageStore class, which as the name suggests, is a message store designed for algorithms where vertices IDs are of type int and messages are of type float. Also in this case, only the part of the implementation relevant to the discussion is presented.

Listing 11-5. IntFloatMessageStore

```
public class IntFloatMessageStore
    implements MessageStore<IntWritable, FloatWritable> {
  /** Map from partition id to map from vertex id to message */
  private final Int2ObjectOpenHashMap<Int2FloatOpenHashMap> map;
  /** Message messageCombiner */
  private final
  MessageCombiner<? super IntWritable, FloatWritable> messageCombiner;
  /** Service worker */
  private final CentralizedServiceWorker<IntWritable, ?, ?> service;

  /**
   * Constructor
   *
   * @param service Service worker
   * @param messageCombiner Message messageCombiner
   */
```

```
  public IntFloatMessageStore(
      CentralizedServiceWorker<IntWritable, Writable, Writable> service,
      MessageCombiner<? super IntWritable, FloatWritable> messageCombiner) {
    this.service = service;
    this.messageCombiner = messageCombiner;

    map = new Int2ObjectOpenHashMap<Int2FloatOpenHashMap>();
    for (int partitionId : service.getPartitionStore().getPartitionIds()) {
      PartitionStore<IntWritable, Writable, Writable> partitionStore =
        service.getPartitionStore();
      Partition<IntWritable, Writable, Writable> partition =
        partitionStore.getOrCreatePartition(partitionId);
      Int2FloatOpenHashMap partitionMap = #1
          new Int2FloatOpenHashMap((int) partition.getVertexCount()); #1
      map.put(partitionId, partitionMap); #1
      partitionStore.putPartition(partition); #1
    }
  }

  @Override
  public void addPartitionMessages(int partitionId,
      VertexIdMessages<IntWritable, FloatWritable> messages) throws
      IOException {
    IntWritable reusableVertexId = new IntWritable(); #2
    FloatWritable reusableMessage = new FloatWritable(); #2
    FloatWritable reusableCurrentMessage = new FloatWritable(); #2

    Int2FloatOpenHashMap partitionMap = map.get(partitionId);
    synchronized (partitionMap) {
      VertexIdMessageIterator<IntWritable, FloatWritable>
          iterator = messages.getVertexIdMessageIterator();
      while (iterator.hasNext()) {
        iterator.next();
        int vertexId = iterator.getCurrentVertexId().get();
        float message = iterator.getCurrentMessage().get();
        if (partitionMap.containsKey(vertexId)) { #3
          reusableVertexId.set(vertexId); #3
          reusableMessage.set(message); #3
          reusableCurrentMessage.set(partitionMap.get(vertexId)); #3
          messageCombiner.combine(reusableVertexId, reusableCurrentMessage,
              reusableMessage); #4
          message = reusableCurrentMessage.get(); #4
        }
        partitionMap.put(vertexId, message); #4
      }
    }
  }
```

```
@Override
public Iterable<FloatWritable> getVertexMessages(
    IntWritable vertexId) throws IOException {
  Int2FloatOpenHashMap partitionMap = getPartitionMap(vertexId);
  if (!partitionMap.containsKey(vertexId.get())) {
    return EmptyIterable.get();
  } else {
    return Collections.singleton(#5
        new FloatWritable(partitionMap.get(vertexId.get())));  #5
  }
}
}
}
```

#1 We create all the partitions at construction for efficiency

#2 We use reusable objects that we reinitialized to save objects

#3 We reinitialized objects with primitive typed data

#4 We use the reusables to combine current and new message

#5 We return a single iterable containing an object created on-the-fly

This time the outer map is a fastutil `Int2ObjectOpenHashMap` class, as keys are the integer partition IDs and the values are the inner map objects, in this case `Int2FloatOpenHashMap` objects, that map vertex IDs to their message. The functioning of the class in both `add PartitionMessages()` and `getVertexMessages()` methods recalls what you have seen in the previous two classes. This time however it was not possible to reuse code coming from the `SimpleMessageStore` class because all the functionality is tailored to the specific primitive types. The last thing to mention is how to specify which class to use as a message store. As with other classes, the message store class is specified with a Giraph parameter. In particular, you must use the `giraph.messageStoreFactoryClass` parameter to specify a *factory* class that generates `MessageStore` implementations. Examples of factory classes come together with the message stores in the Giraph source code; the interested reader is directed to the source code for examples of factory implementations.

Going Out-of-Core

Many times this book has underlined how Giraph was designed to compute graph algorithms in memory, and that this is a key factor in allowing such fast computations on massive graphs. However, this can also be Giraph's weakness, as sometimes data can be excessive and workers can run out of memory. In the previous sections you have seen how Giraph already implements some nice tricks to minimize memory footprint, and how you can tailor your data structures to minimize it even more. Yet, there will be times when all of these efforts won't be enough. In those cases, you want to try the capability of Giraph to spill excessive data to disk. Keep in mind that this capability should be considered a last-resort option because it may degrade performance substantially if used excessively.

Giraph can store both messages and graph (both the topology, and the vertex and edge values) on local disks during the computation of each superstep. The fact that Giraph can make use of disks does not turn it into a database. Its access to disk should not be considered as a layer of persistence, but more of a swap partition in an operating system. It spills to disk the portion of the state and the graph that is excessive to keep in main memory to reach the conclusion of the computation. In other words, from the perspective of the user there is no difference to have Giraph running with or without the out-of-core capability activated, except maybe for performance. The only thing the user is asked to do is to choose whether to spill messages, graph, or both, and how much of them. This section briefly describes how the functionality works, so that you can make your choice and investigate what works best for your application.

Out-of-Core Graph

Each worker stores the vertices assigned to it inside so-called *partitions*. Inside a partition are all the vertices, and their edges, assigned to that partition. During a superstep, a worker processes one partition after the other, or multiple partitions in parallel if multiple compute threads are used. Within partitions, one vertex is processed after the other. The out-of-core graph functionality (or OOCG) allows a worker to keep only a user-defined number of partitions in memory, while the remaining ones are stored on local disk. When a worker is done processing all the partitions that are currently in memory, it spills a number of processed partitions to disk to load unprocessed ones from disk into memory. The worker will not start spilling or loading any vertex until the partition the vertex belongs to has been processed completely. This means that OOCG works at a partition-level granularity. No single vertex (or edge) is spilled or loaded from disk individually.

When OOCG is active by default Giraph keeps in memory N partitions, of which a number is considered *sticky*, typically defined as $N - k$ (where N is the number of partitions to be kept in memory at all times, and k is the number of compute or input threads). Sticky partitions are partitions that are never swapped to disk. You want to keep at least k non sticky partitions when using k compute threads so that each compute thread can swap a partition to disk with an unprocessed one without having to wait. By default, which partitions are sticky is chosen at random at the beginning of the computation and it does not change. Because in most graph algorithms, all partitions need to be processed at each superstep (but not necessarily all vertices in them); keeping a number of sticky partitions in memory at all times minimizes the number of swaps.

However, your partitioning algorithm may allow to split the graph in such a way that only certain partitions contain active vertices to be processed, while other partitions can be ignored for some supersteps. For those cases, Giraph can also ignore stickiness and use a *least-recently-used* (*LRU*) policy when choosing which partition to swap to disk to make space for a new one. In this setting, the partition stores operate practically as a cache. You can activate OOCG with the Giraph `giraph.useOutOfCoreGraph` parameter, you can define the number of partitions to keep in memory with the `giraph.maxPartitionsInMemory` parameter, and you can overwrite the automatic setting of sticky partitions with the `giraph.stickyPartitions` parameter.

The functioning of OOCG is simple. When stored on local disk, each partition is saved in two files, one for vertices and one for edges. The layout of the vertices file is, a part of a short header, a sequence of vertices serialized through the `Writable` interface one after the other. Similarly, the edges file is a sequence of edges, grouped by the source vertex. Spilling a partition to disk is IO-efficient, as it is a sequential write to a disk of the two files that does not involve read or seek operations, hence maximizing the use of IO disk bandwidth. In the same way, reading a partition from disk is also a sequential read of the two same files, which are parsed again through the methods defined by the `Writable` interface. The reason why vertices and edges are stored in two separate files is to minimize IO when possible. Many algorithms act on static graphs, as, for example, PageRank, SSSP, Connected Components, and so forth. Static graphs mean that the edges and vertices are not added or removed, and that edge values (e.g., weights) do not change. By this definition, Label Propagation Algorithm and Stochastic Gradient Descend (both described in Chapter 4) do not operate on static graphs as they mutate the edge values. Because only vertex values change during the computation, Giraph can spill to local disk both vertices and edges only the first time a partition is swapped to disk. The subsequent times, only the changing elements, the vertex values, need to be written to disk. When the partition is read back to memory, the first write of edges and the most recent write of the vertices are used. This saves substantial write IO, considering that edges are by far the largest portion of the graph. The user can specify whether its algorithm acts on a static view of the graph with the `giraph.isStaticGraph` parameter.

Finally, you can specify the directories on local disk used to store out-of-core partitions with the `giraph.partitionsDirectory` parameter. The parameter accepts a comma-separated list of directories. If multiple disks are available, it is convenient to specify one directory on each disk. OOCG spreads partitions across disks, and as compute threads swap partitions from memory to disk, they parallelize IO for faster data transfer, maximizing throughput.

Out-of-Core Messages

You have seen that Giraph stores incoming messages inside of inboxes called message stores, one for each worker. If a message store is filled with more messages than the heap can manage, the worker will run out of memory. For this reason, Giraph allows you to set a threshold to the maximum number of messages that are stored in memory, before messages start being spilled to disk. Every time the message store reaches its maximum capacity, its content is spilled to disk in a file, and a new empty message store is created to accommodate new incoming messages.

More precisely, out-of-core messages (or OOCM) works as follows. First, when OOCM is active, vertices are processed by the compute threads in sorted order, as dictated by their ID. Second, when a message store is written to disk, after the given threshold is reached, messages are stored sorted by destination vertex ID, one message after the other. As a natural effect, messages are grouped together in the file by destination vertex ID, as all messages destined to the same vertex are written together as a result of the sorting. After the messages are sorted in memory, they are written very efficiently as a sequential operation. Third, because messages arrive at different times during a superstep, messages destined to a specific vertex are spread across multiple files. These three elements allow OOCM to serve messages from disk without indices or without seek operations inside of a file. In other words, they are scanned through sequential reads.

When a superstep begins and the message store is opened for reading, the store places a *cursor* at the beginning of each file. When the store is asked for the messages for a specific vertex, it looks at all its cursors and check whether they are pointing to messages destined to that vertex. If a cursor is indeed pointing to such messages, it is used to read the messages (actually literally streamed as they are consumed) from the file. Note that the cursor is advanced automatically, as a side-effect of the reads, to the next group of messages. If the messages pointed by the cursor are destined to a different vertex, the file can be skipped completely for this vertex. In fact, if the file contained messages for that vertex, they would be currently pointed by the cursor. This is a necessary condition due to the fact that messages are stored sorted in the same order they are processed during a superstep. Assuming vertices (and messages) are stored in ascending order, if a cursor is not pointing to messages for the given vertex, it must be pointing to messages destined to a vertex with a bigger ID that still needs to be processed. Note that the message store does the same operations with a cursor pointing to the messages currently stored sorted in-memory. The operations are transparent to where messages are stored, as long as they are sorted. Figure 11-1 shows the layout of OOCM when reading data for vertex v4.

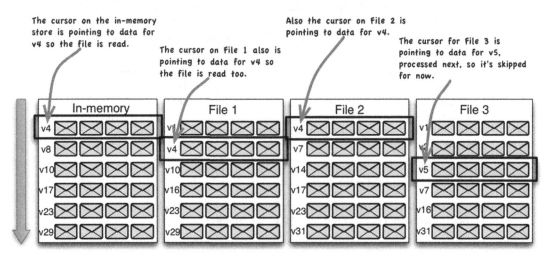

Figure 11-1. *Out-of-core messages reading data for vertex v4*

This technique is very efficient, because it is based only on sequential writes and reads to disk, and it is hence very IO efficient. Yet, it needs to periodically sort (in-memory) portions of messages before they are written to disk, which is expensive if many messages need to be written to disk. Moreover, using a low threshold pushes Giraph to produce many small files. While reads are logically sequential within each file, the disk has to physically seek to different areas to read the content pointed by each cursor. For this reason, one should use OOCM with high thresholds, hence not in case of extremely low memory availability (in which case the more predictable OOCG should be used). OOCM is activated by means of the `giraph.useOutOfCoreMessages` parameter, and the threshold is set via the `giraph.maxMessagesInMemory` parameter. The list of directories on local disk(s) used to store messages is set via the `giraph.messagesDirectory` parameter. As for OOCG, also OOCM can make use of multiple disks, to parallelize disk IO.

Giraph Parameters

Giraph is a large system, and as any large enough system it has an even larger number of parameters that influence its behavior and performance. It is difficult to define how certain parameters characterize precisely the behavior and the performance of a system for all applications. To complicate things up a bit, it's even harder to characterize how multiple parameters interact with each other. The following is a list of parameters with explanations about what they do. Some of them are straightforward; others, like numbers of threads and size of buffers, depend on your application, graph, cluster, and setting. As said at the beginning of this chapter, you'll have to measure yourself how different values impact your applications. Through this section, you should be guided in this exploration.

- `giraph.useBigDataIOForMessages=[true|false]`, default false: BigDataIO is a set of classes that allows you to go beyond Java's limitation of size of byte arrays, by wrapping together multiple ones. It can be used if certain vertices are expected to receive many messages.

- `giraph.jmap.histo.enable=[true|false]`, default false: By enabling this parameter, you request the JVM to print histograms of objects in a worker, possibly also to a remote controller. It is handy to see where your memory is going (as in which objects use it and should be hence optimized).

- `giraph.metrics.enable=[true|false]`, default false: By enabling this parameter you ask Giraph to print additional metrics information, as in where each worker is spending most time. For example, you can measure, per superstep and per worker, the time spent computing vertices, the time spending serving messages, and so forth.

- `giraph.oneToAllMsgSending=[true|false]`, default false: If you activate this parameter you tell Giraph that each vertex sends the same message to all its neighbors. This is the case, for example, of PageRank, but not of weighted SSSP where the message depends on each edge weight. Once Giraph knows this, it can make certain optimizations that minimize the amount of messages it stores in memory.

- `giraph.numInputThreads=integer`, default 1: Input splits can be read concurrently to load the graph faster.

- `giraph.numOutputThreads=integer`, default 1: Multiple threads can go through vertices and output results at the end of the computation.

- `giraph.nettyClientThreads=integer, default 4`: Client threads are responsible of sending data from workers to other workers. You want to increase the number of threads when you have a large number of workers. Having a too large number of threads increases overhead, while a too small number underuses network while buffers fill-up.

- `giraph.nettyServerThreads=integer, default 16`: Server threads are responsible of receiving data from other workers. You want to increase the number of threads when you have a large number of workers. Having a too large number of threads increases overhead, while a too small number underuses network while buffers fill-up.

- `giraph.serverReceiveBufferSize=integer, default 524288`: You want to increase the buffer size when your application sends a lot of data; for example, many and/or large messages. This avoids that the server is constantly moving little pieces of data around.

- `giraph.clientSendBufferSize=integer, default 524288`: Similar to server buffers, you want to increase the buffer size when your application sends a lot of data; for example, many and/or large messages. This avoids the server constantly moving little pieces of data around.

Some of these parameters play a joint role. In particular buffer sizes and number of threads are strongly connected. Increasing the buffer size makes sure that IO costs are amortized. Increasing the number of threads makes sure that data is not stalling in buffers. However, using large buffers for many threads increases the usage of heap with risks of going out of memory.

Summary

Giraph is a large-scale graph processing system that runs your code potentially on thousands of cores across hundreds of machines with little complexity. Its generality, although specific to graph algorithms, leaves some space to optimizations tailored to each use case. In particular, you have seen how:

- The amount of data stored in memory and sent over the network can produce bottlenecks or hinder the ability to conclude the computation

- While running on commodity machines, certain hardware and Hadoop cluster setup choices are more suitable to large-scale graph processing

- Giraph uses particular memory layouts to minimize the impact of the garbage collector in the JVM

- Depending on the access pattern to the edges, different edge stores can be more suitable

- The user can implement tailored data structures based on primitive types for both edges and messages to decrease memory footprint

- When necessary, Giraph can spill excessive data, graph and messages, to disk

A number of parameters can influence the behavior and performance of your applications depending on its usage of resources

CHAPTER 12

■ ■ ■

Giraph in the Cloud

This chapter contains

- A high-level architectural overview of cloud computing

- An analysis of specific requirements graph processing presents in the cloud

- A deep dive into using Giraph on a public cloud provided by Amazon

As you learned in previous chapters, Giraph leverages Hadoop clusters as its main execution framework and can fetch data from multiple different sources, not just HDFS. This makes Giraph applications ideally suited for running on a shared infrastructure provided by various public and private cloud computing platforms. Utilizing cloud computing services lets you focus on graph processing, while the vendors take care of things like deploying and operating scalable Hadoop clusters. This chapter begins by providing an overview of cloud computing services, outlining the ones ideally suited for making Giraph applications run as seamlessly as possible.

Since not all cloud computing vendors go all the way to providing Hadoop clusters as a service, we will briefly overview the more basic, but still useful infrastructure virtualization services. These types of services also come in handy in cases where precise control over the version of Hadoop available to Giraph is required.

Even though there are plenty of public cloud providers available today and the ecosystem of private cloud implementations is as robust as ever, this chapter mostly focuses on Amazon Web Services (AWS). We are focusing on AWS for two reasons. First of all, it is one of the most popular public clouds available. On top of that, having experience in running graph processing jobs on AWS is going to make using other public and private clouds a much more natural experience. Within AWS, we are going to focus on a set of APIs known as Elastic MapReduce, which lets you dynamically spin up an arbitrary number of virtual servers that appear to your Giraph application as a bona fide Hadoop cluster. Even though Giraph doesn't come preinstalled, it is still the easiest way to get your application from a single node execution to as many nodes as you wish (or willing to pay for). All the complexities of installing, configuring, and managing Apache Hadoop are hidden from you.

A Quick Introduction to Cloud Computing

Consider for a moment the amount of effort involved in putting any graph processing application into production use at your company. At the very minimum, you would have to buy and install as much storage and computer hardware as required to accommodate your potential datasets and processing needs. Before you can turn that hardware into a fully functioning Hadoop cluster, you would have to spend time and money on managing power, cooling, and networking. And after that, you would still have to install and configure operating systems and Hadoop on every single node in your cluster. Only then could you start collecting data and running Giraph applications. Of course, to keep everything up and running, you must monitor anything (be it hardware or software) that could go wrong and maintain your cluster accordingly. In short, it means a lot of upfront investment and not much scalability. In other words, when your requirements for the size of datasets change, you are still stuck with the same cluster.

One way to avoid this initial investment and complexity is to run your Giraph applications in a cloud computing environment where someone else is taking care of setting up and managing all the infrastructure.

Simply put, cloud computing is a set of services built around the idea of providing various computing, software, and informational resources in a scalable and elastic fashion over the network. Another key aspect of cloud computing is the concept of self-service: something that allows any customer to request as many resources as an application requires without any human intervention. No longer do you have to wait for IT personnel procurement and deployment of physical hardware in company datacenters: you can simply request resources online via well-defined API calls. A few API calls, for example, can give you a fully functioning Hadoop cluster accessible over the network.

WHAT KIND OF SERVICES CAN I EXPECT OUT OF CLOUD COMPUTING?

All the services provided by cloud computing vendors fall into three broad categories, each of which builds off the previous one:

- **Infrastructure as a Service (IaaS)**: A set of services delivering virtualized instances of the physical infrastructure, such as servers, storage, load balancers, network equipment, and so forth.

- **Platform as a Service (PaaS)**: A set of services delivering fundamental APIs on which applications can be built. Examples include execution runtimes, database instances, web and application servers, and so forth.

- **Software as a Service (SaaS)**: A set of services delivering end-user software applications over the network. The range of applications could be as vast as what is available on a desktop, but the most common examples are e-mail, communications, games, and virtual desktops.

Most cloud providers give you a combination of services from all three categories. For example, Amazon Web Services (AWS) can be utilized either as an IaaS, letting you create virtual datacenters of any size, or as a PaaS, giving you not just the bare servers, but things like scalable databases or Hadoop clusters. Running Giraph in the cloud requires IaaS capabilities, but not necessarily PaaS or SaaS. However, the presence of either PaaS or SaaS capabilities may provide additional sources of data to be analyzed by Giraph.

Cloud computing is typically associated with public clouds: services available to anybody over the public Internet. However, the agility and self-service nature of cloud computing makes it appear in enterprise companies' private datacenters at a rapid pace. It used to be that an enterprise datacenter was considered as racks of servers running various operating systems. Today, chances are that it is considered a cloud computing service providing at least IaaS-level APIs. This is what is known as a *private cloud*.

It may be tempting to think that the APIs offered by different cloud computing solutions are compatible, and anything that you do on one cloud you could do on a different one. The good news is that conceptually this is more or less the case (and this is why this chapter only focuses on AWS). The bad news is that the set of APIs needed to achieve the same goal differs between clouds. Every cloud computing solution (public or private) typically comes with software libraries and tools that let you create and manage various resources.

Giraph on the Amazon Web Services Cloud

One of the primary services that the Amazon Web Services (AWS) cloud provides is the Elastic Compute Cloud (EC2). EC2 lets anyone instantiate any number of virtual servers from static virtual machine (VM) images; it can be considered the fundamental building block of the AWS IaaS layer. The VM images are called Amazon Machine Image (AMI). You can instantiate any number of virtual servers from a single AMI by associating virtual resources (CPU, memory, I/O bandwidth, etc.) with an AMI. Those servers are self-contained, virtualized operating systems that run somewhere in one of the AWS datacenters. They are called *instances*. Your account is charged based on the number of instances that are running each hour and the amount of resources (CPU, memory, local storage, network bandwidth, etc.) that each instance utilizes. The only way to interact with these instances is over the Internet. For example, you can decide to log in to them directly using SSH or run any kind of networked service on them, such as HTTP and so forth. All you need to know is the public IP address of each instance.

EC2 is a very simple, but at the same time an extremely scalable service that you can use for deploying Hadoop clusters in the cloud manually. After all, once EC2 returns a set of IP addresses corresponding to each of the instances you created, you can deploy a Hadoop cluster on them exactly the same way you would do with physical hosts (refer to the Appendix for Hadoop deployment tips). In fact, when AWS first came out in 2006, doing it yourself was the only way of organizing instances into any kind of compute cluster.

Things changed when Amazon introduced a higher-level service called Elastic MapReduce (EMR) into its portfolio of AWS services. EMR was one of the first services that operated more like PaaS, rather than an IaaS-level service. With EMR, the details of managing individual hosts and configuring them into a Hadoop cluster are hidden from you. Still, an EMR cluster is comprised of one or more instances and thus can be considered an IaaS-level offering. Instances that are part of an EMR cluster are derived from specific AMIs maintained by Amazon. Think of it as Amazon's own Hadoop distribution packaged in a form of AMIs. There are multiple versions of that distribution (built on top of different versions of Hadoop) available for use. The entire cluster is instantiated and destroyed as a single action, without the need to track and manage all the nodes. In fact, you can even add nodes to the cluster on the fly to speed up computation or take advantage of Amazon's pricing model.

Before you try to spin up the first EMR cluster, however, you need to do a bit of upfront configuration of the environment that would let you interact with AWS.

Before You Begin

Interacting with AWS requires that you set up an account with Amazon. Detailed instructions on how to do that are available from Amazon in their "Getting Started with AWS" tutorial. Once you have your account set up, you are able to log in to the AWS management console (a web application), use the command-line tools, and start making direct AWS API calls from within your business applications. The AWS management console is the easiest way to get started, since it offers an intuitive Web UI for common operations. That said, once you start running more and more jobs in the Amazon cloud, you find yourself asking for an increased degree of automation and programmatic control. One option allows you to opt out of a language-specific SDK to embed AWS cluster and infrastructure management logic directly into your business application. This offers the most precise level of control over cloud infrastructure usage, but comes at the price of investing in orchestrating all the interactions with AWS. The command-line (CLI) tools provide a convenient middle ground between the ad hoc nature of Web UI and a precise control of a language-specific SDK. Even though there are a few different versions of CLI tools available, for the remainder of this chapter, you are using Amazon AWS CLI implementation.

All programmatic interactions with AWS require user authentication. This is done by providing two pieces of information: access key ID (a string of characters similar to IDAKIAIOSFODNN7EXAMPLE) and a secret access key (a string of characters similar to wJalrXUtnFEMI/K7MDENG/bPxRfiCYEXAMPLEKEY).

Both of these can be requested from the Amazon AWS web console during (or after) the registration process. You can provide both of these to the AWS CLI tools either via environment variables or by running the aws configure command and entering the required information at the prompt.

If you haven't installed AWS CLI tools on your desktop, make sure to read and follow the instructions provided at http://docs.aws.amazon.com/cli/latest/userguide/. Keep in mind that, unlike everything else in this book, AWS CLI is implemented in the Python programming language, which may require you to set up a Python environment. The Amazon user guide provides instructions on how to do that. Once the Python environment is available, installing AWS CLI boils down to a few steps, which are outlined in Listing 12-1.

Listing 12-1. Setting up AWS CLI Tools on Linux or Mac OS

```
$ wget https://s3.amazonaws.com/aws-cli/awscli-bundle.zip   # downloading the bundle
$ unzip awscli-bundle.zip                                    # unzipping the bundle
$ ./awscli-bundle/install -b ~/bin/aws    # installing AWS CLI into user's home directory
$ PATH=~/bin:$PATH                          # making sure aws CLI is available on the PATH
$ aws configure                            # configuring AWS CLI with user credentials
AWS Access Key ID [None]: IDAKIAIOSFODNN7EXAMPLE
AWS Secret Access Key [None]: wJalrXUtnFEMI/K7MDENG/bPxRfiCYEXAMPLEKEY
Default region name [None]: us-east-1
Default output format [None]: json
$ aws ec2 describe-regions   # run simplest EC2 request to make sure everything is working
```

After telling aws configure your access and secret keys, and setting the default AWS datacenter to US EAST, you'll have almost everything set up and configured to instantiate your first cluster on the Amazon cloud. However, to remotely log in to individual nodes of the cluster, you have to set up a Secure Shell (SSH) key pair. This is the last bit of authentication information that you need to set up. Think of it this way: while access key ID and secret access keys are used to request services from AWS, once those services instantiate virtual infrastructure for you, an SSH key pair is needed to remotely log in to that infrastructure. SSH key pairs need unique names. You will call this one giraph-keys. If you have an SSH key pair by than name already, you can skip the steps in Listing 12-2.

Listing 12-2. Setting up a Giraph-Keys SSH Key Pair

```
$ aws ec2 create-key-pair  --key-name giraph-keys --query 'KeyMaterial' \
    --output text > ~/giraph-keys.pem
$ chmod 500 ~/giraph-keys.pem
$ aws emr create-default-roles
```

The first command in Listing 11.2 requests creation of the SSH key pair and outputs the private key material into the file called giraph-keys.pem in the user's home directory. The second command is required to make sure that the permissions on the file only allow reading for the user and no one else on the system. These are the default permissions that SSH expects. Finally, the last command makes sure that the user has rights to work with EMR; this needs to be executed only once for each user account.

One thing that you have probably already noticed is the common pattern of how AWS CLI tools are executed from the command line. First, you specify the subset of Amazon cloud APIs that you would like to operate on (e.g., EC2, EMR, etc.). That becomes your main command. Then you select an action (verb) that you need to perform on that subsystem (e.g., creating a key pair or listing the regions). That becomes your subcommand. Finally, you can specify flags that would affect the semantics of the action via --flag options (e.g., give your key pair a name). The result of this command-line invocation is always one or a few AWS API calls that AWS CLI makes on your behalf and gives you the JSON response back. There is no magic in how AWS CLI calls the APIs; your own application can issue the very same API calls by using various language-specific SDKs.

With that in mind, let's proceed with creating your very first functional Hadoop cluster on the Amazon cloud. Amazingly enough, all it takes is a single execution of the AWS CLI tools.

Creating Your First Cluster on the Amazon Cloud

Most of the actions required to manipulate Hadoop clusters on the Amazon cloud are provided by the emr command. You can run aws emr help to see which subcommands are available. You are using quite a few of them in the following sections, but for now, let's create the most basic Hadoop cluster using the create-cluster subcommand, as shown in Listing 12-3. Note that, once again, you are using backslash characters to break lengthy single command lines for readability purposes.

Listing 12-3. Creating Your First Cluster on the Amazon Cloud

```
$ aws emr create-cluster    \
  --ami-version 2.4.11          \
  --instance-groups            \
    InstanceGroupType=MASTER,InstanceCount=1,InstanceType=m3.xlarge     \
    InstanceGroupType=CORE,InstanceCount=2,InstanceType=m3.xlarge         \
  --no-auto-terminate                                                        \
  --ec2-attributes KeyName=giraph-keys
{
    "ClusterId": "j-DEADBEEF"
}
```

Congratulations! You have just instantiated your first cluster on the Amazon cloud. The cluster consist of one MASTER node and two CORE nodes (three nodes total); it is capable of running MapReduce jobs. The invocation you used to create the cluster is long, but pretty self-explanatory. You had to invoke the emr subcommand of aws and give it a verb asking to create a cluster: create-cluster. That action required you to specify the version of Hadoop that you wanted to use (--ami-version 2.4.11), and then the topology of your cluster (--instance-groups), describing the role of each node, the overall number of nodes assigned to a given role, and an AWS instance type associated with each node. You also asked not to terminate the cluster immediately (--no-auto-terminate) and provided a key pair so that you can log in to the cluster later on (--ec2-attributes KeyName=giraph-keys). This is the minimum amount of information that you need to give to AWS EMR to create a cluster. The result of running this command was a cluster ID (j-DEADBEAF) output formatted as JSON. Keep an eye on that cluster ID. You are using it for a few other things in the chapter.

Now that the Hadoop cluster is up and running in the Amazon cloud, what can you do with it? It would be awesome if at this point you could run Giraph applications without doing much else. Unfortunately, Amazon EMR clusters don't come with Giraph bits preinstalled the same way that they bundle Apache Hive, Pig, and some of the other Hadoop ecosystem projects. Before proceeding to making Giraph available on the cluster, let's first make sure that you can at least run a simple MapReduce job on it.

The way you are going to launch your test MapReduce job on the freshly minted cluster is simply by logging into its gateway node and using a preinstalled hadoop command-line utility to launch a Pi job (Hadoop's answer to HelloWorld in MapReduce), as shown in Listing 12-4. Note that setting up a cluster can take quite a bit of time. Hence, the first command may not give you a login on the cluster for several minutes while waiting for the cluster to be fully online.

Listing 12-4. Logging into the Gateway Node of the EMR Cluster and Running a Pi MapReduce Job

```
$ aws emr ssh                                     \
  --cluster-id j-DEADBEEF                    \
  --key-pair-file ~/giraph-keys.pem
Waiting for the cluster to start.
ssh -i ~/giraph-keys.pem hadoop@ec2-54-82-189-117.compute-1.amazonaws.com

% hostname    # printing the hostname to make sure we are really in the cloud now
ip-172-31-51-237.ec2.internal
% hadoop jar hadoop-*examples.jar pi 10 100   # lets run a test job
...
Estimated value of Pi is 3.14800000000000000000
% exit # exiting a cluster back to our workstation
```

What you have done here is used the `ssh` action from the EMR collection of APIs. You have supplied the cluster ID (`--cluster-id`) from the output given in Listing 12-3, and also the same key pair that you used to create the cluster in the first place (`--key-pair-file`). The result of running this command is a regular SSH session that gives you a shell on the gateway node of the cluster. All the commands prefixed with % are at that point executing on the remote gateway node. The first command you ran on the gateway printed its `hostname` to make sure that you're really in the cloud. After that, you used the `hadoop` command-line utility to run the Pi calculation job, and you received an expected (if somewhat imprecise) result. The cluster definitely looks and feels like a real Hadoop cluster. The last thing you did was exit (`exit` command) the gateway node back to your workstation.

Please keep in mind that at this point, you have three nodes running on the Amazon cloud. Even though they are not doing any useful work, Amazon still charges you for them. Once you are done experimenting, make sure to destroy your cluster(s) and check that there are no leftover nodes running, as shown in Listing 12-5.

Listing 12-5. Finding Leftover Cluster(s) and Terminating Them

```
$ aws emr list-clusters
$ aws emr terminate-clusters --cluster-id j-DEADBEAF
```

The first command gives you a very detailed JSON output that describes all the clusters that Amazon is or has been maintaining on your behalf. You need to look for the ones that are not listed as TERMINATED and terminate them with the `terminate-clusters` action.

The Building Blocks of an EMR Cluster

Most of the actions that happen on the Amazon cloud can be decomposed into operations on virtual servers. Those virtual servers are called instances, and to a user of the Amazon cloud, they are indistinguishable from racks of real servers. Just like any real server, instances have a copy of an operating system running, and you can either get a remote shell into that operating system (via SSH) or interact with other software services (such as a web or a database service) running on these virtual servers. Unlike real servers, though, you can create as many instances as you are willing to pay for.

Each instance running on the Amazon cloud is defined by

- An image of a fully configured operating system. This is known as Amazon Machine Image (AMI) and essentially consists of the content of a disk drive from which an operating system boots.

- The amount of physical resources (e.g., the number of CPUs, RAM, and disk size).

There are thousands of AMIs with all sorts of operating systems available on the Amazon cloud today. Some are maintained by operating system vendors and some are maintained by volunteers. Some are free (like the ones based on Linux) and some cost money to use (like the ones based on Microsoft Windows). And if you don't find an AMI that works for you, Amazon gives you all the tools to maintain your own AMIs. If you are looking for your favorite operating system image, a good place to do it is on the Cloud Market web site at http://thecloudmarket.com. Keep in mind that the decision on which AMI to run is orthogonal to the amount of resources given to it.

While the Amazon cloud doesn't allow you to specify an arbitrary number of resources to give to an instance, you can choose from a preselected set of templates. These templates are called *instance types*. Amazon provides a detailed overview of the different types of resources assigned to each instance type at http://aws.amazon.com/ec2/instance-types/. This is also a good place to visit if you want to understand the billing implications of running different instance types.

In order to facilitate management operations, instances can be grouped into chunks called *reservations* (or instance groups). All instances in a group are identical in terms of what image they use (AMI specification) and the number of resources given to each instance in a group (instance type). This makes it easy to tell the Amazon cloud to always maintain a number of identical servers. Regardless of whether you ask for one instance in a reservation or a hundred, you're still creating a group.

These are the basic building blocks that Amazon's IaaS layer provides you. By using them, you can opt out of rolling your own Hadoop cluster or you can ask Amazon services to do it for you.

The Composition of an EMR Cluster: Instance Groups

Building off the fundamental notion of an instance group, an Amazon EMR cluster consists of a number of nodes belonging to three different instance groups.

- *The MASTER instance group.* This instance group always contains just one node that is configured to run all the centralized services of a Hadoop cluster: HDFS NameNode, YARN ResourceManager, or MapReduce JobTracker. It is configured as a gateway so that you can manually launch ad hoc MapReduce or YARN jobs by logging into it using the aws emr ssh action. Amazon also makes sure that client-side bundled software such as Hive, Pig, and so forth, are available by default. Unfortunately, Giraph is not part of this list and needs to be installed on this node separately. If you are running a cluster with a single node, that node must belong to this group. It will also run all the services from a CORE instance group.

- *The CORE instance group.* This instance group contains one or more nodes that function as Hadoop slave nodes. The two fundamental services that each node in this group runs are HDFS DataNode (for storing data) and YARN NodeManager/ MapReduce TaskTracker. The nodes in this group do all the data processing. Whenever you want to speed up your computation, you can dynamically add nodes to this group, but you cannot dynamically remove them.

- *The TASK instance group.* This instance group is optional and contains nodes that function almost exactly as the nodes in the CORE instance group, with one exception: they do not run HDFS DataNode services and thus are not capable of processing data locally (the data always needs to be fetched from an external location). Because the nodes in this instance group are completely ephemeral (they don't store HDFS blocks), you can increase and decrease the number of nodes in this group. This instance group is convenient when processing data that tends to reside in external repositories, such as Amazon's Simple Storage Service (S3), or when you want to use spot instances to reduce operational costs.

Putting it all together, you can now make sense of the first two options given to the aws emr create-cluster subcommands (repeated here in Listing 12-6).

Listing 12-6. Creating Your First Cluster on the Amazon Cloud

```
$ aws emr create-cluster
  --ami-version 2.4.11
  --instance-groups
    InstanceGroupType=MASTER,InstanceCount=1,InstanceType=m3.xlarge
    InstanceGroupType=CORE,InstanceCount=2,InstanceType=m3.xlarge
```

The first option (`--ami-version`) specifies which AMI to use for spinning up instances in all three groups. This is an indirect specification, since instead of referring to an AMI ID you are actually referring to the version of the EMR stack maintained by Amazon. Think of it as any other Hadoop distribution: you can only select the version of the distribution itself. You don't get to choose versions of Apache Hadoop and its ecosystem components within the distribution. If you are curious about which versions of Amazon EMR stack map to which components, you can visit the web page at `http://docs.aws.amazon.com/ElasticMapReduce/latest/DeveloperGuide/ami-versions-supported.html`. The version that you used in the example corresponds to the Hadoop 1.0.x based stack. It must be noted that there is absolutely nothing mysterious about the AMI that Amazon maintains for you as part of a given Hadoop stack. You can create an instance of that AMI the same way you can create an instance of any other AMI with an arbitrary OS.

The second option that you are giving to the `create-cluster` command (`--instance-groups`) is an exact specification of each instance group's size and type. It consists of one, two, or three lists of key value pairs describing each of the three instance groups that you're adding to the cluster (MASTER is the only mandatory group). The following keys are required.

- `InstanceGroupType` defines the instance group. The values here are MASTER, CORE, or TASK.

- `InstanceCount` defines the number of instances (nodes) in the group.

- `InstanceType` specifies the instance type (the amount of resources each node is given) that every node in the group gets. The values here are any valid instance type specification recognizable by AWS.

The graphical representation of the cluster that you created is shown in Figure 12-1.

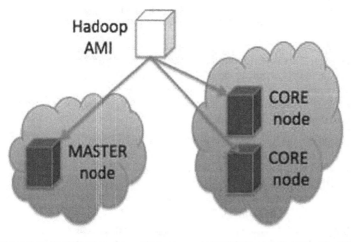

Figure 12-1. *Cluster composed of two instance groups and three nodes*

The white box in Figure 12-1 represents an AMI template from which actual instances are derived and placed into requested instance groups: one into a MASTER instance group and two into a CORE instance group.

Deploying Giraph Applications onto an EMR Cluster

As mentioned, Amazon doesn't bundle Giraph as part of its Hadoop stack. It would be nice if that changes in the future (Amazon is known to listen to its customers' requests), but given that Giraph is primarily a framework, rather than an application, it is not that big of a deal.

From a Hadoop cluster's point of view, any graph processing application built on top of Giraph looks like a custom MapReduce or YARN job. With that in mind, let's create a jar that contains one of the example applications from Chapter 5 and also bundles Giraph and all of its dependencies. This is pretty easy to do with Maven. All you need to do is tell Maven to create a jar-with-dependencies assembly, as outlined in Listing 12-7 (just make sure to go back to the folder you created for the example application in Listing 5-4).

Listing 12-7. Deploying a Giraph Application to an EMR Cluster

```
$ mvn clean compile assembly:assembly -DdescriptorId=jar-with-dependencies
$ aws emr put --cluster-id j-DEADBEEF --key-pair-file ~/giraph-keys.pem \
  --src target/*jar-with-dependencies.jar
```

Your first command instructed Maven to build a self-contained JAR file that you then copied to the gateway node of the EMR cluster using the put action. As usual, you had to specify the cluster ID (--cluster-id) and key pair (--key-pair-file) to allow access to a given cluster. An extra argument that you had to give to the put action was the location of the JAR file to copy (--src).

At this point, you could have logged into the EMR cluster using the aws emr ssh command and executed the graph processing example the same way you executed the Pi estimation job in Listing 11-3. Instead, you will look into scheduling it remotely as one of the steps in EMR cluster processing specification.

EMR Cluster Data Processing Steps

The most common use case for EMR clusters is to create them on demand, keep them running for as long as there are data processing steps to be executed, and then let the Amazon cloud tear the cluster down. Of course, you can request the cluster to stick around and wait for an explicit tear-down API call. This is exactly what you did with the --no-auto-terminate option. Having a cluster available for interactive use is convenient for exploring EMR. However, once you start moving more into full automation of your data pipelines, it becomes more natural to let the Amazon cloud manage the cluster life cycle.

If you want to rely on the Amazon cloud for end-to-end cluster life cycle management, there are a few additional bits of information you need to communicate to AWS when creating your cluster:

- *The set of additional applications to be installed on an EMR cluster.* Currently, you can install Hive, Pig, HBase, Ganglia, Impala, and MapR distribution that way.

- *The set of bootstrapping actions you would like to perform after the instances are created but before the cluster gets started.* This could include things like installing additional software, tweaking clusters, and operating system configuration. Very simply put, bootstrap actions are scripts that reside on externally visible storage (S3, HTTP, etc.) and are run on the cluster nodes.

- *The set of data processing steps you would like to perform once the cluster is up and running.* Currently these steps can include the following:

  - Running a custom MapReduce or YARN job from a self-container jar

  - Running a custom script on the MASTER node

  - Running a Hadoop streaming job

  - Running a Hive, Pig, or Impala query

Using both bootstrapping and data processing actions follows a very similar pattern. Since you don't need any special tweaks on the cluster, we will hold off showing example bootstrapping actions until later and will focus on data processing steps for now. An interesting side note is that while it is not possible to request additional bootstrapping actions for a cluster that is already up and running, it is possible to request additional data processing steps.

When using AWS CLI tools, both bootstrapping actions and data processing steps need to be communicated to the `aws emr create-cluster` command via either JSON specification or a shorthand notation. Listing 12-8 uses JSON specification to add a Giraph data processing step to an already running cluster.

Listing 12-8. Running a Giraph Data Processing Step

```
$ aws emr add-steps --cluster-id j-DEADBEAF
  --steps '[{
"Type":"CUSTOM_JAR",
"Name":"GiraphApp",
"MainClass":"org.apache.giraph.GiraphRunner",
"ActionOnFailure":"CONTINUE",
"Jar":"file:///home/hadoop/book-examples-1.0.0-jar-with-dependencies.jar",
"Args":["GiraphHelloWorld",
        "-vip", "input",
        "-vif", "org.apache.giraph.io.formats.IntIntNullTextInputFormat",
        "-w", "1"]}]'

{ "StepIds": ["s-STEPID"]}
```

The preceding command added a processing step for immediate execution on an already available cluster identified by `--cluster-id`. The description of processing steps is a well-formatted JSON array with one entry per processing step that you wish to run on the cluster. Each entry consists of the following key-value pairs:

- Type: Defines a type of a processing step. In this case, you requested a custom jar execution via Java. There are other types of steps available, including the execution of Pig or Hive scripts, streaming jobs, and other actions. CUSTOM_JAR type is the most flexible.

- Name: A symbolic name identifying the processing step in all future outputs; it can be any string.

- MainClass: For a CUSTOM_JAR type of a step; identifies the name of the entry point class.

- ActionOnFailure: Specifies what to do if the step fails. Valid values include CONTINUE, TERMINATE_CLUSTER, and CANCEL_AND_WAIT.

- Jar: For a CUSTOM_JAR type of a step; specifies location of the JAR file.

- Args: For a CUSTOM_JAR type of a step; specifies arguments to be given to the main method.

The output of the command gives you a step ID to refer to if you want to find out more about the step(s) status or result. Finally, if you are wondering whether you could've avoided the lengthy JSON specification types on the command line, the answer is yes. The `--steps` flag can read JSON from files, provided you give it a file URL instead of an actual JSON like this: `--steps file://./step.json`.

In general, adding data processing steps to an EMR cluster is pretty trivial. In fact, you can specify a bunch of steps when you request the cluster to be created by providing the very same `--steps` JSON specification to the `create-cluster` action. That way, you can have the Amazon cloud spin up a cluster, run a series of data processing steps, and at the end, tear down the cluster, thus minimizing the amount of time you have to pay for EC2 resources. You will try that later in this chapter, but for now, let's see whether your processing step was successful or not. You can do it by running the query command shown in Listing 12-9 and looking for the State key in the JSON output.

Listing 12-9. Running Giraph Data Processing Step

```
$ aws emr describe-step --cluster-id j-DEADBEEF --step-id s-STEPID
{
    "Step": {
        "Status": {
            "Timeline": {
                "EndDateTime": 1440475780.643,
                "CreationDateTime": 1440475759.648,
                "StartDateTime": 1440475780.635
            },
            "State": "FAILED",
            "StateChangeReason": {}
        },
....
```

Believe it or not, the data processing step failed. You now have to find out why this happened.

When Things Go Wrong: Debugging EMR Clusters

When anything goes wrong with your EMR cluster or a job that it was supposed to process, you need to be able to track what went wrong. Typically, your best option is to analyze the log files and try to deduce the source of failure. There are two sources of log files available for EMR clusters: aggregated logs on S3 and local log files on cluster nodes. On a cluster that is still up and running, you can `ssh` into a gateway node and go directly to the log files collected under `/mnt/var/log/`. For example, to figure out why the data processing step from the previous section failed, let's take a look at `/mnt/var/log/hadoop/steps`. That folder contains subfolders corresponding to each data processing step submitted to a cluster. Underneath these subfolders, you see three log files:

- `controller`: A log file containing the exact invocation of a data processing step. In this case, it has a custom jar Hadoop invocation command.

- `stdout`: A log file with the output of whatever controller command was produced on its standard output stream.

- `stderr`: A log file with the output of whatever controller command was produced on its error output stream.

First, let's take a look at what the controller invoked when you submitted the Giraph data processing step via a custom JAR. You see something very similar to Listing 12-10.

Listing 12-10. Contents of /mnt/var/log/hadoop/steps/s-STEPID/controller

```
2014-07-14T19:36:33.251Z INFO startExec 'hadoop jar /home/hadoop/book-examples-1.0.0-jar-
with-dependencies.jar org.apache.giraph.GiraphRunner GiraphHelloWorld -vip inputt -vif org.
apache.giraph.io.formats.IntIntNullTextInputFormat -w 1 -ca giraph.SplitMasterWorker=false'
...
```

This looks like exactly the same command that you would've submitted manually to run a Giraph workflow via the 'hadoop jar ...' command. Perhaps if you take a look at the error execution that command produced, you would have a clue as to why it failed. What you see there is very similar to Listing 12-11.

Listing 12-11. Contents of /mnt/var/log/hadoop/steps/s-STEPID/stderr

```
Exception in thread "main" java.lang.IllegalArgumentException: Invalid vertex input path
(-vip): input
at org.apache.giraph.utils.ConfigurationUtils.populateGiraphConfiguration(ConfigurationUti
ls.java:406)
at org.apache.giraph.utils.ConfigurationUtils.parseArgs(ConfigurationUtils.java:207)
    at org.apache.giraph.GiraphRunner.run(GiraphRunner.java:74)
```

Of course, this makes sense—you tried running a Giraph application on a nonexistent input data set. How to make datasets available to the EMR clusters and get the results is the subject of the next section.

Where's My Stuff? Data Migration to and from EMR Clusters

EMR clusters are an ephemeral collection of nodes. They are created on demand to accomplish a certain sequence of data processing steps; they are typically destroyed immediately afterward. This is a convenient, cloud-friendly model of utilizing resources, but it does require that the initial dataset comes from some permanent storage location and a resulting dataset is somehow captured before the cluster and its HDFS storage layer are gone.

In general, the Amazon cloud relies on S3 as its permanent storage layer. Almost all AWS services support S3 as a data source or data sink. For example, AMI are stored in S3; it can also be used as a target for logging most AWS services. S3 stores all of its objects in buckets. You can think of buckets as top-level file system folders that you manage under your AWS account. Once you create a bucket, you can use it to create file-like objects holding any kind of data. An arbitrary, unique key that is no more than 1024 bytes long references each object. It is convenient (but in no way enforced) to name your objects as though they were files in a filesystem with a traditional Unix path name convention of folders delimited by the / character. Following this convention makes it natural to reference S3 objects using either S3 or HTTP URIs. For example, if you create a bucket with the name giraph.examples and put an object named datasets/1/data.txt in it, you can then reference it as either s3://giraph.examples/datasets/1/data.txt or https://s3.amazonaws.com/giraph.examples/ datasets/1/data.txt.

Keep in mind that bucket names share a global namespace on AWS. This means that you may need to get creative. Chances are, any simple name like mybucket is already taken. With that in mind, the first steps of suppling the Giraph application with data are to create a unique bucket and copy the example input data you used in Chapter 5 under it, as shown in Listing 12-12.

Listing 12-12. Putting Input Data into S3

```
$ aws s3 mb s3://unique.bucket
$ aws s3 cp src/main/resources/1/graph.txt s3://unique.bucket/datasets/1/graph.txt
```

Now that the example graph description is available in S3, you are half way to making it available for Giraph processing. Amazon EMR conveniently offers a custom JAR file that implements efficient data transfer from S3 buckets into HDFS, where Giraph can pick it up. The way that you're going to trigger this transfer is by adding yet another step to the cluster, as shown in Listing 12-13. Note that this time you're using a shorthand notation to specify all the required key-value pairs instead of using a properly formatted JSON object. Either way of doing it is fine.

Listing 12-13. Transferring Data from S3 to HDFS via Custom JAR Step

```
$ aws emr add-steps --cluster-id j-DEADBEAF                                       \
        --steps Type=CUSTOM_JAR,Name=UploadInputData,                            \
          Jar=/home/hadoop/lib/emr-s3distcp-1.0.jar,            \
          Args=--src,s3://unique.bucket/datasets/1/,        \
                       --dest,hdfs:///user/hadoop/input
```

The arguments that you are giving to the custom JAR specify the source S3 bucket via `--src` and destination in HDFS via `--dest`. Of course, this step can work in the other direction as well. If you have any datasets that you need to capture after the cluster is destroyed, you can simply add a step that copies a bunch of files from HDFS into an S3 bucket.

A word of caution: S3 is expensive and it holds your data until you explicitly delete it; use it only for final results. Any data that remains in S3 is charged to your account on a monthly basis based on its size.

At this point you could just rerun the graph processing step from Listing 12-8 to see that this time it runs to completion with an exist status being `COMPLETED`. Instead of doing this, however, let's tie it all together to see how a true ephemeral cluster can be used for data processing.

Putting It All Together: Ephemeral Graph Processing EMR Clusters

At this point, you have everything you need to consider a graph processing job running on the Amazon cloud that is as close as possible to what you would actually use in production. You are sticking with your good, old friend the "Hello World" graph processing example, but you will make sure that the data and custom JAR file are uploaded to the ephemeral EMR cluster from S3 and the resulting data set is recorded in one of the S3 buckets. You will also make sure that all the logs are transferred to S3, in case you need to do any kind of diagnostics after the cluster disappears. Given what you need to do, the definition of the data processing steps gets to be pretty long. Instead of specifying it all on the command line, you're going to use a JSON file, as shown in Listing 12-14.

Listing 12-14. steps.json a JSON Definition of Data Processing Steps

```
[{"Type"            : "CUSTOM_JAR",
   "Name"            : "UploadInputData",
   "ActionOnFailure" : "CANCEL_AND_WAIT",
      "Jar"            : "/home/hadoop/lib/emr-s3distcp-1.0.jar",
      "Args"           : [ "--src", "s3://unique.bucket/datasets/1/",
                          "--dest","hdfs:///user/hadoop/input"]},
```

```
{"Type"            : "CUSTOM_JAR",
 "Name"            : "GiraphJob",
 "MainClass"       : "org.apache.giraph.GiraphRunner",
 "ActionOnFailure" : "CANCEL_AND_WAIT",
 "Jar"             : "s3://unique.bucket/jars/giraph-hello.jar",
 "Args"            : [ "GiraphHelloWorld",
        "-vip", "input",
        "-op", "output",
        "-vif", "org.apache.giraph.io.formats.IntIntNullTextInputFormat",
        "-vof", "org.apache.giraph.io.formats.GraphvizOutputFormat",
        "-w", "1",
        "-ca", "giraph.SplitMasterWorker=false"]},

{"Type"            : "CUSTOM_JAR",
 "Name"            : "DownloadOutputData",
   "ActionOnFailure" : "CANCEL_AND_WAIT",
 "Jar"             : "/home/hadoop/lib/emr-s3distcp-1.0.jar",
 "Args"            : [ "--src", "hdfs:///user/hadoop/output",
                       "--dest","s3://unique.bucket/output/"]}
 ]
```

An interesting side note is that since EMR clusters allow custom jars to come from S3, you don't actually need to invoke an extra put action. All you need to do is to make the GiraphHelloWorld JAR available in S3, as shown in Listing 12-15.

Listing 12-15. Copying GiraphHelloWorld JAR to S3 Bucket

```
$ aws s3 cp target/*-jar-with-dependencies.jar s3://unique.bucket/jars/giraph-hello.jar
```

Now that you have all the required files available to you in S3, the only thing left to do is start up an ephemeral EMR cluster and make it run through the data processing steps specified in steps.json, as shown in Listing 12-16 and using the familiar steps. But you also need to add a location for where to store logs (--log-uri) and request to terminate the cluster once the processing steps are done (--auto-terminate).

Listing 12-16. Starting an Ephemeral Graph Processing Job on EMR

```
$ aws emr create-cluster --ami-version 2.4.11     \
   --instance-groups                                        \
     InstanceGroupType=MASTER,InstanceCount=1,InstanceType=m3.xlarge    \
   --log-uri s3://unique.bucket/logs/ --auto-terminate                            \
   --steps file://./steps.json
{
    "ClusterId": "j-DEADBEEF2"
}
```

As before, the command outputs the cluster ID and you need to wait for this cluster to exit before you can inspect the overall execution logs. It is convenient to monitor the status of the cluster using the command shown in Listing 12-17. This commands prints the status of the cluster based on the JSON output of the describe-cluster action processed by a custom JSON query request, specified via the --query flag. This is a convenient way to cut down on the amount of JSON output you are getting. Refer to aws help for information on how JSON queries can be specified.

Listing 12-17. Monitoring Status of the Cluster

```
$ aws emr describe-cluster --cluster-id j-DEADBEEF2
   --query Cluster.Status.State
"TERMINATED"
```

Once the cluster is terminated, all that you need to do is inspect the logs left in S3, make sure that the processing steps were completed successfully, and that you get the final output of the job. This can be done by a series of s3 commands, as shown in Listing 12-18. The first command copies the output of the Giraph application run to your workstation, under the file name output.txt, and the second is an example of how you can list the contents of a part of the bucket (as though it was a folder) looking for log files that may be of interest.

Listing 12-18. Inspecting Logs and the Output After the Cluster Is Terminated

```
$ aws s3 cp s3://unique.bucket/output/part-m-00000 output.txt
$ aws s3 ls s3://unique.bucket/logs/j-DEADBEEF2/
                        PRE daemons/
                        PRE jobs/
```

At this point, all of your output data (including the logs) is available to you in S3. This is great for centralized storage, but it could end up being quite expensive if you don't purge the logs that you no longer need and prune the output datasets.

Reducing cost while running on EMR is one of the top concerns, and storage costs can contribute greatly to the final bill. Amazon offers various attractive pricing models that make the usage of its EMR clusters cost-effective, however. One such model, Spot Instances, is reviewed in the next section.

Getting the Most Bang for the Buck: Amazon EMR Spot Instances

The pricing model of the Amazon cloud that you have seen so far assumed a fixed billing rate associated with every instance type. It is not, however, the only model. A different way to pay for using compute resources is to bid a certain amount of money on unused capacity within the Amazon cloud. This is applicable to all instance types, and the price per hour typically ends up being less than a flat billing rate of an instance type of the same kind. The downside to spot instances is that Amazon shuts them down when demand increases, and your bidding price falls below the going rate. This is not a big deal for traditional MapReduce workloads (after all, Hadoop was built to withstand node failures) and it gives you a very convenient way to utilize cheap compute resources to speed up data processing.

A typical use for spot instances with EMR clusters is to use them as part of the TASK instance group so that nodes can be added (and removed) at any time without disrupting the cluster operations. In order to indicate that an instance group needs to have a spot instance billing enabled, all you need to do is add the BidPrice property to the instance group specification, as shown in Listing 12-19.

Listing 12-19. Creating a Cluster with a Spot Instance TASK Instance Group

```
$ aws emr create-cluster --ami-version 2.4.11      \
    --instance-groups                                       \
InstanceGroupType=MASTER,InstanceCount=1,InstanceType=m3.xlarge       \
InstanceGroupType=CORE,InstanceCount=2,InstanceType=m3.xlarge          \
InstanceGroupType=TASK,InstanceCount=3,InstanceType=m3.xlarge,BidPrice=0.10 \
    --log-uri s3://giraph.examples/logs/ --auto-terminate     \
    --steps file://./steps.json
```

If you run the preceding command, the Amazon cloud immediately creates three nodes for you: one in the MASTER instance group and two in the CORE instance group. These nodes are used to manage storage (HDFS) on the cluster and provide a minimum guaranteed amount of compute resources available to the graph processing application. You may also expect an additional three nodes from the TASK instance group to be added to your cluster at any time (and also taken away from you at any time). For a traditional MapReduce application, these additional nodes give a temporary performance advantage. Giraph applications, however, present a slight complication to this model.

Most traditional MapReduce workloads don't have hard requirements on the overall number of compute nodes available to mappers. If some of the nodes fail, most likely, it will result in a slower overall execution, but your job will finish anyway. Giraph uses mappers as parallel workers and expects a certain number of them (the value specified via the --w option) to be available at all times. If a node fails, Giraph expects Hadoop to reschedule a mapper on a different node, which is only possible if the overall number of compute resources available to a Giraph application is more than the number of workers it expects. This complicates spot instance usage with Giraph applications. After all, the whole point of spot instances is to provide a temporary boost to the performance of your cluster when the price is right and to scale back when the price increases.

Using spot instances with Giraph is considered an experimental feature; it is only recommended to advanced users. If you are interested in playing with it, make sure to pay attention to the following configuration properties that you would have to explicitly specify in your configuration (either via the --ca command-line option or giraph-site.xml).

- giraph.maxWorkers: The total number of workers that Giraph can utilize on a cluster. Experiment with setting it according to the maximum number of nodes you expect Amazon to give your cluster (CORE + TASK).

- giraph.minWorkers: The minimum number of workers that Giraph needs to proceed to the next superstep. Experiment with setting it according to the guaranteed number of nodes you expect Amazon to give to your cluster (CORE).

- giraph.minPercentResponded: The minimum percentage of healthy workers needed to proceed to the next superstep. Experiment with setting it to about $100 \times$ giraph.minWorkers ÷ giraph.maxWorkers.

- giraph.checkpointFrequency: The number of consecutive supersteps between checkpoints of the worker state. Experiment with setting it to 1 and increasing according to the performance characteristics of your application.

At this point, you have seen all the different ways to run vanilla EMR clusters. But what if you need to optimize price, performance, or both to best suit your graph processing application?

One Size Doesn't Fit All: Fine-Tuning Your EMR Clusters

So far you have been using a vanilla configuration of an EMR cluster without paying too much attention to how well it is optimized to run your graph processing jobs. This is a fine first step (and these simple examples don't require much else anyway), but for any real workload, some level of tuning is required. There are two related reasons why you want to tune your EMR cluster: performance and cost.

Every Giraph application is different, but given its flexible architecture, any application utilizing Giraph can stress all the three basic resources: CPU, memory, and I/O. All the performance tuning advice that you saw in the previous chapter still applies to running Giraph in the cloud. Of course, you need to view it through the lens of actually being charged for all resources usage. Thus, identifying the minimum number of resources that still gives you adequate performance becomes the key.

The most rewarding choice you can make on EMR is the right instance size for your job. As mentioned, EMR offers dozens of different instance sizes, and guessing the right one can be intimidating. The best advice here is to start on the small side, deploy performance monitoring tools such as Ganglia, and then experiment with different instance sizes to find the best trade-off of cost vs. performance. Keep the instance type reference table at `http://aws.amazon.com/ec2/instance-types/` handy and do a few experiments to see whether you're on the right track.

Some workloads exhibit the best performance when run on instances with a lot of RAM. If you fall in that category, make sure to tell EMR to configure your cluster for such an environment. In general, any kind of cluster-specific tweaks can be achieved as part of the bootstrapping actions for the cluster. Bootstrapping actions are nothing but scripts that are executed as part of the cluster bring up and affect Hadoop configuration files. For example, the easiest way to configure Hadoop for a memory-intensive workload is to add a bootstrap action referencing Amazon's own script, as shown in Listing 12-20.

Listing 12-20. Configuring Hadoop for Memory-Intensive Workloads

```
$ aws emr create-cluster -- bootstrap-actions                                    \
    Path=s3n://elasticmapreduce/bootstrap-actions/configurations/latest/memory-intensive, \
    Name=mem-config                                                             \
    Args=string1,string2                                                        \
  --ami-version 2.4.6 ...
```

Since bootstrapping actions are the main tool for tweaking Hadoop configuration, it is recommended that you download the `memory-intensive` script mentioned in Listing 2-20 and use it as a reference for creating your own bootstrapping actions.

Another fundamental variable you need to consider is the number of instances in CORE and TASK instance groups. Once again, EMR makes it easy to start small and dynamically adjust the size of the TASK instance group to see if it helps speed up the processing.

Finally, when working with any AWS service, make sure that everything that you do is confined to the same AWS region; and better yet, the same availability zone. Cross-region data transfers are costly and provide much less bandwidth compared to the I/O within the same region. The best-case scenario is for all the elements of your data processing cluster to be part of the same availability zone.

Summary

Running Giraph in the cloud provides a scalable, elastic, and easy-to-use way of executing graph processing applications, without worrying about managing the infrastructure. All the cluster configuration is done for you, yet leaving enough flexibility to do all the needed tweaks. Not all clouds are created equal. While this chapter only talked about the Amazon cloud, the same core principles are applicable to other public and private clouds. If your cloud provider is not Amazon, at least you now know what services to look for.

APPENDIX A

■ ■ ■

Install and Configure Giraph and Hadoop

This chapter covers

- System requirements for running Giraph and Hadoop
- Installation methods for Giraph and Hadoop
- Different modes for running Hadoop for Giraph applications
- Configuring Hadoop for all three different running modes

Throughout this book, it was assumed that you had a working Hadoop installation available to run examples and experiment with Giraph. Since Giraph runs on top of Hadoop, having a working Hadoop environment is a fundamental prerequisite. This appendix begins by looking into system dependencies for Hadoop and Giraph deployments. It proceeds to describe various methods of installing Hadoop and Giraph, showing their relative strengths and weaknesses. Next, it looks into the different types of Hadoop deployments and how Giraph deals with the different versions of Hadoop ecosystem projects that it needs to leverage. Finally, this appendix outlines the basics of configuration management for both Hadoop and Giraph and discusses which Hadoop configuration is required for running it in different execution modes.

System Requirements

Both Giraph and Hadoop are implemented in the Java programming language and can run on Unix-based, Max OS X, and Windows systems. Giraph requires Oracle JDK version 7 or higher. Since Giraph jobs are executed by the same JVM that runs the Hadoop framework, this puts a lower bound on the JDK version that can be used for Hadoop deployment. What this means is that if you have an existing Hadoop cluster that you need to use for running Giraph applications, you have to make sure that it was deployed using JDK 7 or above.

You also need JDK installed on the host where Giraph applications will be launched and on the host that is going to be used for Giraph application development. Both Giraph and Hadoop have been wildly tested on Oracle's JDK, with OpenJDK (an open source, community-driven version of Oracle's JDK) a close second choice for deployment. If you don't have JDK installed and you want to use Oracle's version, go to www.java.com/jdk and follow the installation procedures for your operating system. If, on the other hand, you decide to go the OpenJDK route, you may find it bundled for you operating system by a vendor.

Regardless of how you install JDK, make sure that the location of installation tree is available to all your applications via environment variable JAVA_HOME. If you want that location to also supply the binaries for all Java command-line utilities (including launching JMV itself), you may want to update your PATH with that setting. On Unix platforms, it is often convenient to set up those values in a global shell startup file such as /etc/profile, ~/.bashrc or ~/.bash_profile, similar to what is shown in Listing A-1.

Listing A-1. Setting up JAVA_HOME and Adding Java Binaries to Your PATH

```
JAVA_HOME=/usr/lib/jvm/default-java
export JAVA_HOME

PATH=$JAVA_HOME/bin:$PATH
```

■ **Note**   When it comes to the operating systems, various flavors of Linux are the most widely tested deployment platforms. While it is possible to run Giraph and Hadoop on Mac OS X and even Windows, these platforms are mostly used for development.

Hadoop Installation

Unless you already have a Hadoop cluster available to run Giraph applications on (either on-premises or as a cloud-based service offering), your first step is to decide which major version of Hadoop you would like to use. Currently, there are two major versions available: Hadoop 1.x and Hadoop 2.x. Both of these are considered stable and can be safely used in production. The difference between the two boils down to Hadoop 1.x slowly transitioning into a maintenance mode with very little development activity going on. In contrast to that, Hadoop 2.x development activity remains very high, with bug fixes and new feature development progressing at a brisk pace. Another fundamental difference between these two versions is the architecture of the MapReduce framework. While Hadoop 1.x offers a faithful implementation of MapReduce framework as it is described in the original Google paper, Hadoop 2.x takes it one step further, essentially providing a MapReduce v2 implementation as an application sitting on top of a general-purpose resource scheduler called YARN.

What was wrong with MapReduce v1?   The original implementation of MapReduce (now known as MapReduce v1) made an architectural decision of conflating MapReduce-specific logic with lower-level cluster resource management and scheduling. While this provided the fastest route to a functional implementation (and paved the way for Hadoop's world domination), it also suffered from a number of technical limitations: scalability concerns, failure tolerance, and difficulty in running non-MapReduce frameworks on the same clusters among the top issues. Indeed, as you saw in Chapter 6, Giraph implementation has to trick MapReduce v1 into thinking that it is running as a generic map-reduce application, while in reality, Giraph applications don't map at all into the generic MapReduce model.

Hadoop 2.x tries to fix these limitations by providing low-level cluster resource management and scheduling capabilities as an independent layer called YARN (Yet Another Resource Negotiator) and running the MapReduce framework on top of YARN. This architecture makes it possible for other distributed frameworks to run side by side with MapReduce applications, without the need to pretend to map into the MapReduce model. It is worth repeating that this is purely an implementation change; all of the existing MapReduce v1 applications remain compatible with MapReduce v2 APIs. Even though MapReduce workloads still dominate Hadoop 2.x clusters today, there's a robust interest in porting other distributed computation frameworks to run on top of YARN: Giraph, Apache Spark (in-memory engine for data processing), and Hamster (OpenMPI) are just a few examples.

Throughout this book, you used Hadoop 1.x implementation to run examples via map-only MapReduce jobs. The very same workflow still applies to running on top of Hadoop 2.x MapReduce implementation. Even though you haven't explored running on YARN (since it is still considered somewhat experimental), if you decide on the Hadoop 2.x installation, you can experiment with Giraph as a YARN client in addition to running Giraph via MapReduce. Regardless of which version of Hadoop you choose, your next step is installing the binary Hadoop distribution on your host(s).

Unless you have to worry about installing Hadoop on a large cluster of Linux hosts, the easiest way to get the binary distribution of Hadoop is to download the stable release packaged as a gzipped tar file from the Apache Software Foundation Hadoop release page at `http://hadoop.apache.org/releases.html`.

Unpack the resulting file in a subdirectory somewhere in your filesystem (you will use that same subdirectory later for installing Giraph), as shown in Listing A-2.

Listing A-2. Installing Hadoop 1.2.1 on a Unix or Max OS X Workstation

```
$ mkdir ~/dist
$ cd ~/dist
$ tar xzf hadoop-1.2.1.tar.gz
```

At this point, you have a binary installation of Hadoop and you need to expose the location of that installation to all the other command-line utilities that you are going to run. This is achieved by setting a few environment variables, as shown in Listing A-3 (don't forget that you can add them to your profile files the same way you could have added JAVA_HOME earlier).

Listing A-3. Setting up an Environment to Run Hadoop Applications

```
$ HADOOP_HOME=~/dist/hadoop-1.2.1
$ export HADOOP_HOME
$ PATH=$HADOOP_HOME/bin:$PATH
$ hadoop version
Hadoop 1.2.1
Subversion https://svn.apache.org/repos/asf/hadoop/common/branches/branch-1.2
Compiled by mattf on Mon Jul 22 15:23:09 PDT 2013
From source with checksum 6923c86528809c4e7e6f493b6b413a9a
This command was run using /Users/shapor/dist/hadoop-1.2.1/hadoop-core-1.2.1.jar
```

The last command in Listing A-3 proves that the Hadoop installation was successful by running the hadoop command-line utility and seeing the expected output. This is the easiest way to make sure that your Hadoop was installed correctly and can find Java on your system.

▓ **Note** Install Hadoop as described on all machines in the cluster.

Even though the preceding method of installing Hadoop is extremely easy and it should work on most operating systems, you may want to install from a binary distribution using a package manager. Doing so will guarantee that both Giraph and Hadoop binaries are coming from the same distribution and that they were integrated and tested together to work side by side. The following sections provide additional information on what it takes to go this route.

Giraph Installation

Now that you've completed your Hadoop installation, it may be tempting to think that installing Giraph on your systems should be even simpler. The good news is that since Giraph happens to be a client-side-only application, it doesn't need to be installed on all the hosts of your cluster. It is sufficient to make Giraph installation available on a machine in a cluster from which MapReduce or YARN jobs are typically submitted (these machines are known as *gateway* or *edge nodes*). The bad news is that unlike with Hadoop, choosing the ready-made Giraph binary distribution that is compatible with the rest of the Hadoop ecosystem deployed on your cluster may be non-trivial.

The thorny issue here is dependencies. At the very minimum, Giraph has a fundamental dependency on Hadoop; but realistically, its set of dependencies is at least as big as the number of Hadoop ecosystem projects Giraph I/O formats have to interface with. Apache Hadoop ecosystem projects are fortunate enough to have many contributions to their code base from different stakeholders in the open source and business communities. The high pace of project development resulted in quite a few major refactoring cycles of the code base and also produced quite a few commercial offerings based on different points in the Hadoop evolution history.

In general, the Hadoop ecosystem development community needs to be praised for paying a lot of attention to backward compatibility at the public API level. Just like the Linux kernel made it taboo to break user-land application by introducing changes to the public APIs, Hadoop has had a good track record in not overly upsetting the writers of the applications. Modulo bugs, a MapReduce application written against Hadoop 1.x (and not using any private or evolving APIs) should be able to run on Hadoop 2.x unmodified. The catch, however, is that it requires a recompilation of application Java code.

What this means for binary releases of Giraph is that at build-time they need to target the exact same version of Hadoop (and Hadoop ecosystem projects) that will be deployed on the cluster at runtime. At the very minimum, Giraph has to publish two binary releases: one built against Hadoop 1.x and another build against Hadoop 2.x. That, however, still doesn't account for differences in other dependencies, such as Hive, HBase, and so forth.

This particular issue is not specific to Giraph; it is known as *combinatorial explosion of dependencies*: the number of permutations of various versions of dependencies grow exponentially with the number of dependencies.

So far, the software engineering community has developed two different strategies for dealing with the combinatorial explosion of dependencies:

- Building binaries from source code as part of the software installation process

- Releasing complete stacks (or binary software distributions) of tightly integrated components instead of providing independent binary artifacts of individual components and expecting any combination of versions to work with each other

Before you deep-dive into the detailed descriptions of these strategies, let's consider the fact that the first strategy is really a subset of the second one. The decision of building your Giraph installation from source effectively means that you're embarking on a mission of producing your own binary software distribution with exact versions of components deployed to your cluster.

Software projects vs. software stacks   It is interesting to note that, in general, Apache Software Foundation tries to stay away from binary distributions of its projects, leaving this responsibility to downstream packagers. For the established projects (e.g., Apache HTTP server), these packagers are typically distributors of the various operating systems (Linux, etc.). OS distributors make sure that all of the various ASF software projects that end up in their particular version of an OS distribution work smoothly with each other. Hadoop and its ecosystem projects, however, haven't been on the agenda of operating system packagers.

To fill this void, various commercial vendors of Hadoop distributions began offering fully integrated sets of packages that don't require recompilation and are known to work well with each other. Most of these commercial distributions are based on the work done at Apache Bigtop: a 100% open source, community-driven Big Data management distribution of Apache Hadoop. For most of the users of Hadoop and its ecosystem projects, installing a fully integrated distribution from either Apache Bigtop or a commercial vendor is the easiest way to get started. For those with existing clusters, compiling Giraph from source code to match the exact versions of dependencies deployed on the cluster may be the only option.

Installing the Binary Release of Giraph

As mentioned, installing a binary release is the easiest option, but also the most limiting one. If you decide to go this route, you are essentially making the Giraph binary release dictate which versions of Hadoop and the Hadoop ecosystem components you'll be using. While it is unlikely to be useful for practical work—outside of quickly setting up an environment for running examples in this book, it is simple and very similar to the Hadoop installation process described at the beginning of this appendix.

As with Hadoop, there are two binary releases of Giraph available: one built against the latest version of Hadoop 1.x and the other against the latest version of Hadoop 2.x. And just like a binary release of Hadoop, binary releases of Giraph are packaged as gzipped tar files. Each released version of Giraph includes two binary artifacts: a Giraph binary release that is targeting Hadoop 1.x (available for download as `giraph-dist-X.Y.Z-bin.tar.gz`) and a Giraph binary release targeting Hadoop 2.x (available for download as `giraph-dist-X.Y.Z-hadoop2-bin.tar.gz`). Both of these gzipped tar files are available for download from the Giraph project website at `http://giraph.apache.org/releases.html`.

Because this book has been using examples of Giraph 1.1.0 running on Hadoop 1.2.1, the binary you need to grab is `giraph-dist-1.1.0-bin.tar.gz`. Once you download the binary, make sure to unpack it on the same workstation that you previously installed Hadoop. Note how the top-level folder that is created after you unpack the archive has an exact version of Hadoop 1.x embedded in its name, which looks like `giraph-1.1.0-for-hadoop-1.2.1`. Also, remember that on a real cluster, this is a gateway node. If you don't know where your gateway node is, ask your Hadoop administrator. Regardless of whether you are installing on your laptop or a gateway node, you have to go through the series of steps shown in Listing A-4.

Listing A-4. Installing Giraph 1.1.0 Built to Run on Hadoop 1.2.1

```
$ cd ~/dist
$ tar xzf giraph-dist-1.1.0-bin.tar.gz
$ GIRAPH_HOME=~/dist/giraph-1.1.0-for-hadoop-1.2.1
$ export GIRAPH_HOME
$ PATH=$GIRAPH_HOME/bin:$PATH
$ giraph
   Usage: giraph [-D<Hadoop property>] <jar containing vertex>
                 <parameters to jar>
   At a minimum one must provide a path to the jar containing the vertex to be executed.
```

As you can see, Listing A-4 is very similar to what you have done with Hadoop (see Listing A-3), and as with Hadoop, the last line in the listing proves that Giraph is installed and ready to go.

Installing Giraph As Part of a Packaged Hadoop Distribution

Almost all real-world deployments of Hadoop clusters happen on the Linux OS in a form of native Linux packages (DEB or RPM) that come from a binary software distribution of Hadoop. There are two sources for those packages: commercial vendors of Hadoop distributions and Apache Bigtop. Given that almost all commercial Hadoop distributions are derived from Apache Bigtop, you will focus on it in this section. The difference between Bigtop and commercial distributions boils down to support (you cannot buy support for Bigtop) and how quickly new versions of Hadoop ecosystem components are incorporated into the packaging (Bigtop tends to run ahead of the commercial distributions).

Installing Hadoop and its ecosystem projects as part of a packaged Hadoop distribution on Linux guarantees that every component has been integrated and is known to work with every other component coming from the same distribution. Not only that, but the packages are also guaranteed to be well-integrated with the underlying Linux OS by following the packaging guidelines of a given flavor of Linux. The end result is that working with Hadoop installed this way is no different from working with any piece of system software that came bundled with the Linux OS. The only downside to this method of installation is the fact that it is limited to Linux (although the Max OS X brew port is in the works) and that it requires elevated superuser privileges. Make sure to talk to your system administrator (or consult your Linux OS documentation) so that your account can run commands under elevated privileges using sudo(8).

The first step in enabling Hadoop installation from Apache Bigtop distribution consists of telling your Linux repository manager the URL where the packages can be found. Navigate to http://archive.apache.org/dist/bigtop/stable/repos/. Make sure that you find the folder corresponding to your Linux OS flavor and download the repository definition file named bigtop.XXX (where XXX is an extension specific to the Linux flavor that you are using). Once you get the file, make sure to copy it to the location where repository manager of your Linux OS looks for definitions of external repositories (don't forget to use sudo(8) so that you can copy into the system location). Table A-1 summarizes the location of repository definition files for various Linux flavors.

Table A-1. Locations of Repository Definitions

Linux Flavor	Folder Where Repo File Needs to Be Copied
Debian and Ubuntu	/etc/apt/sources.list.d
CentOS, RHEL and Fedora	/etc/yum/repos.d
SUSE and OpenSUSE	/etc/zypp/repos.d

Once you add the Bigtop repository definition file to one of the folders listed in Table A-1, the next step is to import a repository key. Importing the key allows the package manager to make sure it is installing genuine packages. Since the repository key establishes trust, always make sure to download it from the secure https://dist.apache.org/repos/dist/release/bigtop/KEYS and store it in the current directory. Once that is done, all that is left is adding the key and refreshing the repository definition cached locally. After that, you can install Hadoop and Giraph using the usual means of package installation on Linux. The required steps are summarized in the Table A-2, with the rows again corresponding to different flavors of Linux.

Table A-2. *Installing Giraph via Linux Binary Packages*

Debian and Ubuntu	CentOS, RHEL and Fedora	SUSE and OpenSUSE
`$ sudo apt-key add < KEYS`	`$ sudo rpm --import KEYS`	`$ sudo rpm --import KEYS`
`$ sudo apt-get update`	`$ sudo yum clean metadata`	`$ sudo zypper update`
`$ sudo apt-get install giraph`	`$ sudo yum install giraph`	`$ sudo install giraph`

An interesting side effect of running that last command is that the dependency between Giraph and Hadoop is properly recognized and the right Hadoop package is implicitly installed on your system. In a way, once you decide to install Giraph via Linux packages, your best (and easiest!) option for installing Hadoop is to install it in a packaged form as well. In fact, you don't even have to execute that step explicitly: the correct package of Hadoop is fetched.

In case you are wondering what you need to do to set environment variables HADOOP_HOME and GIRAPH_HOME to after installing Giraph via Linux packages, the good news is that you don't have to worry about those anymore. You can simply run the hadoop and giraph command-line utilities the same way you would run any executable on your Linux system: just type its name.

There is no denying that installing from Linux packages is by far the easiest way to install both Giraph and Hadoop. If you choose one of the commercial vendors, you can sign up for professional support. The downside, however, is that by installing the prepackaged bits, you are giving up your right to have a precise combination of Giraph and Hadoop versions that are unique to your environment. If you find yourself needing to match Giraph to the custom versions of Hadoop and Hadoop ecosystems components, installing Giraph by building from source code may be your only option.

Installing Giraph by Building from Source Code

The source code of Giraph is packaged as a gzipped tar file and available for at the same project web site that you used for downloading binaries: http://giraph.apache.org/releases.html. For the Giraph version 1.1.0, get the file named giraph-dist-1.1.0-src.tar.gz and unpack it somewhere on your development workstation.

The Giraph build infrastructure is managed by Apache Maven. If you decide to install Giraph by building it from the source code, you have to make sure that Maven version 3 or higher is available in your environment. If you don't have Maven available, make sure to follow installation instructions provided by the project's web site at http://maven.apache.org. The rest of this appendix assumes that you have set up Maven and that you can successfully run the mvn command-line utility.

■ **Note**   You will install Giraph by building it from source code in situations where you have a preexisting Hadoop cluster with the versions of dependencies (including Hadoop itself) not matching those of packaged Giraph binaries. Although building Giraph against arbitrary versions of dependencies is possible, be warned that it may very well be that such a combination has never been tested.

Giraph's build infrastructure provides a convenient way for specifying exact version of dependencies that Giraph needs to be built for. You don't need to download any of these dependencies or otherwise make them available in your environment—Maven does it for you. You are expected to use Maven build profiles for specifying the major version of Hadoop and when you want to build Giraph as a YARN client rather than a map-only MapReduce application. The most commonly used build profiles are summarized in Table A-3.

Table A-3. *Commonly Used Maven Build Profiles*

Profile Name	Profile Effect
hadoop_1	Produces a build that is compatible with Hadoop 1.x
hadoop_2	Produces a build that is compatible with Hadoop 2.x
hadoop_yarn	Produces a YARN-based Giraph build (assumes Hadoop 2.x)

Using a Maven build profile lets you preset the majority of properties to the desired values, while still allowing surgical overrides to match minor and micro versions of dependencies. Whereas there are dozens of properties that you can tweak while building Giraph from source code, the most commonly used ones are summarized in Table A-4.

Table A-4. *Commonly Used Properties for Specifying Exact Versions of Giraph Dependencies*

Property Name	Property Effect
hadoop.version	Used for specifying the exact version of Hadoop dependency
dep.accumulo.version	Used for specifying the exact version of Accumulo dependency
dep.hbase.version	Used for specifying the exact version of HBase dependency
dep.hcatalog.version	Used for specifying the exact version of HCatalog dependency
dep.hive.version	Used for specifying the exact version of Hive dependency
dep.zookeeper.version	Used for specifying the exact version of ZooKeeper dependency

Starting with specifying a profile and then all the desired version properties allows you to have very fine-grained control over the resulting Giraph binary. For example, suppose your cluster was running Hadoop 2.7.1 and HBase 0.98. The way to build YARN-aware Giraph compatible with your cluster dependencies is outlined in Listing A-5. Before running this command, however, keep in mind that a particular set of dependencies that you are requesting may not have been tested. The implications of this could range from build failure, unit test failures, or runtime failures. Once you deviate from the beaten path, you are on your own, although it may just work. Thus, don't be dismayed if unit tests fail. Some of those unit tests happen to be sensitive to the versions of dependencies they have been developed against; they can be turned off by passing –DskipTests to the Maven build.

Listing A-5. An Example of Building Giraph Targeting Hadoop 2.7.1 and HBase 0.98

```
$ mvn –Phadoop_yarn –Dhadoop.version=2.7.1 –Ddep.hbase.version=0.98 -DskipTests
```

After the build is done, you find the binary artifact very similar to the one you downloaded in previous chapters under the `giraph-dist/target` folder. Look for the gzipped tar file there and follow the steps described in the previous section to untar and install that custom version of Giraph on your gateway node. Don't forget to set up `GIRAPH_HOME` and to add Giraph binaries to your $PATH variable the same way that you did in Listing A-4.

As you have seen, so far there are a number of different ways to install Hadoop and Giraph on a system. Of course, after you install them, you then have to configure them. You can't really use the bits for much of anything before you configure. This is the subject of the next section.

Fundamentals of Hadoop and Hadoop Ecosystem Projects Configuration

Almost all members of the Hadoop ecosystem (including Hadoop itself) share the common configuration management system that is based on flat XML files encoding a set of key-value property settings. The folders where these configuration files are located happen to be project and installation method specific. For any project installed via Linux packages (DEB or RPM), the location of configuration files is under /etc/<project name>/conf. If, however, the installation was manual, the same configuration files are located under the project's HOME folder in the conf subfolder. Where you will find configuration files for Hadoop and Giraph depend on the installation method, which is summarized in Table A-5.

Table A-5. *Location of Configuration Files by Installation Method*

Installation Method	Hadoop Configuration Folder	Giraph Configuration Folder
Linux packages	/etc/hadoop/conf	/etc/giraph/conf
Manual	$HADOOP_HOME/conf	$GIRAPH_HOME/conf

Inside of these folders you will find one or more XML files, similar to the example outlined in Listing A-6. Note that whereas Giraph requires just a single configuration file called giraph-site.xml, Hadoop spreads its configuration over at least three different files—core-site.xml, hdfs-site.xml, and mapred-site.xml—with yarn-site.xml being an additional file for configuring YARN as part of Hadoop 2.x.

Listing A-6. Example Hadoop Configuration File

```
<configuration>
  <property>
   <name>prop.name</name>
   <value>prop.value</value>
   <description>Free form description of what this property does</description>
  </property>
</configuration>
```

Configuring Giraph

Strictly speaking, by default, Giraph doesn't actually require any settings in its XML configuration file. Anything that may be a required configuration option (such as specifying the number of workers) can be passed to Giraph via its command-line utility. Pretty soon, though, constantly typing all of the required options on the command line gets frustrating, at which point you can start leveraging giraph-site.xml to store the ones specific to your environment (just make sure to put it in the folder listed in Table A-5). For example, if you find yourself running Giraph applications on top of Hadoop configured in local mode, you may find it useful to have a Giraph configuration file similar to the one in Listing A-7.

Listing A-7. Example Giraph Configuration File giraph-site.xml

```
<configuration>
  <property>
    <name>giraph.SplitMasterWorker</name>
    <value>false</value>
    <description>This lets Giraph app run in a single task</description>
  </property>
</configuration>
```

Even though Giraph has its own configuration file, it still depends on the Hadoop client to be correctly configured via Hadoop-specific configuration files. Configuring Hadoop and matching the Hadoop client is the subject of the next section.

Configuring Hadoop

Fundamentally, Hadoop jobs can run in one of three modes:

- On a fully distributed Hadoop cluster with a client configured on the gateway node

- On a pseudo-distributed Hadoop cluster with a client co-located

- In a local Hadoop configuration with the client mocking the cluster-specific Hadoop machinery

The vast majority of examples in this book assumed you were running the Giraph application on top of Hadoop configured in local mode. Local mode is the simplest one to set up. It doesn't require any active processes running anywhere and it makes sure that the Giraph job runs within a single JVM with everything else. Instead of using HDFS, local mode reads and writes files to a local filesystem, which makes it even easier to use. You don't even have to configure anything to enable local mode, since it also happens to be the default (thus no XML configuration files or empty files would enable it).

Local mode is great for development and debugging your Giraph application, but it also has a few limitations. First of all, it can only run one task at a time, which requires you to limit the number of Giraph workers to one, and to combine master and worker into the same task. Perhaps an even bigger limitation is that local mode execution doesn't hit the same code path that is normally triggered when running on a real Hadoop cluster. Keep this in mind when debugging Giraph applications—you may see differences in behavior. Finally, even though local mode is supposed to be the fastest (at least on tiny datasets), there's still quite a bit of work that Hadoop has to do to bootstrap the MapReduce framework. This can lead up to a 20-second delay before Giraph code starts to execute.

The total opposite of a local mode is, of course, a fully distributed Hadoop cluster. This is the usual way of running a Giraph application in production; it assumes that a fully distributed Hadoop cluster is available. Giraph acts as a pure client and you need to submit Giraph jobs from a cluster gateway node. The Hadoop configuration available on a gateway node needs to match the configuration that is being used by Hadoop running on all the other nodes of the cluster. This is typically done by a Hadoop cluster administrator; it doesn't require any explicit configuration by the users of the cluster. All the administrator needs to make sure of is that the contents of the Hadoop configuration folder is kept in sync on all the nodes in the cluster, including the gateway node.

Configuring Hadoop in Pseudo-Distributed Mode

A really nice compromise between a somewhat limiting local mode and a heavyweight, fully distributed mode requiring a lot of setup is Hadoop's pseudo-distributed configuration. In this mode, you can run your jobs in an environment as close to the real cluster as possible, but without the hassle of setting up a multinode cluster (although as you've seen in Chapter 12, that hassle can be mitigated by using Hadoop cloud services). Pseudo-distributed mode runs all the processes (daemons) that a real cluster would have on the same host. These processes run in different JVMs and communicate over the loopback network interface. Other than that, the configuration is identical to what you would see on a real cluster. Pseudo-distributed mode is a nice middle ground between local and fully distributed modes.

The easiest way to enable a pseudo-distributed mode is by installing Hadoop on Linux from packages. There is a special package called hadoop-conf-pseudo that pulls all the right dependencies and presets the configuration parameters needed for pseudo distributed mode in /etc/hadoop/conf. Simply installing that package is all that is needed.

In the situations where installing Hadoop from Linux packages is not an option, an alternative is to manually install the Hadoop binary under the $HADOOP_HOME folder (the way that it was described earlier in this appendix), and make sure that the properties summarized in Table A-6 are set in the appropriate configuration files.

Table A-6. *Hadoop Configuration Required for Pseudo-Distributed Mode*

Configuration File	Property Name	Property Value	Property Effect
core-site.xml	fs.default.name	hdfs://localhost/	Sets up a URL for the pseudo-distributed HDFS filesystem
hdfs-site.xml	dfs.replication	1	Sets up a replication factor of 1, since in pseudo-distributed mode, you only have a single data node
mapred-site.xml	mapred.job.tracker	localhost:8021	Sets up a URL for the pseudo-distributed Job Tracker
yarn-site.xml	yarn.nodemanager. aux-services	mapreduce_shuffle	Makes YARN aware of an extra service required by the MapReduce framework
yarn-site.xml	yarn.nodemanager. aux-services. mapreduce_shuffle. class	org.apache. hadoop. mapred. ShuffleHandler	Specifies the name of the class implementing the mapreduce_ shuffle service

Since pseudo-distributed mode required a few processes to run, the final step after configuration is to launch the necessary services and initialize their state. First, you need to start HDFS. Use the instructions provided in Table A-7, depending on whether you installed from Linux packages or from a binary gzipped tar file. Of course, if you didn't install Hadoop from packages, don't forget to add $HADOOP_HOME to your $PATH, as was shown in Listing A-3.

Table A-7. *Starting HDFS in Pseudo-Distributed Mode*

Hadoop Installed from Linux Packages	Hadoop Installed from Binary gzipped tar File

$ sudo service hadoop-hdfs-namenode format
$ sudo service hadoop-hdfs-namenode start
$ sudo service hadoop-hdfs-datanode start
$ sudo -u hdfs hadoop fs -chmod 777 /
	$ hadoop namenode -format
	$ hadoop-daemon.sh start namenode
	$ hadoop-daemon.sh start datanode
	$ hadoop fs -chmod 777 /

■ **Note** The last command makes the root subdirectory of HDFS readable and writable by anyone. This would be a huge security issue on a real, fully distributed, multitenant cluster. On a pseudo-distributed setup, however, it allows you to shortcut a few setup steps without compromising the functionality of HDFS.

At this point, you should have a fully functional HDFS. You can verify that it is up and running by issuing the commands shown in Listing A-8.

Listing A-8. Verifying That You Can Copy Files into and out of Pseudo-Distributed HDFS

```
$ hadoop fs -put /etc/hosts /test.file
$ hadoop fs -cat /test.file
$ hadoop fs -rm /test.file
```

As long as all the commands ran without an error, you can be sure your pseudo-distributed HDFS setup is fine. The final step in setting up your pseudo-distributed Hadoop cluster is starting MapReduce services (for Hadoop 1.x) or YARN services (for Hadoop 2.x). The commands needed are summarized in Tables A-8 and A-9, respectively.

Table A-8. *Starting MapReduce in Pseudo-Distributed Mode*

Hadoop Installed from Linux packages	Hadoop Installed from Binary gzipped tar File

`$ sudo service hadoop-mapred-jobtracker start`
`$ sudo service hadoop-mapred-tasktracker start`
`$ cd /usr/lib/hadoop`
	`$ hadoop-daemon.sh start jobtracker`
	`$ hadoop-daemon.sh start tasktracker`
	`$ cd $HADOOP_HOME`

Table A-9. *Starting YARN in Pseudo-Distributed Mode*

Hadoop Installed from Linux Packages	Hadoop Installed from Binary gzipped tar File

`$ sudo service hadoop-yarn-resourcemanager start`
`$ sudo service hadoop-yarn-nodemanager start`
`$ cd /usr/lib/hadoop-mapreduce`
	`$ hadoop-daemon.sh start resourcemanager`
	`$ hadoop-daemon.sh start nodemanager`
	`$ cd $HADOOP_HOME/share/hadoop/mapreduce`

Regardless of whether you are setting up Hadoop 1.x or Hadoop 2.x, once you are done with these commands, make sure to run a test MapReduce job while you are still located in the current working directory (where the last cd command put you). If the command shown in Listing A-9 runs successfully to completion, it means your pseudo-distributed Hadoop cluster is fully set and ready for your Giraph applications.

Listing A-9. Verifying That You Run MapReduce Jobs on Pseudo-Distributed Hadoop Cluster

```
$ hadoop jar *examples*jar pi 10 1000
....
Job Finished in 35.513 seconds
Estimated value of Pi is 3.14080000000000000000
```

Summary

You can install Hadoop and Giraph in a number of different ways. The key issue to keep in mind is that whichever version of Hadoop you install, it has to match the version Giraph binary that you will be using. If you plan to use I/O formats connecting Giraph to other members of the Hadoop ecosystem (such as Hive, HBase, etc.), those versions have to match as well. After installation is done, the last step is to configure both. Giraph doesn't require any configuration by default. Hadoop has to be configured to run in one of the three modes: local, fully distributed, or pseudo-distributed. In this appendix you looked at the following topics:

- System requirements for running Hadoop and Giraph: JDK 7+ on Linux or Mac OS X

- The steps involved in installing Hadoop using two main methods: a binary gzipped tar file or Linux binary packages

- The steps involved in installing Giraph using three different methods of installation: a binary gzipped tar file, Linux binary packages, or building from source code

- Configuring Hadoop and Giraph using local flat XML configuration files

- Executing simple command lines to make sure that your Giraph and Hadoop deployments are functioning correctly

Although this appendix offered a number of alternative options for accomplishing two basic tasks—installation and configuration, if you're installing everything from scratch, you can simply choose whichever is the easiest for your environment. If, however, you are dealing with existing Hadoop clusters, you may find some of the alternative methods helpful.

Summary

Index

Get the eBook for only $5!

Why limit yourself?

Now you can take the weightless companion with you wherever you go and access your content on your PC, phone, tablet, or reader.

Since you've purchased this print book, we're happy to offer you the eBook in all 3 formats for just $5.

Convenient and fully searchable, the PDF version enables you to easily find and copy code—or perform examples by quickly toggling between instructions and applications. The MOBI format is ideal for your Kindle, while the ePUB can be utilized on a variety of mobile devices.

To learn more, go to www.apress.com/companion or contact support@apress.com.

Printed in the United States
By Bookmasters